Cooperative Management

The Book Series on Cooperative Management provides an invaluable forum for creative and scholarship work on cooperative economics, organizational, financial and marketing aspects of business cooperatives and development of cooperative communities throughout the Mediterranean region and worldwide. The main objectives of this book series are to advance knowledge related to cooperative entrepreneurship as well as to generate theoretical knowledge aiming to promoting research within various sectors wherein cooperatives operate (agriculture, banking, real estate, insurance, and other forms). Scholarly edited volumes and monographs should relate to one of these areas, should have a theoretical and/or empirical problem orientation, and should demonstrate innovation in theoretical and empirical analyses, methodologies, and applications. Analyses of cooperative economic problems and phenomena pertinent to managerial research, extension, and teaching (e.g., case studies) regarding cooperative entrepreneurship are equally encouraged.

More information about this series at http://www.springer.com/series/11891

L. Jan Slikkerveer · George Baourakis · Kurniawan Saefullah
Editors

Integrated Community-Managed Development

Strategizing Indigenous Knowledge and Institutions for Poverty Reduction and Sustainable Community Development in Indonesia

A Community-Based Contribution to the United Nations 2030 Agenda for Sustainable Development

Editors
L. Jan Slikkerveer
Leiden Ethnosystems and Development Programme (LEAD)
Leiden University
Leiden, The Netherlands

George Baourakis
Department of Business Economics and Management
Mediterranean Agronomic Institute of Chania
Chania, Greece

Kurniawan Saefullah
Faculty of Economics and Business
Padjadjaran University
Bandung, Indonesia

ISSN 2364-401X ISSN 2364-4028 (electronic)
Cooperative Management
ISBN 978-3-030-05422-9 ISBN 978-3-030-05423-6 (eBook)
https://doi.org/10.1007/978-3-030-05423-6

Library of Congress Control Number: 2018964258

This Springer imprint is published by the registered company Springer Nature Switzerland AG
The registered company address is: Gewerbestrasse 11, 6330 Cham, Switzerland

This Volume is dedicated with affection, respect, and gratitude to

Alkinoos Nikolaidis (1945–2016)

in memory of his extraordinary accomplishments as an inspiring mentor and pioneer, rendering under his Directorship for more than 25 years the Mediterranean Agronomic Institute of Chania (CIHEAM-MAICh) a highly acclaimed international research and training institute

Foreword I

In certain parts of the world, poverty has been historically accepted by some as inevitable, as non-industrialised economies provided insufficient goods and services to serve the entire population with an adequate standard of living, rendering poverty the trademark and curse of humanity since its birth. On the contrary, all that one has to do is to look around and study the dreary statistics. The large majority of the world's 7 billion people survives on just one meagre meal a day. According to a recent *United Nations Development Report*, in Indonesia, almost 20% or about 50 million people earn less than $1.25 per day, while the renowned international journal *The Economist* indicates that 100 million Indonesians have to survive on only $2 dollar a day or less. Throughout history, various efforts have been exerted by individuals and organizations to improve the situation of the underprivileged through the alleviation of poverty. King Hammurabi of Babylon and the Pharisees in ancient Jerusalem, the Templar Knights throughout Europe and the Mormons in the USA, for the good of humanity or their own salvation, all tried in their own ways to help the destitute. However, all these efforts were merely disparate activities which eventually bore relatively few results.

It was not until the 1970s when in Bangladesh Muhammad Yunus of the Grameen Bank started the microcredit scheme to provide small loans with subsidised low interest rates to the poor on a group basis, repayable through group efforts. This scheme soon exploded into a worldwide movement which helped some 20 million poverty-stricken people, especially women, who borrowed money in order to start micro-enterprises. Later on, microcredit evolved into the microfinance stage, where borrowers were provided not only with credit, but increasingly with broader services, such as insurance and deposit banking, albeit on a more commercial basis of higher market-based interest rates. Despite the initially significant efforts through microcredit to help the poor, the later microfinance schemes often led to over-indebtedness and continued dependence of the poorer segment of the population on MFIs, village banks and local resident credit organisations.

However, since the beginning of this century, international organisations and agencies started to pay more attention to economic and human development programmes with a view to alleviating global poverty within the context of sustainable development, but the results remained largely disappointing as these programmes did not reach, nor did they involve the target population, in order to make a positive impact on the position of the poor at the community level.

A recent breakthrough in the alleviation of poverty emerged in the form of the alternative approaches of *Integrated Microfinance Management* (*IMM*) and *Integrated Community-Managed Development* (ICMD), generated by the Leiden Ethnosystems and Development Programme (LEAD) of Leiden University under the auspices of Prof. Dr. L. Jan Slikkerveer, which not only provides for the prospect of obtaining credit, but also for a comprehensive and inclusive strategy of solving the problem of the poor to attain a better standard of living.

While the previous IMM approach sought to alleviate poverty through the combined provision of not only microfinancial services, but also health, education, communication, social and cultural services, the newly introduced ICMD approach focuses on the integration of global and local knowledge systems, embodied in the indigenous institutions which are functionalised as the people-centred locus of participatory decision-making and the implementation of programmes of poverty reduction and sustainable development at the grassroots level. The novel idea here is to strive for sustainable development through the indigenous institutions of mutual aid and communal action which epitomise the indigenous systems of knowledge, beliefs and practices among fellow members of the community with a view to achieving well-being in a comprehensive array of needs, such as health, education, socio-economic and financial services, ecological preservation and the conservation of the bio-cultural resources.

The key significance of the new approach is the integration of the peoples' cosmologies underlying the indigenous knowledge systems, exemplified in Indonesia in the indigenous worldviews of *Tri Hita Karana* in Bali and *Tri Tangtu* in the Sunda region of Java, rendering their participation optimal in sustainable development. The expected results are not only to improve local peoples' welfare, but also to realise self-determination, empowerment and sustainability of the local communities.

The ICMD approach is a formidable breakthrough concept in human development which places emphasis on indigenous systems of knowledge and institutions, encompassing collective action, mutual aid and self-help in order to achieve communal prosperity for all members of the community. In Indonesia, the indigenous knowledge, practices and institutions of communal action, including *i.a. Gotong Royong, Metulung, Gintingan* and *Subak*, have been functioning over many generations, which can support and make *Integrated Community-Managed Development* a new reality.

Needless to say, some seed money is needed to further develop this new approach both for advanced education and training of new cadres of *Integrated Community Managers* and for the implementation of *Integrated Community-Managed Development Programmes* in the field, as a nationwide endeavour to alleviate poverty in Indonesia for the coming years.

Now is a most opportune time to make it happen, as globally, *macroeconomics* has currently evolved into *microeconomics*, and national governments are striving to focus on satisfying the needs of smaller units of the economy such as districts, villages and families, as opposed to that of the entire nation. Such is especially the case in Indonesia, where the government recently initiated a programme whereby each village is given *one billion rupiah* every year in order to realise their own priorities. Additionally, the democratic climate, which promotes equality and freedom of speech, facilitates such current 'bottom-up' aspirations. The new approach can foster the development of community empowerment and self-help improvement, which in time will make the cost of national welfare and prosperity less costly and more effective. The promising experience with strategizing indigenous knowledge and institutions for poverty reduction and sustainable community development, currently accumulating in training and in the field in Indonesia, will certainly provide a community-based contribution to the United Nations' *2030 Agenda for Sustainable Development* and the *Sustainable Development Goals.*

Given these golden opportunities, the new concept of Integrated Community-Managed Development has the immense potential to make a major difference in the global fight against poverty by mustering the inherent strengths of individual groups and traditional institutions, and harking back to their indigenous knowledge and practices through mutual assistance in order to achieve sustainable prosperity. For this grand prospect, we have to thank Prof. Jan Slikkerveer and his two co-editors for this significant volume. It is now up to all of us to contribute in whatever way we can to make this concept and its aspirations come true.

The destitution of the past compels us to change the future and history for the better of all humankind.

Jakarta, Indonesia Prof. Dr. Anak Agung Gde Agung

Anak Agung Gde Agung is Prince of Gianyar, a descendant of Balinese Royalty, and Former Cabinet Minister of Social Affairs of the Republic of Indonesia. He is currently Professor at the Trisakti School of Tourism, Jakarta. Educated in International Law and Diplomacy at Harvard University (USA), he received his Ph.D. degree in Ethnoscience from Leiden University in The Netherlands in 2005. An expert in Balinese cosmology, he has published several articles on the concept of *Tri Hita Karana and Sustainable Development.*

Foreword II

The origin of the newly developing approach of *Integrated Community-Managed Development* (ICMD), introduced as an innovative human development strategy to reduce poverty and realise sustainable development, is located in the joint academic endeavours among the editors of this impressive Volume as Representatives of the collaborating academic institutions of higher education in The Netherlands, Indonesia and Greece, going back to the early 1990s of the previous century: Leiden University (LEAD/UL), Universitas Padjadjaran (FEB/UNPAD) and the Mediterranean Agronomic Institute of Chania (CIHEAM-MAICh).

The educational dimensions of the development of this new ICMD approach followed the ongoing training and research programmes in the multidisciplinary field of *Indigenous Knowledge Systems and Development* at the counterpart institutions in different fields, such as health, agriculture, economics and communication, which pertained in 2009 to the previous joint Project on the Development of a new Master Course on *Integrated Microfinance Management* (IMM) at Universitas Padjadjaran in Bandung, Indonesia, which received substantial support for three years from the Indonesian Facility of the Netherlands Ministry of Economic Affairs in The Hague.

This unique programme, in which the three counterparts were joined by the Indonesian Movement for Microfinance (Gema PKM) and advised by the Martha Tilaar Foundation (MTF), has set the tone for a renewed focus on the societal relevance of scientific endeavours particularly for sustainable development and change in developing nations. As the Former President of Leiden University, I have had the privilege to witness and support the innovative scientific orientation and attitude towards indigenous peoples and their knowledge systems, pioneered by the Leiden Ethnosystems and Development Programme (LEAD), which had evolved since the late 1980s to study, document, analyse and functionalise indigenous knowledge and practices within the dynamic context of socio-economic development. The breakthrough in such scientific reassessment of the so-called 'ethnoscientific' knowledge systems became particularly manifest in the transition from monodisciplinary to multidisciplinary approaches as a means of finding more adequate solutions for increasingly multifaceted problems in the present-day societies.

Indeed, the traditional academic view on science has mostly been monodisciplinary in orientation. Although there are some exceptions, for instance in medicine, in most fields of study there has been a primary drive towards the implementation of a monodisciplinary approach, such as in astronomy, physics, chemistry, biology, and economics. Nowadays, however, we have to acknowledge that it is not sufficient anymore to explain complicated phenomena and processes only in science, but also in society. Such new developments need students, scientists and scholars who are open in their orientation, who are qualified to go beyond the boundary of their monodisciplinary field, and above all, who are prepared to work with colleagues in a team.

As an example of such re-orientation in applied science towards the multidisciplinary education and research of indigenous knowledge systems within the context of socio-economic development has been the introduction of the above-mentioned Master Course of *Integrated Microfinance Management* (IMM) in 2012 in Indonesia, which since then has successfully delivered highly skilled Integrated Microfinance Managers capable of managing financial and non-financial services in the communities, and as such, provides a strong contribution to poverty alleviation and sustainable development in the rural areas of the country.

While the IMM approach is focused on the provision of not only financial, but also non-financial services to the poor and low-income families, the new ICMD approach has emerged from the field, where a need has been identified for more human poverty alleviation programmes for the rural poor segment of the population, which remains largely excluded from participation not only in the largely commercial activities of MFIs, NGOs and village banks, but also in the recent activities of the Government of Indonesia in financial economic programmes for poverty reduction, largely supported by the World Bank.

Since we have learned that such participation by community members is crucial for the long-term success of eventually realising the objective of poverty reduction and can only be achieved by programmes which are engendering the indigenous peoples' local knowledge, values and institutions which have been guiding their own processes of decision-making, management and development for many centuries, the renewed 'bottom-up' ICMD approach has been designed and tested by the counterparts. Although this ICMD approach is directly linked to the previous IMM approach, it embarks on the specific integration of local and global knowledge systems and on strategizing the related indigenous institutions in the policy planning and implementation process at the community level.

Therefore, I believe that this interesting Volume in the *Cooperation Management Series* on the new ICMD approach not only links up very well with the present scientific educational demands from the intelligentsia in Indonesia, but also responds directly to the urgent need from the society for well-trained and skilled *Integrated Community Managers*, capable of reducing the unacceptably high rates of poverty in

Indonesia, and as such, providing a constructive contribution to the realisation of the Sustainable Development Goals (SDGs) of the United Nations in the South-East Asian Region and the rest of the world.

The Hague, The Netherlands Loek E. H. Vredevoogd, M.Sc.

Loek E. H. Vredevoogd is the Former President of Leiden University in The Netherlands and Former Chairman of the Netherlands—Flemish Accreditation Commission for Higher Education. He holds a M.Sc. degree in Economics, is Senior Advisor of the LEAD Programme of Leiden University and a Guest Senior Lecturer in the Course on *Integrated Microfinance Management* (IMM).

Acknowledgements

The editors would like to acknowledge the great input of all the individuals who made this Volume possible, particularly the local people in the rural communities of Indonesia who kindly shared their knowledge, experience, practices and institutions, as well as the cosmologies of their universe and their vision on processes of poverty reduction and sustainable community development which have been affecting their way of life and livelihood over so many generations.

Similarly, the editors wish to express their high appreciation for the kind support of Mrs. Maria Verivaki MA to provide the excellent editing work and styling of the entire manuscript, rendering this Volume a valuable contribution to the Cooperative Management Series of Springer Academic Publications.

Furthermore, our gratitude goes to those individuals who have contributed substantially to facilitate the development of the new concepts pertaining to the processes of poverty reduction and sustainable development in Indonesia and beyond, including the former Rectors of Universitas Padjadjaran in Bandung, Prof. Dr. Himendra Wargahadibrata and Prof. Dr. Ganjar Kurnia, as well as the current Rector Prof. Dr. Tri Hanggono Achmad and the Coordinator of the new M.IMM Master Course Dr. Mokhamad Anwar.

Finally, the editors wish to thank Mrs. Mady Slikkerveer MA for her unflagging support and encouragement in the development of the *Integrated Microfinance Management* (IMM) Programme since 2009.

Contents

Editors and Contributors

About the Editors

L. Jan Slikkerveer is the Director of the *Leiden Ethnosystems and Development Programme* (LEAD) and Professor of Applied Ethnoscience in the Faculty of Science of Leiden University in The Netherlands. He received his Ph.D. degree from Leiden University in 1983 and an Honorary Degree from Universitas Padjadjaran in 2005. He conceptualised both the IMM and ICMD approaches since 2008 and is Senior Lecturer in the Course on *Integrated Microfinance Management* (IMM).

George Baourakis is the Director of the Mediterranean Agronomic Institute of Chania (CIHEAM-MAICh) in Greece and the Research Coordinator of the Business Economics and Management Department of CIHEAM-MAICh. He holds a Ph.D. in Food Marketing and is an expert in behavioural economics. He is also Affiliate Professor at Nyenrode University in The Netherlands and has lectured in the Course on *Integrated Microfinance Management* (IMM), Bandung.

Kurniawan Saefullah is Lecturer at the Faculty of Economics and Business of Universitas Padjadjaran, Bandung, Indonesia. He received his M.Sc. degree from the International Islamic University, Malaysia and is currently finalising his Ph.D. research on the concept of IMM with special attention to the local cosmology and institutions in Subang, West Java, at Leiden University. He is also Senior Lecturer in the Course on *Integrated Microfinance Management* (IMM), Bandung.

Contributors

Prihatini Ambaretnani FISIP, Universitas Padjadjaran, Bandung, Indonesia

George Baourakis CIHEAM-MAICh, Chania, Greece

Nury Effendi FEB, Universitas Padjadjaran, Bandung, Indonesia

J. C. M. (Hans) de Bekker LEAD, Leiden University, Leiden, The Netherlands

Bambang Ismawan Bina Swadaya Indonesia, Jakarta, Indonesia

Tati S. Joesron FEB, Universitas Padjadjaran, Bandung, Indonesia

Benito Lopulalan Sinergi Indonesia, Depok, Indonesia

Asep Mulyana Department of Management of the Faculty of Economics and Business, Universitas Padjadjaran, Bandung, Indonesia

Dimitrios Niklis CIHEAM-MAICh, Chania, Greece

Alkinoos Nikolaidis CIHEAM-MAICh, Chania, Greece

Kurniawan Saefullah LEAD, Leiden University, Leiden, The Netherlands; FEB, Universitas Padjadjaran, Bandung, Indonesia

Adiatma Y. M. Siregar FISIP, Universitas Padjadjaran, Bandung, Indonesia

L. Jan Slikkerveer LEAD, Leiden University, Leiden, The Netherlands

Martha Tilaar Martha Tilaar Group, Jakarta, Indonesia

Constantin Zopounidis CIHEAM-MAICh, Chania, Greece

Prologue

The purpose of this Volume on *Integrated Community-Managed Development* (ICMD) is to provide a timely follow-up to the *Handbook for Lecturers and Tutors of the new Master Course on Integrated Microfinance Management for Poverty Reduction and Sustainable Development in Indonesia* (IMM), edited by Slikkerveer (2012), focusing on an *Indigenous Knowledge Systems*-oriented contribution to poverty reduction as the primary objective of the recent *2030 Agenda for Sustainable Development* of the United Nations (2015).

The Volume links up well with the *Cooperative Management Series* of Springer as it highlights the importance of community-managed development of the activities of all stakeholders concerned, to realise the objective of poverty reduction and sustainable community development through the integration of global and local knowledge systems. Moreover, it prioritises the so far largely ignored managerial role of *indigenous institutions* in achieving sustainable community development as they are informal—sometimes invisible to the outsider—institutions which are rooted in the history of the community and based on strong local cosmological principles of cooperation, mutual aid and communal action, which are all crucial to achieve community participation. The indigenous institutions are crucial for sustainable development as they have guided and sanctioned the interests, resources and capacities of the community members in a complex of norms, values and behaviours over many generations, with a view to serving socially valued goals in order to achieve common goods and services for the entire community in a non-commercial way. The indigenous institutions of collective action in Indonesia include *i.a. Gotong Royong, Metulung, Gintingan* and *Subak* (cf. Agung 2005; Saefullah 2019).

While the newly developed *Integrated Community-Managed Development* (ICMD) approach is representing a largely human development strategy to attain poverty reduction and sustainable development through the integration of local and global systems of knowledge and technology at the community level, the previously developed *Integrated Microfinance Management* (IMM) approach is representing a largely economic development strategy to attain poverty reduction and sustainable development through the management of a comprehensive set of integrated community-based services, encompassing not only inclusive financial, but also health, education, communication and socio-cultural services. Both IKS-oriente`d approaches are rather innovative as they are based on the *Indigenous Knowledge Systems Integration Model* (IKSIM) and embark on strategising the often disregarded indigenous institutions in order to strengthen their shared objectives of reaching poverty reduction for sustainable community development through a holistic framework of increased processes of knowledge integration, communication, decision-making, participation, cooperation and capacity building at the community level.

The ICMD approach has been developed by the LEAD Programme of Leiden University since the late 2010s on the basis of its prolonged experience in *ethnoscience* as part of the shift in the development paradigm of the 1990s towards a 'bottom-up' approach to achieve sustainable community development, by embarking on the integration of indigenous knowledge systems as part of the underlying world vision or cosmology of the target population, known as *The Cultural Dimension of Development* (cf. Warren et al. 1995; Slikkerveer 1999, 2012).

As the previous community development approaches of *Community-Based Development* (CBD) of the 1960s and *Integrated Rural Development* (IRD) of the 1990s had been introduced in a predominant 'top-down' mode in sectors such as agriculture, health and family planning, largely implemented and coordinated by external organisations, they have not only suffered from the ignorance of the potentially significant role of indigenous knowledge and institutions in the process of development, but also been confronted by limited community participation, eventually resulting in the decline of both community development approaches by the turn of the century (cf. Jones and Wiggle 1987; De Berge 2011).

Following the disillusions with these 'top-down' approaches, some international organisations and donor agencies active in financial economic development programmes led by the World Bank have recently turned to the community-driven approaches to development, such as *Community-Driven Development* (CDD) and *Community-Led Development* (CLD).

The CLD approach emerged from a new wave of collaborative, place-based action projects which began emerging in the early 2000s in New Zealand, Australia and Canada, partly as a response to the decline of communities as a result of government-led economic and social restructuring programmes (cf. MacLennan et al. 2015; Aimers and Walker 2016). After more than a decade, the CLD approach has emerged as one of the fastest growing investments by NGOs, aid organisations and multilateral development banks, mostly driven by the demand from donor agencies and developing countries for large-scale, bottom-up and demand-driven poverty reduction programmes (cf. Wong 2012). Recent estimates of the World Bank's funding for these projects vary between a rise from $325 million in 1996 to $2 billion in 2003, and from $2 billion in 1996 to $7 billion in 2003 (cf. Mansuri and Rao 2004). In general, however, the CLD approach suffers from its vulnerability to capital capture by local elites, limited participation and its failure to reach the extreme poor in developing countries (cf. Mansuri and Rao 2004).

Meanwhile, a few development organisations have extended the renewed community development movement, backed by the World Bank, to launch *Integrated Community Development* (ICD), such as the *Asian Productivity Organisation* (APO) and the *Pyxera Global Programme* (cf. Dhamotharan 2000; JIVA 2012).

However, as regards the general disillusion which is currently transpiring through a growing number of case studies of the effectiveness of microfinance for the reduction of poverty, Slikkerveer (2012) points to the fundamental delusion in the current financial economic approaches to poverty reduction by drawing attention to the growing distance they are creating between, on the one hand, microfinance as a multi-million dollar industry merely interested in investing shareholders' capital for profit in middle-class enterprises and business companies and, on the other hand, the poor and extremely poor who are largely excluded from benefitting from these kinds of services as the 'non-bankable' segment of the population. In his view, the failure to alleviate poverty stems from the basic incompatibility between the neoliberal ideology and the humanitarian solidarity movement, and could only be bridged by the transformation towards a new form of solidarity economy, based on integrated approaches to increase peoples' quality of life mainly through humanitarian not-for-profit policies. As is further elaborated in the following chapters, in this discussion the definition of the concept of poverty and the methods to measure and analyse poverty are crucial for assessing the effectiveness of current policies and contributing to the improvement of the position of the poor and underprivileged at the community level.

In Indonesia, the Government, in conjunction with the World Bank and the Islamic Development Bank Group Indonesia, recently implemented the *Integrated Community-Driven Development* (ICDD) approach in the PNPM Mandiri Programme, focused on a gradual shift from *community* empowerment to *economic* empowerment (cf. PNPM Mandiri 2016). Not surprisingly, however, the financial economic integrated development approaches are similarly faced with capital capture by elites, and failure to reach the extreme poor in the rural areas. Moreover, they encounter limited participation as they continue to ignore the significant role and potential of indigenous knowledge systems (IKSs) as a major component of the indigenous cosmovision as the base for local decision-making of the population in their efforts to realise effective poverty reduction and sustainable development.

Considering the growing need for more effective, participatory and human-centered approaches to reduce poverty and achieve sustainable community development, in which the interaction and integration of local and global systems of knowledge and technology are operationalised through indigenous institutions at the village level in order to respond to local problems and needs, LEAD further developed an innovative *Integrated Community-Managed Development* (ICMD) strategy on the basis of its experience with the newly designed *IKS Integration Model* (IKSIM) of holistic, 'bottom-up' approaches towards poverty reduction and sustainable community development.

Given the importance of the IKS-based IMM and ICMD strategies, which are implemented by current and future generations through the existing indigenous institutions, both approaches embark on the advanced education and training of special cadres of facilitators/managers, who are able to manage and coordinate these 'bottom-up' approaches at the community level. In this context, the newly designed and implemented *Master Course on Integrated Microfinance Management* (IMM) at the Faculty of Economics and Business of Universitas Padjadjaran (FEB/UNPAD) in Bandung, Indonesia, is also described as an innovative academic contribution to achieve poverty reduction and sustainable development through the integration of financial and non-financial services throughout Indonesia (cf. Slikkerveer 2012).

In addition, the ground-breaking strategy of *Integrated Community-Managed Development* (ICMD) is further elaborated as a complementing contribution to current 'human development approaches' of the United Nations towards poverty reduction programmes worldwide through the integration of local and global systems of knowledge and technology with a focus on the instrumental role of indigenous institutions for increased community participation, and to provide a community-based contribution to the realisation of the *2030 Agenda for Sustainable Development Goals* (SDG) of the United Nations (2015).

While a general consensus has emerged, for which the previous *Millennium Development Goals* (MDG) of the United Nations (2000) have already given a major impetus towards the reduction of poverty and hunger globally, progress has been severely uneven and unbalanced as too many people are still poor, extremely poor and undernourished, especially in the rural areas of developing countries, posing the greatest challenge to humanity today in poverty reduction and sustainable community development for the years to come. The situation of the poor has recently been aggravated by the rising economic inequality, widening the gap between the rich and the rest of the world, where since 2015 the richest 1% has owned more wealth than the rest of the planet (cf. Hardoon et al. 2016).

In Indonesia, where poverty rates of well above 10% of the population still continue to remain, unacceptably high for the South-East Asian Region, the experience of ICDD projects such as the *Kecamatan Development Project* (KPD), initiated in 2002 by the Government with the support of the World Bank and several national banks to alleviate rural poverty, has shown disappointing results in terms of local participation, procurement, elite capture, financial management and project sustainability. Unfortunately, since once again indigenous knowledge systems and institutions have largely been underutilised and ignored in this country-wide programme, the integrative ICMD approach would enable a more participatory, effective and sustainable outcome of the future implementation of such programmes for the rural population.

As described in this Volume, the new approaches such as IMM and ICMD, locally managed by newly trained facilitators/managers and operationalised within a neo-endogenous, i.e. a mixed exogenous/endogenous development framework, are indeed showing to be rather effective in poverty reduction for sustainable community development by the implementation of an integrated strategy of local and global knowledge systems through the indigenous institutions at the community level. Rather than the narrow financially based 'economic development approach' of the World Bank, the newly introduced *Integrated Community-Managed Development*

(ICDM) strategy seeks to contribute to the broader people-centered 'human development approach' of the United Nations, established in the 17 Sustainable Development Goals (SDGs) of the *Post-2015 Agenda for Sustainable Development* to be realised in 2030.

References

Agung, A. A. G. (2005). *Bali endangered paradise? Tri Hita Karana and the conservation of the Island's biocultural diversity*. Ph.D. Dissertation. Leiden Ethnosystems and Development Programme. LEAD Studies No 1. Leiden University. xxv + ill., pp. 463.

Aimers, J., & Walker, P. (2016). Can community development practice survive neoliberalism in Aotearoa New Zealand? *Oxford University Press and Community Development Journal*, 1–18.

De Berge, S. (2011). Subsistence to sustainability: An integrated approach to community development via education, diversification and accompaniment. *Changemakers*, 23 May 2011.

Dhamotharan, M. (2000). *Handbook on integrated community development—Seven D approach to community capacity development.* Tokyo: APO.

Hardoon, D., Fuentes-Nieva, R., & Ayele, S. (2016). *An Economy for the 1%: How privilege and power in the economy drive extreme inequality and how this can be stopped.* Briefing Paper, 18 January 2016. Oxfam International

JIVA. (2012). *Joint initiative for village advancement in India.* Washington DC: Pyxera Global.

Jones, J., & Wiggle, I. (1987). The concept and politics of 'Integrated Community Development'. *Community Development Journal*, *22*(2), 107–119.

MacLennan, B., Bijoux, D., & Courtney, M. (2015). Inspiring communities. March, 2015. http://www.inspiringcommunities.org.nz.

Mansuri, G., & Rao, V. (2004). Community-based and driven development: A critical view. *The World Bank Research Observer*, *19*(1).

Oxfam. (2017). *An economy for the 99%—It's time to build a human economy that benefits everyone, not just the privileged few*. Oxford, UK: Oxfam.

PNPM Mandiri. (2016). Indonesia Integrated Community-Driven Development (ICDD). Phase III Project. http://isdb-indonesia.org/.

Saefullah, K. (2019). *Gintingan in Subang: An Indigenous Institution for Sustainable Community-Based Development in the Sunda Region of West Java, Indonesia*. Ph.D. Dissertation. Leiden Ethnosystems and Development Programme. LEAD Studies No. 11. Leiden University.

Slikkerveer, L. J. (1999). Ethnoscience, Traditional Ecological Knowledge (TEK) and its application to conservation. In D. A. Posey (Ed.), *Cultural and spiritual values of biodiversity: A complementary contribution to the global biodiversity assessment*. Nairobi/London: UNEP/ITP.

Slikkerveer, L. J. (2012). *Handbook for lecturers and tutors of the new master course on integrated microfinance management for poverty reduction and sustainable development in indonesia* (IMM). Leiden/Bandung: LEAD-UL/FEB-UNPAD.

United Nations. (2015). *2030 agenda for sustainable development* (*SDG*). New York: United Nations.

Warren, D. M., Slikkerveer, L. J., & Brokensha, D. (1995). The cultural dimension of development. Indigenous knowledge systems. *IT Studies on Indigenous Knowledge and Development.* London: Intermediate Technology Publications Ltd.

Wong, S. (2012). *What have been the impacts of World Bank Community-Driven Development Programs?* CDD impact evaluation review and operational and research implications. Social Development Department. Sustainable Development Network, Washington, DC: The World Bank.

Chapter 1
Introduction

L. Jan Slikkerveer, George Baourakis and Kurniawan Saefullah

> *A large part of the difficulty in all of this is that a significant majority of the world's people are living below the poverty line, and if you don't have a security shelter, health and nutrition, the notion of environmental management is certainly off-center.*
> Leakey (2014)

1.1 The Global Poverty Crisis at the Beginning of the 21st Century

Despite major achievements in human development, the beginning of the 21st Century witnessed a dramatic situation among the poor and underprivileged peoples around the globe. The World Bank (2015) estimates that people living in extreme poverty, i.e. on less than $1.25 a day, were still numbering around 1 billion in 2011, representing 14.5% of the entire global population, establishing the global poverty crisis with the highest number in human history (cf. Food and Agricultural Organisation 2009). In spite of the contribution of the previous United Nations *Millennium Development Goals* (MDGs) (2000) to a certain reduction of poverty at their conclusion in 2015, the continuing overall poverty predicament has now

L. J. Slikkerveer (✉)
Leiden, The Netherlands
e-mail: l.j.slikkerveer@gmail.com

G. Baourakis
Chania, Greece
e-mail: baouraki@maich.gr

K. Saefullah
Bandung, Indonesia
e-mail: kurniawan.saefullah@gmail.com

L. J. Slikkerveer et al. (eds.), *Integrated Community-Managed Development*, Cooperative Management, https://doi.org/10.1007/978-3-030-05423-6_1

turned into a global development crisis, prompting international organisations such as the World Bank (WB) and the United Nations (UN) to target poverty reduction as the highest priority for the *2030 Agenda for Sustainable Development and the Sustainable Development Goals* (SDGs) of the United Nations (2015a). Poverty reduction, or poverty alleviation, generally refers to a combination of largely economic or humanitarian measures, which are intended to lift people permanently out of poverty.

While the World Bank (WB) and the International Monetary Fund (IMF) seek to attain poverty reduction by the implementation of their financial-economic 'World Development Approach' of mainly monetary activities by providing loans, the United Nations (UN) and its Agencies focus on realising poverty reduction through their 'Human Development Approach' of people-centered support, as transpiring through, respectively, the successive *World Development Reports* of the World Bank (1997–2016) and the *Human Development Reports* of the United Nations (1990–2015).

The recent experience of the financial crisis and its lasting impact on the poor, however, calls for the design and implementation of alternative, more effective policies of poverty reduction, which reach beyond the conventional but insufficient financially-based *economic development approach* of the World Bank, and encompass an extension of a broader people-centered *human development approach* as implemented by the *United Nations.*

Although the concept of 'poverty' has initially been described in strict monetary terms as a human condition where people lack money or income to meet their basic needs such as food, clothing and shelter, recent reorientation towards the conceptualisation, measurement and comparison of poverty has led to a more realistic definition so as to encompass a broader persistent condition of a general lack of access to proper services of health, education, justice, employment and freedom, causing poverty to become a wider, more dramatic complex problem in the society. Indeed, people who are categorised as belonging to poor families are suffering from increased vulnerability, running daily risks in their struggle for their livelihood and survival, which further tends to marginalise them from socio-economic life in their communities. Moreover, it is not only a lack of individual responsibility, but also of public conditions of social exclusion, injustice, abuse, over-exploitation and failing socio-economic policies for the poor, who over the past centuries have largely been marginalised and excluded by external forces from access to their own resources and wealth. A major concern has emerged that the poorest people worldwide suffer from limited access to health, education, communication and other social services, impairing their fragile condition in many communities by hunger, malnutrition and disease. In addition, the poorest tend to have little representation or voice in public debates, rendering it even harder for them to make adequate decisions and escape from poverty.

1.2 Poverty Reduction and Community Development Strategies

The overall picture of poverty remains critical, and projections made on the current economic GDP-based model even estimate that it would take 100 years to bring the world's poorest up to the minimum of the standard poverty line of $1.25 a day (cf. Hickel 2015). With the continuation of these disturbing numbers, it is not surprising that a growing number of organisations as well as experts and scientists involved in international development are considering that after the prolonged experience with overall disappointing community development programmes since the 1960s, alternative approaches are now needed to achieve a true reduction of poverty among the poor in an effort to attain genuine, sustainable development in the near future.

Basically, three strategies have been designed and implemented worldwide to reduce poverty to a minimum standard of living of the poor: a humanitarian, a financial-economic and a human development approach. Apart from social considerations of empathy and compassion with the poor who from a humanitarian perspective would deserve better, there is also an interest in reducing poverty from national governments, not only out of embarrassment, but also because of their economic concern as the increased costs of public services which would provide support to the poor tend to have a negative impact on the national economy. International organisations and national governments have become aware of the fact that a worldwide decrease in poverty would play a contributive role in achieving sustainable development throughout the planet, rendering poverty reduction a primary goal in the international development policies and strategies of the early 21st Century.

History shows that in the practical setting of the earliest approaches of development of local, often disadvantaged groups in communities which were adopted in Kenya and British East Africa during the 1930s, the concept of *community development* was introduced, and has over the years further been elaborated and spread into numerous programmes and projects with the aim of improving the position of disadvantaged and poor people in developing countries, largely through the mobilisation of local people's power to bring about social change. The conceptual framework of community development, recently defined by the United Nations (2014) as: "*a process where community members come together to take collective action and generate solutions to common problems*", is generally based on seven strategies, which had previously been developed by Blanchard (1988). These strategies include: comprehensive community participation, motivating local communities, expanding learning opportunities, improving local resource management, replicating human development, increasing communication and interchange, and localising financial access.

In the course of the 1990s, a promising aspect of the practice of community development emerged from the work of the *National Training Organisation for Community Learning and Development* (PAULO), responsible for setting professional training standards for education and development practitioners working within local communities in the UK. The newly-adopted term "*community learning and development*" underscored the fact that most occupations worked primarily within local communities, and that this important education and training component encompassed a concern for the wider socio-economic, environmental, cultural and political development of those communities (cf. McConnell 2002).

The introduction of the important principle of *Community Learning and Development* (CLD) has added an essential aspect to the concept of community development, known as *Community Capacity Building* (CCB). In 1996, the United Nations Development Program (UNDP) (1996) defined CCB as: "*a process and means through which national governments and local communities develop the necessary skills and expertise to manage their environment and natural resources in a sustainable manner within their daily activities.*"

The major principle underlying this rather successful approach of CCB is its new emphasis on the need to build social capital through experimentation and learning, and on developing the skills and performance of both individuals and institutions at the community level. This principle is encouraging the community development approach to focus on the training and education of participants in order to realise its objectives of sustainable development. Although the community development approach was incorporated into a 'new' strategy of *Integrated Rural Development* (IRD) for developing countries by the World Bank and several United Nations Agencies, it led to disappointing results, largely because of their continuing 'top-down' approach. In reaction, a growing number of scientists active in the field of international development research at the community level started to document and highlight the efficacy of the 'bottom-up' model in community development (cf. Rubin and Babbie 1993; Midgley 1993; David 1993; Slikkerveer 1999a, b). In this context, Macdonald (1995) points out that it is particularly important that the community members themselves perceive the existence of the problem and share their willingness to participate in the community development programme's process with a view to jointly finding practical solutions.

The experience in Indonesia shows that renewed efforts to apply *Community Capacity Building* (CCB) along the lines of the more successful participatory approach of training and education 'from the bottom' started to enable a more effective implementation of the principles of *Community Learning and Development* (CLD). In this way, several promising outcomes of pilot studies illustrate how local people have further developed their skills and capabilities to manage their own resources as part of their community life in a sustainable way. Hence, the community capacity building activities at the institutional level proved rather successful, as their aim was to strengthen institutions at the community level. As both the educational and societal needs assessments recently carried out in Indonesia indicate, there is a particular demand for advanced training and education in the area of community capacity building to be initiated at the higher education centres throughout the country (cf. Slikkerveer and Ambaretnani 2010a, b).

1.3 Current International Approaches to Reduce Global Poverty

Although over the past decades, several poverty reduction strategies have been designed and implemented, the results remain disappointing as today, about half the world—over three billion people—continue to survive on less than $2.50 a day, and at least 80% of humanity lives on less than $10 a day. A recent study of the World Bank (2016) shows that, according to the most recent estimates of the situation in 2012, not less than 12.7% of the world's population lived at or below $1.90 a day, encompassing nearly 1 billion (896 million) people living on the planet.

In line with the above-mentioned humanitarian strategy to reduce poverty to a minimum standard of living of the poor, largely implemented by non-profit foundations, organisations and NGOs in developing countries, spontaneous public campaigns have recently emerged as part of the *Global Call to Action Against Poverty* in several countries. *Make Poverty History* (2005), a coalition of aid and development agencies, is working towards raising awareness of global poverty and promoting radical policy changes by the government. Similarly, numerous development organisations and experts have launched a whole array of poverty reduction approaches, varying from development aid, welfare and debt relief to the generation of employment, capital investment, microcredit, microfinance and neo-liberalisation. International organisations such as the World Bank (WB) and the United Nations (UN) have recently put more effort into the study and analysis of the wider concept of poverty with a view to attaining reduction, focusing on its dimensions and indicators on a globally comparable scale. As it is now generally acknowledged that a reduction of poverty will contribute substantially to the overall progress of sustainable development on a global scale, these organisations have begun to address poverty as *the* most important factor which needs to be eradicated in order to realise the timely United Nations (2015a) *Sustainable Development Goals* (SDGs).

The World Bank as one of the leading international organisations in the predominantly financial-economic 'world development approach' to poverty reduction has recently broadened its thinking about the concept of poverty and its causal framework and structure, and widened its strategy of the 1990s to focus on labour-intensive growth, socio-economic sector investment, empowerment and security of the poor (cf. World Bank 2000). To this end, the World Bank has identified several socio-economic indicators of poverty to measure people's access to services of education and health, vulnerability and social exclusion, and also proposed two goals to measure the success of promoting sustainable economic development to monitor its own efficacy in delivering results. The *Global Monitoring Report* (World Bank 2014) describes the first goal as ending extreme poverty by reducing the share of people living on less than $1.25 per day to less than 3% of the global population by the new target year of 2030 [3].

The second goal was launched as promoting shared prosperity by improving the living standards of the bottom 40% of the population in every country. Although the Report indicates that global poverty has declined over the last few decades, as the number of people living on less than $1.25 per day (referred to as 'extreme poverty') has halved since 1990, around 1 billion people in 2011, representing 14.5% of the entire global population, were still living in extreme poverty. These disturbing numbers prompted the Senior Vice-President and Chief Economist of the World Bank, Kaushik Basu (World Bank 2014) to conclude: "*If it is shocking to have a poverty line as low as $1.25 per day, it is even more shocking that 1/7th of the world's population lives below this line. The levels of inequality and poverty that prevail in the world today are totally unacceptable*".

Meanwhile, the World Bank had recalculated the poverty line of $1.25 to $1.90 per day in order to introduce its new international poverty line. It followed the advice of the *International Comparison Program* (ICP), which provided the most recent estimates in 2012, amounting to 12.7% of the world's population living at or below $1.90 a day. However, in spite of the introduction of this new international poverty line of $1.90 a day in 2011, the overall level of global poverty has remained basically unchanged, largely because of the lower price levels in poor countries relative to those in the US, than those used for the 2005 purchasing power parity exchange rates (PPPs) [4]. Soon thereafter, the World Bank (2013) announced its two overarching goals: the end of chronic extreme poverty by 2030, and the promotion of shared prosperity, defined in terms of economic growth of the poorest segments of the society. In its recent assessment of the state of affairs, the *Global Monitoring Report* of the World Bank (2015) elaborates further on its twin goals of ending extreme poverty by 2030 and promoting shared prosperity, measured as income growth of the bottom 40%. Reaffirming the centrality of growth for development, the Report underscores that in turn, growth is more effective in reducing poverty and promoting shared prosperity, if it is inclusive and sustainable. Three key elements are considered to be of particular importance: greater investment in human capital, judicious use of safety nets, and steps to ensure the environmental sustainability of development. However, since it is clear that in terms of living standards, the bottom 40% in the developing world are still much worse off when it comes to access to education, health, and sanitation, the World Bank (2015: 1) duly concludes that: "*much more work needs to be done to end poverty and close the gap in living standards between those in the bottom 40% and the top 60% of the population around the world.*"

Similarly, the United Nations (2000) and its agencies, as a leading international organisation in the people-centered 'human development approach' to poverty reduction, had launched its eight *Millennium Development Goals* (MDGs) at the beginning of the 21st Century, including the objective of halving extreme poverty and providing universal primary education by the year 2015. Although the *Millennium Development Goals* (MDGs) have certainly reached a milestone in joint global and national development efforts, promoting a framework for international development priorities, not all of the eight goals—including poverty reduction—have, however, been completely achieved by the conclusion of 2015. Despite achieving remarkable gains, progress has been unbalanced and the reduction of poverty remains uneven within and across countries, leaving the number of people living in extreme poverty disappointingly high [2].

As a follow-up, a new set of *Sustainable Development Goals* (SDGs) has been committed by the United Nations (2015a) in its *Post-2015 Agenda for Sustainable Development*, as laid down in the 'Declaration of the Agenda': "*to build upon the achievements of the Millennium Development Goals and seek to address their unfinished business*". In accordance with its 'Human Development Approach', the United Nations declared the eradication of poverty by 2030 as its primary development goal. The United Nations Development Programme (UNDP) (2014), as one of the United Nations agencies engaged in the poverty reduction conundrum, contends that the advancement in poverty reduction is continuing, but it has to admit that more than 2.2 billion people are either near or living in 'multidimensional poverty', as categorised by the *Multidimensional Poverty Index* (MPI). This index measures deprivation in the three dimensions of the *Human Development Index* (HDI), i.e. health, education and living standards [5].

As the indigenous peoples make up about 5% of the world's population, they account for some 15% of the world's poor, with as many as one third of them living in extreme rural poverty. Embarking on the overarching concern that poverty eradication is the greatest global challenge facing the world today and an indispensable requirement for sustainable development, the United Nations (2015a, b) *Post-2015 Agenda for Sustainable Development* formulates not less than 17 *Sustainable Development Goals* (SDGs) for its Agenda for 2030, where Goal 1 "*End poverty in all its forms everywhere*" is sub-divided into 7 sub-goals. In general, these sub-goals are focused on poverty eradication, changing unsustainable and promoting sustainable patterns of consumption and production, and protecting and managing the natural resource base of economic and social development, all regarded as essential requirements for attaining sustainable development by 2030.

However, despite the reported achievements over the past one-and-a-half decades, that in most developing regions the extreme poverty rates have declined and that globally, the prime target of poverty reduction of the United Nations (2000) *Millennium Development Goals* (MDGs) and the *Twin Goals* of the World Bank (2014) on the proportion of people living in extreme poverty had been reached in 2015, the reality of the most recent estimates of 2012 show that not less than 12.7% of the world's population lived at or below $1.90 a day, encompassing nearly 1 billion (896 million) people living on the planet.

1.4 Microfinance: Disillusions of the Financial Development Approach

In the worldwide struggle against poverty, several humanitarian, financial-economic development approaches and strategies have been designed and implemented in the past in an effort to permanently lift people out of poverty. Back in the course of the 1880s, George's (1879) popular book on *Progress and Poverty:*

An Inquiry into the Cause of Industrial Depressions and of Increase of Want with Increase of Wealth: The Remedy initiated the progressive era of his ideology for a worldwide social reform, known as "Georgism". His main remedy was the introduction of a land value tax on the annual value of land held as private property which eventually would raise the level of wages to ensure that nobody would need to suffer from poverty.

Later on, efforts to reduce poverty were introduced within the overall framework of socio-economic growth, where several models such as economic liberalisation of extending property rights, regional autonomy and financial services to the poor have complemented development aid and government support for human and physical capital, including health, education and infrastructure. Until the mid-1990s, however, poverty reduction strategies remained part of socio-economic policies, and were left out of most development plans (cf. Pallier and Prévost 2007).

Following such capital-based efforts to increase employment, raise farm income and extend welfare to the poor, a new financial concept had emerged in Asia in the course of the 1970s with a view to assisting the poor, stemming from the pioneering work of Mohammad Yunus and the Grameen Bank which he had established in Bangladesh. Albeit that the concept of small-scale finance known as '*microcredit*' embarked on the narrow monetary definition of poverty, it heralded the '*first stage*' in providing finance to the poor (cf. Yunus 1999). The approach included mainly distributing small, largely subsidised loans with low interest rates to small groups of borrowers and collecting obligatory savings for the poor. The good intentions and initial success of the Grameen Bank of the 1970s and 1980s led to a growing support of the international development community, and microcredit rapidly developed into a popular poverty reduction tool which came to dominate many programmes and projects in developing countries. Soon thereafter, the development of microcredit soon started to attract the interest of many international donors, investors and NGOs to become an international financial-economic development approach to reduce poverty by the end of the 1980s.

In the 1990s, the success of the '*microcredit model*' originally designed for helping the poor led to the introduction of extended financial services, such as micro-insurance, micro-savings and micro-deposits, provided by a growing number of microfinance institutions (MFIs). This process of the '*second stage*' in the provision of finance to the poor adopted the term of '*microfinance*' covering a reference to a complex process of distributing small-scale finance with the objective to make the poor 'bankable'. The MFIs began to re-organise and manifest themselves to become NGOs, initially with a non-profit status, still depending on subsidies for low-interest loans from either national government agencies or international donor organisations. Supported by external funding, these MFIs were in the beginning successful to maintain these low interest rates for loans to their poor clients, making an effort to avoid the same label of the local moneylenders and loan sharks who are generally regarded as local institutions lending money at higher interest rates with an increased number of commercial unsubsidised loans and services.

However, the shift in the 1990s from *microcredit* to *microfinance*, supported by the World Bank (WB) and the International Monetary Fund (IMF), introduced a new era of extension of the commercialisation of various financial and banking institutions, often at the cost of their low-income clients. The process eventually even weakened the position of the poor, who soon started to refer to microfinance as the 'poverty trap'.

Notwithstanding, glamorous initiatives were launched by Hollywood stars, business tycoons, European and Middle-Eastern royalty and politicians committed to promoting and lobbying for microfinance, all pertaining to the designation by the United Nations of the year 2005 as the *International Year of Microcredit*, and the subsequent awarding of the *Nobel Peace Prize* in 2006 to Muhammad Yunus and the Grameen Bank.

As the result of growing disillusions stemming from the approach in microfinance, an international debate developed between the defenders of the *financial systems* approach of microfinance and those who support the *poverty lending* approach of microcredit. While the first approach advocates the neo-liberal notion, that the capacity of MFIs to achieve self-sustainability in terms of providing operational costs of staff, offices and institutes can only be generated from income from lending money to clients, the *poverty lending* approach, however, emphasises the importance of providing credit with subsidised, low interest rates to help overcome poverty as the poor cannot afford high interest rates (cf. Robinson 2001; Hermes and Lensink 2011; Slikkerveer 2012). With the increase of mixed feelings about the continuation of the commercialisation of microfinance, the rather critical interest among wider circles of academicians, experts and policy makers focused on the high-profile 'new wave' MFIs, as many foundations which were dedicated to 'helping the poor' in fact started to capitalise the MFIs in developing countries with investments for supposed poverty reduction by the improvement of the health and well-being of the poor. Such interest increased in particular among the general public after the worldwide exposure in 2007 of the mismanagement, greed and inefficiency of Mexico's largest microfinance bank *Compartamos*, prompting the question of the legitimacy and course of the "new wave" microfinance institutions around the globe (cf. IPO 2007).

Following the *Compartamos* drama, where it became public that microfinance managers and directors paid themselves excessive salaries and bonuses at the cost of their poor clients who were charged sky-high interest rates, even amounting to 100%, a number of independent studies started to appear assessing the impact of microfinance on the socio-economic situation of the poor in many developing nations.

Although the overall impact-oriented research in microfinance tends to show on average mixed results, the number of more outspoken studies can broadly be categorised as either providing evidence for the support of a positive impact of 'new wave' microfinance on poverty reduction and sustainable development, or revealing the failures and incapacity of microfinance to improve the situation of the poor, criticising MFIs as undermining and even obstructing sustainable development.

At the same time, some less-conclusive results continue to show the yet inadequate data and incomplete methods of comparison and analytical analysis of the impact studies of microfinance in different parts of the world.

In his well-documented publication *Why Doesn't Microfinance Work? The Destructive Rise of Local Neoliberalism*, Bateman (2010) presents an excellent analysis of the microfinance myths and realities, which shows that a growing number of studies reveal that the main public justification of microfinance as a means of helping the poor to generate income in order to escape from poverty is largely a misconception as microloans are commonly not used for production, but for the most part for consumption. Roodman (2011) extended his thorough analysis of microfinance in his subsequent study *Due Dilligence: An Impertinent Inquiry into Microfinance*, providing a critical assessment of the history and effects of microfinance as an institution, concentrating on microcredit, microinsurance and microsavings. He concluded that: "*The best estimate we have of the impact of microcredit on poverty among clients is zero.*"

As regards the usefulness of microfinance in economic development, Roodman (2011) warns, however, against too much external funding, particularly against the practice of MFIs using Western donor funding for specific loans, instead of building their local infrastructure and capacity as financial institutions.

Recent approaches trying to improve the deteriorating reputation of microfinance as practiced today by MFIs to 'help the poor' include the introduction of new concepts such as 'social business' (SB), 'inclusive finance' (IF) and 'corporate social responsibility' (CSR), which, however, in the practice of poverty reduction largely remain limited to a form of PR which seeks to create a favourable image among the general public. The concept of 'Social Business' (SB) was recently introduced by Yunus (2007) to show up the poverty reduction aspect of microfinance, while the *United Nations Capital Development Fund* (UNCDF) (2013) introduced the concept of 'Inclusive Finance' (IF) to further promote the provision of extended microfinance to the poor by savings, deposits, credit, insurance, and payment services. In addition, 'Corporate Social Responsibility' (CSR) is one of the latest management strategies of MFIs trying to suggest a positive impact on the society while doing business. As a variant of 'social responsibility' (SR) which offers an ethical framework to organisations to act for the benefit of society at large, 'Corporate Social Responsibility' (CSR) has been criticised as nothing more than superficial window-dressing, or 'green-washing' (cf. Alejos Góngora 2013).

Although these predominantly financial-economic approaches to poverty reduction supported by the World Bank (WB) are meant to contribute to economic growth, perceived of as one of the major sources to attain sustainable development, the growing disillusions with the current microfinance model in its various forms to provide a useful tool for poverty reduction call for the exploration of alternative, community-managed approaches based on the integration of indigenous knowledge, beliefs and practices embodied in institutions as the cultural heritage of the local people themselves, including the poor and marginalised families as the prime participants in such an endeavour at the community level.

1.5 IKS: The Input of the Cultural Dimension to the 'Human Development Approach'

The recent interest and orientation towards indigenous peoples and their knowledge systems in many sectors of the community has resulted in a number of studies which provide a growing body of information of indigenous knowledge, beliefs and practices within the context of development. In their publication *Indigenous Knowledge Systems: The Cultural Dimension of Development*, Warren et al. (1995) further extended the significance of applied ethnoscience into a broader, multidisciplinary field of study in relation with the processes of development and change in various sectors, also known as the 'new' ethnoscience or 'neo-ethnoscience', which transpired through their more dynamic definition of an indigenous knowledge system as: "*a system of knowledge, beliefs and practices which has evolved in a particular area or region, often outside universities and laboratories, and transferred over many generations, and as such forming the base for local-level decision-making processes and which has been recorded in several sectors of the society, such as in animal and human health, environment, natural resources management, agriculture, fisheries, forestry, bio-cultural diversity conservation, and local economies.*"

As the subject matter of the new field of *applied ethnoscience*, indigenous knowledge soon became an interesting new territory for the applied sciences within the framework of socio-economic development. Because of the dynamic setting of development, in which the study of 'Indigenous Knowledge Systems' (IKS) emerged, they quickly provided a promising point of embarkation for thriving socio-economic development which focused on community-level structures as the stepping stone for 'bottom-up' policies (cf. Richards et al. 1983; Richards 1985; Warren et al. 1995; Slikkerveer and Slikkerveer 1995; Slikkerveer 1999b; Gibson et al. 2005; Easterly 2006). In a comparative study between the implementation of a 'top-down' and a 'bottom-up' community development programme in Mexico, Larrison (1999) also uncovered a significant fact where a large portion of the success of the 'bottom-up' programme was related to its provision of local services which were perceived as contributing to the local economy.

Soon, the promising approach of *Indigenous Knowledge Systems and Development* (IKS&D) started to highlight the potential role of local peoples' knowledge, beliefs and practices, guiding their own decisions in the socio-economic development process at the community level, prompting the paradigm shift from 'top-down' to 'bottom-up' strategies in international development cooperation towards developing countries of the 1990s. Previously regarded as superstitious, backward and remote, indigenous knowledge systems have now shown to be risk-minimising, adaptable to change, and as such sustainable. Moreover, they heralded a renewed feeling of pride for the local culture among the participants.

During the first Summit of the *World Culture Forum*, held by the Government of Indonesia under the patronage of the United Nations Educational, Scientific and

Cultural Organisation (UNESCO) in Bali in November 2013, the importance of culture for development is emphasised, particularly in the United Nations *Post-2015 Agenda of Sustainable Development.* The Summit featured the following six significant themes: Holistic Approaches to Culture in Development; Civil Society and Cultural Democracy; Creativity and Cultural Economics; Culture in Environmental Sustainability; Sustainable Urban Development; and Inter-Faith Dialogue and Community Building. The *World Culture Forum* (2013) concluded with the adoption of the Bali Promise, which calls for a measurable and effective role, as well as the integration of culture in development at all levels in the Post-2015 Development Agenda of the United Nations.

As the paradigm shift initiated by 'neo-ethnoscientists', who evoked not only a growing interest in indigenous knowledge systems as the prime expression of the cultural dimension within the context of development and change, but also provided the basis for the implementation of related community-level decision-making processes, follow-up research has later been conducted to include the closely related indigenous institutions, and the role these could play in 'bottom-up' sustainable development policies and programmes. While institutions are generally referred to as regularised practices or patterns of behaviour structured by rules and norms of the society which are widely used, they are either formal or informal. Formal institutions are rules, laws and constitutions legalised by members of the society, while informal institutions include social norms of behaviour, and conventions which prohibit or permit individuals to undertake certain activities within their social settings (cf. Metha et al. 1999; Hembram 2007; Slikkerveer 2012).

Such a definition had further been elaborated by Blunt and Warren (1996: x) in their timely publication on *Indigenous Organisations and Development* which links up with the growing interest in indigenous knowledge, extending the development literature: "*It is expected to complement the recently published book entitled 'The Cultural Dimension of Development: Indigenous Knowledge Systems'*" (cf. Warren et al. 1995). As Watson (2003: 288) contends, the convergence of this institutional approach with the interest in indigenous knowledge and practices generated new expectations among development agencies working in natural resources management: "*if the indigenous NRM institutions can be identified and harnessed, it is thought that this strengthened NRM will be achieved, that it will have the 'added value' of drawing on local expertise, and that the projects will be participatory*".

The interest in the role of these indigenous institutions and organisations in development had initially emerged from works of scholars concerned about the rather limited success of international development programmes in imposing their modern, western-based norms, rules and practices on local systems of knowledge and use of resources, such as De Boef et al. (1993), Uphoff (1996), and Bruce and Migot-Adholla (1994). Metha et al. (1999) locate these indigenous institutions as central to the concept of access, particularly access to natural resources, where efforts made in strengthening local resource management institutions would improve environmental management and social welfare.

By consequence, the use of the term *indigenous institution* in this Volume refers specifically to those local-level institutions—informal and sometimes invisible to the outsider—rooted in the history of the community, which embody the local systems of knowledge, beliefs, practices, values and norms, and are based on strong communal principles of mutual aid, neighbourhood cooperation and collective action, where the interests, resources and capacities of many community members are structurally joined together in order to achieve common goods and services for the entire community in a non-commercial way. Examples of such *indigenous institutions* include local kinship groups, clans, farmers' associations, traditional legal bodies, village elders' councils, communal mutual aid and reciprocity alliances, traditional cooperative associations, collective action groups, neighbourhood groups, community water management groups and women's groups.

1.6 The LEAD Perspective of Integrated Community-Managed Development (ICMD)

The *Leiden Ethnosystems and Development Programme* (LEAD) of Leiden University in The Netherlands, established in 1987 by Prof. Jan Slikkerveer and his team to further develop the academic field of *neo-ethnoscience* with a focus on the dynamic role of *Indigenous Knowledge Systems* (IKS) in development, was invited to co-establish the *Global Network for Indigenous Knowledge and Development* in collaboration with Prof. Mike Warren (CIKARD, Iowa State University, USA), Prof. Richard Leakey (National Museums of Kenya, Nairobi) and Prof. Kusnaka Adimihardja (Universitas Padjadjaran, Bandung, Indonesia).

In 1990, LEAD and CIKARD joined with CIRAN/NUFFIC to further extend the global network and publish the *IKS Monitor*, concentrated on the theory and practice of Indigenous Knowledge Systems and Development in various sectors, covering about 35 IK-Centers worldwide. Following the growing number of specialisations in applied ethnoscience in the decades thereafter, LEAD initially focused on the sub-discipline of *ethno-medicine* in education and research, and promoted the further integration of traditional medicine into primary health care development programmes in East Africa (NMK) and South-East Asia (UNPAD). Similarly, research and post-graduate training in ethno-medicine soon evolved in the Mediterranean Region, particularly in Crete, Greece in cooperation with the University of Crete.

As a result of the growing interest worldwide for *ethno-botany,* LEAD soon became involved in the training and research of *Ethnobotanical Knowledge Systems* (EKS) with the National Herbarium of The Netherlands at Leiden University, after which it soon entered into collaboration with the Mediterranean Agronomic Institute of Chania (CIHEAM-MAICh) in the Medusa Network, followed by the joint LEAD/MAICH/UNPAD/NMK Project on Indigenous Agricultural Knowledge Systems (INDAKS) in East Africa and Southeast Asia, supported by the European Commission in Brussels. Because of its pioneering work to further substantiate the role of IKS as the cultural dimension of development, LEAD was recognised in

1994 by UNESCO in Paris as an International Programme of the prestigious *World Decade of Culture and Development 1994–2004*.

The subsequent relocation of the LEAD Programme in 1999 to the Faculty of Science as the appropriate academic environment for the development of *ethnoscience* as the complementary discipline of *science,* the multidisciplinary LEAD Programme further evolved and specialised into several challenging sub-disciplines as a follow-up to *ethno-medicine*, including *ethno-botany/pharmacy, ethno-communication/history* and *ethno-economics,* and, more recently, *ethno-mathematics.*

Embarking on the research-based experience of the LEAD Programme in these various sub-disciplines of *applied ethnoscience* in its three geographical 'target regions' of South-East Asia, East Africa and the Mediterranean Region, a specific research methodology has been developed to document, analyse and explain indigenous knowledge systems within the dynamic context of development, known as the '*Leiden Ethnosystems Approach*' forming the base for a regional comparative research methodology, both qualitative and quantitative. The specific IK-oriented research methodology encompasses three principles: the '*Historical Dimension*' (HD), the '*Field of Ethnographic Studies*' (FES) and the '*Participant's View*' (PV). The related bivariate, multivariate and multiple regression analyses of quantitative research data of indigenous knowledge systems in relation with patterns of human behaviour have pertained to the design and development of an applied-oriented *IKS integration model* (IKSIM) to provide a comprehensive, evidence-based approach towards poverty reduction through indigenous institutions, in which local and global knowledge systems are integrated to achieve reduction of poverty and sustainable development at the community level. The integrated strategies to sustainable development in several sectors at the community level, including health care, education, bio-cultural diversity conservation, communication and management, have been encapsulated in the concept of 'sustainable community development'. As Toledo (2001) contends, this concept allows for a broader and more integrative approach to define 'sustainable community development' as: "*an endogenous mechanism that permits a community to take (or retake) control of the processes that affect it*" (cf. Toledo 1997).

The United Nations Development Programme (UNDP 1991) had defined community capacity building—also known as capacity development—as: "*…a long-term, continuing process, in which all stakeholders participate (ministries, local authorities, non-governmental organisations and water user groups, professional associations, academics and others).*" As a response to the growing demand for support for more effective strategies at national and sub-national levels, the United Nations Development Programme (UNDP 2008) later developed a *Strategic Plan 2008–2011* with an extended conceptual framework and a methodological approach, in which capacity development is conceptualized as a process through which individuals, organisations and societies obtain, strengthen and maintain their capabilities in order to achieve their own development objectives over time. This capacity development process consists of five steps which are embedded into a particular policy advisory analysis and programming process: (1) engage stakeholders on capacity development; (2) assess capacity assets and needs; (3) formulate a capacity development response; (4) implement the response; and (5) evaluate capacity development.

While, at the individual level, community capacity-building requires the development of conditions which allow individual participants to build and enhance knowledge and skills, at the institutional level, it should not involve creating new institutions, but rather functionalise existing institutions, supporting them in forming sound policies, organisational structures, and effective methods of management and revenue control. Community capacity building at the societal level should support the establishment of a more interactive public administration and be used to develop public administrators that are responsive and accountable.

An important aspect of the potential of indigenous knowledge systems for development is formed by the rich, locally available but often yet untapped resources of indigenous knowledge, practices and experience, which provide new opportunities for sustainable community development through capacity building activities at the community level. As such, *indigenous capacity* refers to the collective ability of community members to meet the needs of the community by drawing on local capital. *Capacity development*, as defined by the United Nations Office for Disaster Risk Reduction (UNISDR) (2016), broadens the term *capacity building* to encompass all aspects of creating and sustaining capacity growth over time: "*It involves learning and various types of training, but also continuous efforts to develop institutions, political awareness, financial resources, technology systems, and the wider social and cultural enabling environment. It encompasses a process by which people, organisations and society stimulate and develop their capability over time in a systematic way in order to achieve social and economic goals.*"

Rethinking its IKS-based perspective on development, LEAD, however, has recently taken one further step in its approach to attain sustainable community development by designing the *IKS-based Integration Model* (IKSIM), which explicitly embarks on the integration and interaction of local and global systems of knowledge, practices and institutions as a stepping stone for the newly-developing concept of *Integrated Community-Managed Development* (ICMD). In this approach, the indigenous institutions are the implementing bodies which are strategised at the intermediate level to attain integrated sustainable community development in the various sectors of the community. Following several IKS-oriented studies implementing various adaptations of the IKSIM model in respectively ethno-medicine by Ambaretnani (2012) and Chirangi (2013), ethno-botany by Ibui (2007) and Aiglsperger (2014), bio-cultural diversity conservation by Agung (2005) and Leurs (2010), ethno-communication by Djen Amar (2010) and Erwina (2019), microcredit management by Gheneti (2007), and ethno-economics by Saefullah (2019), a new IKS-based approach has particularly been developed in *ethno-economics* to seek contributions to worldwide efforts to alleviate persistent poverty in the rural communities in the three target areas of the LEAD Programme (cf. Slikkerveer 2009).

Indeed, the related perspective of the LEAD Programme on poverty reduction substantiates its IKS-based orientation within the wider rural development framework with a focus on the role of indigenous institutions, where it links up with the neo-endogenous approach in the current debate on exogenous, endogenous and neo-endogenous strategies in rural development. As described in the previous Paragraph, the neo-endogenous perspective provides a useful level of integration between two types of knowledge systems which tend to play an important role in this integration process between the so-called external, *expert knowledge*, largely contributed by experts and representatives of the institutions, and *indigenous knowledge*, contributed by members of the local community and based on generations-long experience and tradition.

In Indonesia, for example, the operationalisation of the above-mentioned IKSIM model within a neo-endogenous framework at the community level has led to the introduction of two novel approaches respectively of *Integrated Microfinance Management* (IMM) and *Integrated Community-Managed Development* (ICMD) which are both implemented by the existing indigenous institutions. While the IMM approach focuses primarily on the integration of financial and non-financial services to attain poverty reduction and empowerment as a means to realise sustainable development, the ISCD approach focuses on the integration of local and global knowledge systems and institutions to reach sustainable development as an overall objective to reduce poverty and promote equality. In the *Integrated Microfinance Management* (IMM) approach, the indigenous institutions are managed by newly-trained *Integrated Microfinance Managers* to provide and coordinate a comprehensive set of community-based services, encompassing not only inclusive finance, but also health, education, communication and socio-cultural services (cf. Slikkerveer 2012).

Likewise, in the approach of *Integrated Community-Managed Development* (ICMD), the indigenous institutions are managed by newly-trained *Integrated Community Development Managers* to implement and coordinate the interaction between local and global systems of knowledge and technology in several sectors through capacity-building activities at the community level (cf. Dhamotharan 2000; Pawar and Torres 2011; Slikkerveer 2015). While the IMM approach has been developed by the LEAD Programme of Leiden University in collaboration with Universitas Padjadjaran and the Mediterranean Agronomic Institute of Chania since the late 2000s, the ICD approach has previously been introduced in the field of Family Planning, Area Development, Community Investment, and Agricultural and Health Sector Development, albeit that such approaches so far have been implemented through outside interventions, often missing an overall holistic strategy to integrate local and global knowledge in several sectors through indigenous institutions (cf. Jones and Wiggle 1987; de Berge 2011; PROGRESS 2013).

Most of the long-standing indigenous institutions are generally embodying local peoples' knowledge, beliefs, practices, norms and values aimed at communal, non-commercial objectives, and based on community principles of cooperation, solidarity, reciprocity and mutual aid with a moral dimension without personal material benefits. Although these principles are also reflected in the definition of modern cooperatives as noticed by the *International Cooperative Alliance* (2006), stating that: "*Co-operatives are based on the values of self-help, self-responsibility, democracy, equality, equity and solidarity*", these modern cooperatives recently introduced in rural communities from outside depend predominantly on formal membership and focus mainly on profit-making with material benefits for the members.

In their largely commercial strategies, many modern cooperatives tend to fail to achieve the social goals of democracy, equality, equity and solidarity, often embodied in their principles. Such a trend is particularly manifest in cases where newly-established cooperatives among indigenous communities are involved in international alliances in order to enter global markets with a view to opening up new economic opportunities, but where hardly any form of social development or cooperativism has been realised (cf. Burke 2006).

While criticism has increased concerning the fact that most of these modern cooperatives established in developing countries fail to contribute to sustainable development, and in some cases even create and institutionalise new inequalities and dependencies, the contribution from indigenous institutions to communal cooperative management strategies is evidently demonstrated in the sustainable methods and practices of indigenous management systems of natural resources at the community level, as well as in the supporting age-long philosophies of nature and the environment (cf. Slikkerveer 1999a, b; Agung 2005; Saefullah 2019). The perspective of indigenous institutions contributing to communal cooperative management strategies of resources at the community level opens up a new direction for *integrated community-managed development* with a view to attaining the reduction of poverty and the related achievement of sustainable development through the integration of their local functions of providing not only financial but also non-financial, i.e. medical, educational, communicational, social and cultural services to all community members, particularly the poor and low-income families.

1.7 Indonesia: The Mission of IMM and ICMD to Realise Poverty Reduction

Indonesia, with its population of 257.6 million inhabitants in 2015, a GDP growth rate of 5.1% and an inflation rate of 6.4%, forms the fourth largest country in the world, a newly-developing nation in South-East Asia which has maintained certain political stability over the past decades. The country boasts the largest economy in the *Association of Southeast Asian Nations* (ASEAN), and is the 16th largest worldwide. The complex development process of Indonesia has witnessed a rapidly evolving middle-income country, where the emphasis on agricultural development of the 1970s has shifted to energy, urban infrastructure and education in the 1980s, followed by the focus on the financial sector reform and resilience since the late 1990s.

In 1997, a World Bank-funded programme of community-driven development (CDD) was launched in collaboration with *Bappenas* (Ministry of Planning) with the aim to reduce poverty in Indonesia through community empowerment in the rural areas. As Wong (2003) documents, the *Kecamatan Development Program* (KDP) (1998) became one of the largest World Bank-financed CDD projects, with a budget of some $700 million for its first two phases (1998–2006) based on a revolving-loan scheme of block grants to members of local communities, who became responsible for managing these funds. The *Kecamatan Development Program* (KDP) was able to transfer funds directly, not to the chiefs but to the villagers, enabling them to monitor and control the use of the resources. Although the financial crisis of 1998 forced the devaluation of the country's currency, the rupiah, Bappenas was able to expand the programme to reach tens of thousands of villages over the archipelago's 17,000 islands, but became confronted by new challenges including the creation of new 'elites' at the community level including widespread corruption.

A decade later, the Government of Indonesia launched its *Program Nasional Pemberdayaan Masyarakat Mandiri* (PNPM-Mandiri 2007) ("National Community Empowerment in Rural Areas") which includes a national poverty reduction strategy with special attention for community empowerment. PNPM-Mandiri comprises two major programmes: PNPM-Rural and PNPM-Urban, both of which are based on community-driven development. PNPM Mandiri reflected the government's formal acceptance of community-driven development as its national strategy for poverty reduction. Following the Indonesian national elections of 2014, the key stakeholders under the PNPM-Mandiri programme formulated the PNPM Road Map as the policy framework for sustaining community empowerment and related poverty reduction in Indonesia. The supporters of community empowerment have strongly promoted the new Indonesian Law No. 6/2014 (2014) known as the 'Village Law' (*Undang-Undang Tentang Desa*) as a means of institutionalising community-driven development as a national policy. The 'Village Law' on regional government was introduced, stating that: "*the village is the unity of the legal community who have boundaries that are authorised to regulate and manage the interests of the local community, based on their origin and local customs, which are recognised and respected in the unitary system of government of the Republic of Indonesia.*" (cf. Indonesian Law No. 6/2014).

Although at the deadline in 2015 of the finalisation of the *Millennium Development Goals* of the United Nations, Indonesia had shown a largely on-track realisation of many of the MDGs, progress towards the MDGs has been uneven. In particular, Poverty Reduction (MDGs 1) remains a great challenge in Indonesia despite the impressive economic growth in recent decades. According to the United Nations Central Statistics Agency (UNCSA) (2015), almost 30 million people, or 12.36%, in Indonesia live below the national poverty line. In addition, certain regions of Indonesia are also poorer than others, with poverty rates in regions like Papua and West Papua even more than double that of the national average rates.

The same year of 2015 has not only marked the first year of the actual implementation of the 'Village Law', but also the beginning of the transition of community-driven development in Indonesia from a development programme to a legal institution. One of the major changes in this transition process refers to the annual block-grant funding to villages under the PNPM-Rural programme which will be replaced by the transfer of funds as mandated by the 'Village Law'. This new strategy certainly contributes to the increase of financial resources at the community level, especially to the creation of new employment opportunities and equal distribution of resources, prerequisites for poverty reduction and empowerment.

On the other hand, however, among the main challenges to the successful implementation of the 'Village Law' is the creation of new opportunities for increased corruption, collusion, and nepotism, obstructing the aim of poverty reduction among the non-elitist poor in the communities. In addition, sustainable community development has been based on principles of participation, empowerment, accountability and capacity building transferred over many generations through the pre-existing indigenous institutions, which have been marginalised in the past, and which seem to be ignored or replaced by government-related institutions from outside.

The overall emphasis is on the transfer of large funds (*dana desa*) with a view to reducing poverty, but the understanding of the causes of rural poverty is lacking, rendering the increased funding for villages highly controversial. Since poverty itself has been shown to encompass many non-financial aspects, measures to reduce poverty should indeed go beyond monetary solutions to integrated similar non-financial approaches. In most of the Indonesian rural areas, the communities have been managed for many generations by indigenous institutions, such as *Gotong Royong, Metulung, Kekeluargaan*, and *Rukun*, epitomising local knowledge, beliefs, values, norms and principles which form the foundation of the customary community of Indonesia, or *Masyarakat Adat*.

The various case studies in Indonesia presented in this Volume document the pivotal role of these indigenous institutions encompassing the underlying principles of mutual aid, reciprocity, solidarity and traditional cooperation for the process of sustainable development. Given the present interaction—and confrontation—between global and local systems of knowledge and technology, an interesting progress is described in the implementation and success of two IKS-based approaches of *Integrated Microfinance Management* (IMM) and *Integrated Community-Managed Development* (ICMD), which seek to make a contribution by strategising indigenous institutions for poverty reduction and sustainable development among the rural communities in different settings of the country.

While the *Integrated Microfinance Management* (IMM) approach is largely focused on poverty reduction through the provision of a comprehensive set of integrated community-based services, encompassing not only inclusive finance, but also health, education, communication and socio-cultural services, the approach of *Integrated Community-Managed Development* (ICMD) is dedicated to attaining sustainable development through the integration of local and global systems of knowledge and technology among all sectors of the community through capacity building and participation at the village level. Both IKS-oriented approaches are rather innovative as they embark on strategising the often disregarded indigenous institutions in order to strengthen their shared objectives of community participation, capacity building and local governance (cf. Slikkerveer 2016).

The IMM approach has been developed by the LEAD Programme of Leiden University since the late 2000s on the basis of the role of indigenous knowledge systems and institutions in development with a view to reducing poverty and empowering local communities. As the ICD approach has previously been introduced in a predominant top-down mode in sectors such as family planning, area development, community investment, and agricultural and health sector development, implemented and coordinated by external institutions, they have not only often been confronted by limited community participation, but also suffered from the ignorance of the key role of indigenous institutions in development (cf. Jones and Wiggle 1987; de Berge 2011; PROGRESS 2013).

More recently, a few development organisations have revitalised the community development movement through the adoption of integrated approaches, such as the *Asian Productivity Organisation* (APO), the *Australian Commonwealth Department of Families and Community Services* (DFACS) and the *Pyxera Global Programme* (cf. Dhamotharan 2000; Pawar and Torres 2011; JIVA 2012). In these programmes, however, neither the integration of knowledge systems nor the input of the indigenous institutions in development have been functionalised to achieve sustainable development.

Given the importance of the IMM and ICMD strategies, which have to be implemented by the current and future generations through the existing indigenous institutions, both approaches also need the appropriate education and training of facilitators/managers. In this context, the newly-designed and implemented Master Course on *Integrated Microfinance Management* IMM at the Faculty of Economics and Business of Universitas Padjadjaran (FEB/UNPAD) in Bandung, Indonesia will also be described as an academic contribution to achieve poverty reduction and empowerment through the integration of the provision of financial and non-financial services throughout the country. In addition, the innovative strategy of *Integrated Community–Managed Development* (ICMD) will be further elaborated as a similar contribution to poverty reduction through the integration of local and global systems of knowledge and technology with a focus on the role of indigenous institutions in the capacity-building process at the community level.

The experience so far underscores that integrated approaches within a neo-endogenous framework can indeed be successful by implementing a strategising method of community-based development of financial and non-financial services and capacity building, based on local people's indigenous knowledge systems, embodied and retained in their indigenous institutions.

As 17 October 2015 marked the first commemoration of the *International Day for the Eradication of Poverty* (IDEP) since the launching of the United Nations

Fig. 1.1 Logo of the United Nations' International Day for the eradication of poverty. *Source* United Nations 2016

(2015) *Post-2015 Agenda for Sustainable Development Goals*, particularly its Number One goal to: "*End poverty in all its forms everywhere*", the timely consideration of the United Nations (2016) puts the record straight: "*Recent estimates show that despite significant gains since 2002 - the number of people living below the poverty line dropped by half - 1 in 8 people still live in extreme poverty, including 800 million people who do not have enough to eat*" (cf. Fig. 1.1). In line with the United Nations commitment to fight against poverty, this Volume seeks to respond from the academic i.e. education and training perspective to IDEP's opportunity not only to acknowledge and support the efforts and struggles of people living in poverty worldwide, but also to promote the participation of those living in poverty by means of the integration of their indigenous knowledge and institutions through the strategy of *Integrated Community-Managed Development* (ICMD) for poverty reduction and sustainable community development in the years to come.

1.8 Structure of the Book: The Sequence of Parts and Chapters

This Volume is made up of five parts, which seek to reflect recent approaches and principles in international development cooperation which are pertaining to alternative strategies of integrated sustainable development with a focus on poverty reduction as the prime global objective of international and national organisations around the world.

The discourse begins with a section on the recently changed development paradigm from 'development from the top' to 'development from the bottom'. Starting with a *Prologue* and Introduction (*Editors*), the book is divided into five parts. Part I 'Ethnoscience and the Paradigm Shift in Global Development Cooperation' encompasses the following Chapters: (2) 'The Indigenous Knowledge Systems' Perspective on Sustainable Development' (*L. Jan Slikkerveer*); (3) 'The Failure of Financial-Economic Policies to Reduce Global Poverty' (*Kurniawan*

Saefullah); and (4) 'Towards a Model of Integrated Community-Managed Development' (*L. Jan Slikkerveer*). The next section focuses on the dynamic interaction between global and local processes: Part II 'Global Versus Local Economic Development Processes', encompassing the following Chapters: (5) 'Macro- and, Microeconomics and Social Marketing' (*Tati S. Joesron & Alkinoos Nikolaidis* [†]); (6) 'Microfinancial Sector Assessment and Development' (*Dimitrios Niklis, George Baourakis & Constantin Zopounidis*); and (7) 'Integrated Poverty Impact Analysis (IPIA): A New Methodology for IKS-Based Integration Models' (*L. Jan Slikkerveer*).

The following sections describe the current situation in Indonesia with regard to the role of various agencies, organisations and institutions involved in poverty reduction and sustainable development, highlighted by several examples from the field: Part III 'Indonesia: Transitional Development Organisations and Institutions', encompassing the following Chapters: (8) 'Governance, Policies and Regulations in Indonesia' (*J. C. M. [Hans] de Bekker & Kurniawan Saefullah*); (9) '*Bank Rakyat Indonesia*: The First Village Bank System in Indonesia' (*Kurniawan Saefullah and Asep Mulyana*); (10) '*Bina Swadaya*: A Community-Based Organisation' (*Bambang Ismawan*); (11) 'Recent Government Policies of Poverty Reduction: KDP, UPP and PNPM' (*L. Jan Slikkerveer & Kurniawan Saefullah*); (12) 'Satu Desa Satu Milyar': 'Village Law' No. 6/2014 as a Rural Financial Development Programme' (*Benito Lopulalan*); and (13) '*Kopontren* and *Baitul Maal wat Tamwiil*: Islamic Cooperative Institutions in Indonesia' (*Kurniawan Saefullah & Nury Effendi*). Part IV 'Indonesia: Indigenous Institutions for Integrated Community-Managed Development' encompasses the following Chapters: (14) '*Gotong Royong*: An Indigenous Institution of Communality and Mutual Assistance in Indonesia' (*L. Jan Slikkerveer*); (15) '*Gintingan*: An Indigenous Socio-Cultural Institution in Subang' (*Kurniawan Saefullah*); (16) '*Grebeg Air Gajah Wong*: An Integrated Community River Management in Java' (*Martha Tilaar*); and (17) '*Warga Peduli AIDS*: The IMM Approach to HIV/AIDS-Related Poverty Alleviation in Bandung, West Java' (*Prihatini Ambaretnani & Adiatma Y. Siregar*). The concluding section, Part V 'The New Paradigm of Integrated Community-Managed Development', includes the following Chapters: (18) 'Strategic Management Development in Indonesia' (*L. Jan Slikkerveer*); (19) 'Advanced Curriculum Development for Integrated Microfinance Management' (*Nury Effendi, Asep Mulyana & Kurniawan Saefullah*). The Volume ends with an *Epilogue*.

This section elaborates on the operationalisation of the new paradigm of integrated approaches towards achieving poverty reduction and sustainable development through the advanced education and training in *Integrated Community-Managed Development* (ICMD) of new cadres of Integrated Community Managers (ICMs), capable of revitalising, functionalising and managing indigenous institutions in the capacity-building processes for poverty reduction and sustainable community development.

Following the introduction of integrated strategic management development in indigenous institutions for achieving sustainable development at the community level, the implementation of the integration of interacting local and global knowledge systems is illustrated in recent initiatives in post-graduate training in Indonesia.

Following the successful development and implementation of the Master Course on *Integrated Microfinance Management* (IMM) since 2012 by the LEAD Programme of Leiden University in The Netherlands in collaboration with the Faculty of Economics and Business of Universitas Padjadjaran (FEB/UNPAD) in Bandung, Indonesia and the Mediterranean Agronomnic Institute of Chania (CIHEAM-MAICh) in Crete, Greece, in which the integration of the provision of financial and non-financial services is implemented by *Integrated Microfinance Managers* (IMMs), the newly-developing post-graduate curriculum of *Integrated Community-Managed Development* (ICMD) is focused on the education and training of *Integrated Community Development Managers* (ICMDs), well-equipped to implement the integration of global and local knowledge systems in order to realise poverty reduction and sustainable development in Indonesia and beyond.

Notes

(1) The International Monetary Fund (IMF) and the World Bank (WB) are international institutions in the United Nations system, sharing the goal of raising living standards in their member countries. Their largely financial approaches to reach this goal are different but complementary. The World Bank focuses on long-term economic development and poverty reduction by providing technical and financial support to help countries to reform particular sectors or implement specific projects, such as building roads, schools and hospitals and providing water and electricity services. The International Monetary Fund promotes short- and medium-term financial support and provides policy advice and technical assistance to help countries build their economies by granting loans on affordable terms to solve balance of payments problems, mainly funded by quota contributions of its members.

(2) The United Nations *Millennium Development Goals* (2000) include: (1) eradicate extreme poverty and hunger; (2) achieve universal primary education; (3) promote gender equality and empower women; (4) reduce child mortality; (5) improve maternal health; (6) combat HIV/AIDS, malaria, and other diseases; (7) ensure environmental sustainability; and (8) develop a global partnership for development.

(3) The World Bank (WB) comprises two institutions: the *International Bank for Reconstruction and Development* (IBRD), and the *International Development Association* (IDA). Although the World Bank's official goal is the reduction of poverty, as stated in its Articles of Agreement, however, all its decisions have to be guided by a commitment to the promotion of foreign investment and to the facilitation of capital investment (cf. World Bank 2011).

(4) The $1.25 per day poverty line of 2005, used until recently by the Word Bank, represents the average of the national poverty lines of the 15 poorest developing countries. Later, in 2011, the poverty line was recalculated to also include actual purchasing power parity exchange rates (PPPs) provided by the International Comparison Program (ICP), converging the line into the currencies of each developing country. In this way, the upwards adaptation of the line to $1.90 established the World Bank's new international poverty line in use since 2011.
(5) The *Multidimensional Poverty Index* (MPI) measures deprivation in the three dimensions of the *Human Development Index* (HDI), i.e. health, education and living standards (cf. UNDP 1996).

References

Agung, A. A. G. (2005). *Bali endangered paradise? Tri Hita Karana and the conservation of the Island's biocultural diversity*. Ph.D. Dissertation. Leiden Ethnosystems and Development Programme. LEAD Studies No. 1. Leiden University. xxv + ill., p. 463.

Aiglsperger, J. (2014). *'Yiatrosofia yia ton Anthropo': Indigenous knowledge and utilisation of MAC plants in Pirgos and Pretoria, Rural Crete: A community perspective on the plural medical system in Greece*. Ph.D. Dissertation. Leiden Ethnosystems and Development Programme. LEAD Studies No. 9. Leiden University. xxv + ill., p. 336.

Alejos Góngora, C. L. (2013). Greenwashing: Only the appearance of sustainability. *IESE Insight Magazine*.

Ambaretnani, P. (2012). *Paraji and Bidan in Rancaekek: Integrated medicine for advanced partnerships among traditional birth attendants and community midwives in the Sunda Area of West Java, Indonesia*. Ph.D. dissertation. Leiden Ethnosystems and Development Programme (LEAD) Studies, No. 7 (xx + ill., 265 pp). Leiden: Leiden University.

Bateman, M. (2010). *Why doesn't microfinance work? The destructive rise of local neoliberalism*. London/New York: Zed Books.

Blanchard, D. (1988). Empirical strategies of bottom-up development. In *ICA International IERD Regional Development Symposia* (pp. 318–338).

Blunt, P., & Warren, D. M. (1996). *Indigenous organisations and development*. Rugby, UK: Practical Action Publishing.

Bruce, J. W., & Migot-Adholla, S. E. (1994). *Searching for land tenure security in Africa*. Dubuque, IA: Kendall Hunt Publishing Company.

Burke, B. J. (2006). *Brazil's extractive frontier: Contemporary and historical globalizations*. M.A. thesis. Graduate College, The University of Arizona.

Chirangi, M. M. (2013). *'Afya Jumuishi': Towards interprofessional collaboration between traditional & modern medical practitioners in the Mara region of Tanzania*. Ph.D. Dissertation. Leiden Ethnosystems and Development Programme. LEAD Studies No. 8. Leiden University. xxvi + ill., 235 pp.

David, G. (1993). Strategies for grass roots human development. *Social Development Issues, 12* (2), 1–13.

de Berge, S. (2011, May 23). Subsistence to sustainability: An integrated approach to community development via education, diversification & accompaniment. *Changemakers*.

de Boef, W., Manor, K. A., Wellard, K., & Bebbington, A. (Eds.). (1993). *Cultivating knowledge. Genetic diversity, farmer experimentation and crop research*. London: IT Publications Ltd.

Dhamotharan, M. (2000). *Handbook on integrated community development—Seven D approach to community capacity development*. Tokyo: APO.

Djen Amar, S. C. (2010). *Gunem Catur in the Sunda Region of West Java: Indigenous communication on the MAC plant knowledge and practice within the Arisan in Lembang*. Ph.D. Dissertation. Leiden Ethnosystems and Development Programme. LEAD Studies No. 6. Leiden University. xx + ill., 218 pp.

Easterly, W. (2006, April 2). Why doesn't aid work? *CATO Unbound: A Journal of Debate*.

Erwina, W. (2019). *Iber Kasehatan in Sukamiskin: Utilisation of the Plural Health Information & Communication System in the Sunda Region of West Java, Indonesia*. Ph.D. Dissertation. Leiden Ethnosystems and Development Programme. LEAD Studies No. 10. Leiden University.

Food and Agricultural Organisation (FAO). (2009). *The state of food insecurity in the world. Economic crises—Impacts and lessons learned*. Rome: FAO.

George, H. (1879). *Progress and poverty: An inquiry into the cause of industrial depressions and of increase of want with increase of wealth: The remedy*. New York: Appleton & Company.

Gheneti, Y. (2007). *Microcredit management in Ghana: Development of co-operative credit unions among the Dagaaba*. Ph.D. Dissertation. Leiden Ethnosystems and Development Programme. LEAD Studies No. 2. Leiden University. xvii + ill., 330 pp.

Gibson, C. C., Andersson, K. E., Ostrom, E., & Shivakumar, S. (2005). *The Samaritan's dilemma: The political economy of development aid*. Oxford, UK: Oxford University Press.

Hembram, D. (2007). *Use of non-timber forest product by households in the Daniel Boone National Forest Region*. Ph.D. Dissertation. West Lafayette, Indiana, USA: Purdue University Graduate School.

Hermes, N., & Lensink, R. (2011, June). Microfinance: Its impact, outreach, and sustainability. *World Development, 39*(6), 875–881.

Hickel, J. (2015, June 10). The microfinance delusion: Who really wins? *The Guardian*.

Ibui, A. K. (2007). *Indigenous knowledge, belief and practice of wild plants among the Meru of Kenya*. Ph.D. Dissertation. Leiden Ethnosystems and Development Programme. LEAD Studies No. 3. Leiden University. xxv + ill., 327 pp.

IPO. (2007). *Compartamos: Prompting to question the legitimacy and course of the "New Wave" microfinance*. IPO.

Indonesian Law No. 6/2014. (2014). *Undang-Undang Republik Indonesia tentang Desa tahun* 2014 [Indonesian Law on the Village]. Jakarta: SEKNEG.

JIVA. (2012). *Joint initiative for village advancement in India*. Washington, D.C.: Pyxera.

Jones, J., & Wiggle, I. (1987). The concept and politics of integrated community development. *Community Development Journal, 22*(2), 107–119.

Kecamatan Development Project (KDP). (1998–2006). *Indonesia—Kecamatan Development Project, Jakarta*. Ministry of Home Affairs. Community Development Office (PMD).

Larrison, C. R. (1999). *A comparison of top-down and bottom-up community development interventions in rural Mexico: Practical and theoretical implications for community development programs*. Athens GA: University of Georgia, School of Social Work.

Leakey, R. E. (2014, April). Julie Gichuru: Interview with Dr. Richard Leakey. *Inside Africa: Leadership Dialogues*. https://www.youtube.com/watch?v=JnQL1d9CDz8.

Leurs, L. N. (2010). *Medicinal, aromatic and cosmetic (MAC) plants for community health and bio-cultural diversity conservation in Bali, Indonesia*. Ph.D. Dissertation. Leiden Ethnosystems and Development Programme. LEAD Studies No. 5. Leiden University. xx + ill., pp. 343.

Macdonald, L. (1995). NGOs and the problematic discourse of participation: 75 cases from Costa Rica. In D. B. Moore & G. J. Schmitz (Eds.), *Debating development discourse: Institutional and popular perspectives* (pp. 201–229). New York: St. Martin's Press.

Make Poverty History. (2005). *Get involved*. UK: Makepovertyhistory.org. 2005-06-26. Retrieved December 1, 2016.

McConnell, D. (2002). Action research and distributed problem-based learning in continuing professional education. *Distance Education, 23*(1), 59–83.

Metha, L., Leach, M., Newell, P., Scoones, I., Sivaramakrishnan, K., & Way, S. A. (1999). Exploring understandings of institutions and uncertainty: New directions in natural resource management. *IDS Discussion Paper*, No. 372.

Midgley, J. (1993). Ideological roots of social development strategies. *Social Development Issues, 12*(1), 1–13.

Pallier, J., & Prévost, B. (2007). Vulnérabilité et gestion des risques potentialités et limites de la microfinance. *Mondes en Développement, 2*(138).

Pawar, M., & Torres, R. (2011). Integrated Community development through dialogue, capacity-building and partnership in an Australian town. *Journal of Comparative Social Welfare, 27*(3), 253–268.

PNPM-Mandiri. (2007). *National community empowerment in rural areas (PNPM)*. Jakarta: Directorate General of Rural and Community Empowerment (PMD), Ministry of the Interior Affairs.

PROGRESS. (2013). *Integrating family planning into other development sectors.* FHI 360 Project. Durham, NC: FHI 360.

Richards, P. (1985). *Indigenous agricultural revolution: Ecology and food production in West Africa* (p. 192p). London: Hutchinson Education.

Richards, P., Slikkerveer, L. J., & Phillips, A. O. (1983). *Indigenous knowledge systems for agriculture and rural development: The CIKARD inaugural lectures*. Studies in Technology and Social Change (Vol. 13, 41p). Ames, Iowa, U.S.A.: Iowa State University.

Robinson, M. S. (2001). *The microfinance revolution: Sustainable finance for the poor. Lessons from Indonesia. The emerging industry.* Washington, D.C.: World Bank.

Roodman, D. (2011). *Due dilligence: An impertinent inquiry into microfinance*. Washington, D.C.: Center for Global Development.

Rubin, A., & Babbi, E. R. (1993). Research methods for social work. *Brooks/Cole Empowerment Series.* Belmont, CA.

Saefullah, K. (2019). *Gintingan in Subang: An Indigenous Institution for Sustainable Community-Based Development in the Sunda Region of West Java, Indonesia*. Ph.D. Dissertation. Leiden Ethnosystems and Development Programme. LEAD Studies No. 11. Leiden University.

Slikkerveer, L. J. (1999a). Ethnoscience, TEK and its application to conservation. In D. A. Posey (Ed.), *Cultural and spiritual values of biodiversity: A complementary contribution to the global biodiversity assessment.* London/Nairobi: Intermediate Technology Publications Ltd./United Nations Environment Programme (UNEP).

Slikkerveer, L. J. (1999b). Traditional ecological knowledge (TEK): Practical implications from the African experience. In K. Adimihardja & M. Clement (Eds.), *Indigenous knowledge systems research & development studies* (Vol. 3). Bandung: UPT.

Slikkerveer, L. J. (2009). *Applied ethnoscience in the three target areas of the LEAD programme.* LEAD, Leiden University.

Slikkerveer, L. J. (Ed.). (2012). *Handbook for lecturers and tutors of the new master course on integrated microfinance management for poverty reduction and sustainable development in Indonesia (IMM).* Leiden/Bandung: LEAD/UL/UNPAD/MAICH/GEMA PKM.

Slikkerveer, L. J. (2015). *Integrated Microfinance Management for Poverty Reduction and Sustainable Development in Indonesia.* Paper presented at the International Workshop in Integrated Microfinance Management. Bandung Universitas Padjadjaran.

Slikkerveer, L. J. (2016). *LEAD: Leiden Ethnosystems and Development Programme. Verleden, Heden en Toekomst van het Leidse Speerpunt. Lokale Kennis Systemen en Ontwikkeling. 1993-2016-2020.* Internal Report. LEAD, Leiden University.

Slikkerveer, L. J., & Ambaretnani, P. (2010a). *Societal needs assessment for IMM in Indonesia.* LEAD, Leiden University.

Slikkerveer, L. J., & Ambaretnani, P. (2010b). *Educational needs assessment for IMM in Indonesia*. LEAD, Leiden University.
Slikkerveer, L. J., & Slikkerveer, M. (1995). Tanaman Obat Keluarga (TOGA). In D. M. Warren, L. J. Slikkerveer, & D. Brokensha (Eds.), *The cultural dimension of development: indigenous knowledge systems*. London: Intermediate Technology Publications Ltd.
Toledo, V. M. (1997). Sustainable development at the community level: A third world perspective. In E. Smith (Ed.), *Environmental sustainability*. Baco Raton, Fla: St. Lucie Press.
Toledo, V. M. (2001). Biocultural diversity and local power in Mexico. In L. Maffi (Ed.), *On biological diversity: Linking language, knowledge, and the environment*. New York: Smithsonian Institution Press.
United Nations (UN). (1990–2015). *Human development reports*. New York: United Nations.
United Nations (UN). (2000). *Millennium development goals (MGDs)*. New York: United Nations.
United Nations (UN). (2014). *Global sustainable development report*. New York: Department of Economic and Social Affairs (UNDESA).
United Nations (UN). (2015a). *The 2030 agenda for sustainable development and the sustainable development goals (SDGs)*. New York: United Nations.
United Nations (UN). (2015b). *Transforming our World: The 2030 agenda for sustainable development*. Resolution 70/1 of the General Assembly on 25 September, 2015. New York: United Nations.
United Nations (UN). (2016). *International day for the eradication of poverty (IDEP)*. New York: UN Department of Public Information.
United Nations Capital Development Fund (UNCDF). (2013). *Global programmes: UN capital development fund—A summary*. New York: United Nations.
United Nations Central Statistics Agency (UNCSA). (2015). *World statistics pocketbook. 2015 edition*. Series V, No. 39. New York: United Nations.
United Nations Development Programme (UNDP). (1991). *Human development report*. New York: United Nations.
United Nations Development Programme (UNDP). (1996). *Economic growth and human development*. New York: United Nations.
United Nations Development Programme (UNDP). (2008). *Supporting capacity development: The UNDP approach*. New York: United Nations.
United Nations Development Programme (UNDP). (2014). *Sustaining human progress*. New York: United Nations.
United Nations Office for Disaster Risk Reduction (UNISDR). (2016). *Terminology. United Nations Office for disaster risk reduction*. Retrieved March 31, 2016.
Uphoff, N. (1996). *Local institutional development: An analytical sourcebook with cases*. West Hartford: Kumarian Press.
Warren, D. M., Slikkerveer, L. J., & Brokensha, D. (Eds.). (1995). *The cultural dimension of development: Indigenous knowledge systems. IT studies on indigenous knowledge and development*. London: Intermediate Technology Publications Ltd.
Watson, E. E. (2003). Examining the potential of indigenous institutions for development: A perspective from Borana, Ethiopia. *Development and Change, 34*(2), 287–309.
Wong, S. (2003). *Indonesia Kecamatan development program: Building a monitoring and evaluation system for a large-scale community driven development program*. Working Paper. Washington, D.C.: World Bank.
World Bank (WB). (1997–2016). *World development reports*. Washington, D.C.: World Bank.
World Bank (WB). (2000). *World development report 2000/2001. Attacking poverty*. Washington, D.C.: World Bank.
World Bank (WB). (2011). *About us—The World Bank*. Washington, D.C.: World Bank.
World Bank (WB). (2013). *Poverty and prosperity*. Washington, D.C.: World Bank.
World Bank (WB). (2014). *Global monitoring report 2014*. Washington, D.C.: World Bank.

World Bank (WB). (2015). *Global monitoring report 2015*. Washington, D.C.: World Bank.
World Bank (WB). (2016, October). *Indonesia economic quarterly*. Pressures Easing.
World Culture Forum (WCF). (2013). *The Bali promise of the World culture forum*. Paris: UNESCO.
Yunus, M. (1999). *Banker to the poor: Micro-lending and the battle against world poverty*. New York: Public Affairs.
Yunus, M. (2007). *Creating a world without poverty: Social business and the future of capitalism*. New York: Public Affairs.

L. Jan Slikkerveer is Professor of Applied Ethnoscience and Director of the Leiden Ethnosystems and Development Programme, Faculty of Science, Leiden University, Leiden, The Netherlands. He received his Ph.D. on his fieldwork in the Horn of Africa from Leiden University in 1983 and an Honorary Degree from the Faculty of Medicine of Universitas Padjadjaran in Bandung, Indonesia in 2005. He further extended advanced training and research in the newly-developing field of applied ethnoscience in several sub-disciplines, including ethno-economics, ethno-medicine, ethno-biology/botany, ethno-pharmacy, ethno-communication, ethno-agriculture and ethno-mathematics in combination with a focus on three target regions of South-East Asia, East-Africa and the Mediterranean Region. He is the supervisor of 25 Ph.D. students at Leiden University and has published more than 100 books and 300 articles on the subject. While he has also received a number of substantial subsidies from the European Union in Brussels for international development programmes in South-East Asia, East-Africa and the Mediterranean Region, he also conceptualised both the IMM & ICMD approaches in Indonesia, and is Senior Lecturer in the International Master Course on Integrated Microfinance Management (IMM) at the Faculty of Economics and Business of Universitas Padjadjaran in Bandung, Indonesia.

George Baourakis is Director of the Mediterranean Agronomic Institute of Chania (CIHEAM-MAICh) in Crete, Greece, and the Studies and Research Coordinator of the Business Economics and Management Department of CIHEAM-MAICh. He holds a Ph.D. in Food Marketing and is an expert in Behavioural Economics with more than three decades' experience in the management and coordination of educational and research activities, including coordination of the M.Sc. Programme of CIHEAM-MAICh, and organization of international seminars and training courses both at CIHEAM-MAICh and in many other Mediterranean countries. He has co-ordinated a large number of EU projects (FP 4th, 5th, 6th and 7th, INTERREG I, II and III-Archimed, MED, Tempus, Phare, Life, Lifelong Learning, Leonardo Da Vinci, European Social Fund/Operational Sectoral Programme), as well as international and national-regional research projects, and has also supervised their scientific, economic and administrative management. He is a Research Fellow at the Centre of Entrepreneurship, Nyenrode University, The Netherlands Business School, and Senior Lecturer in the International Master Course on Integrated Microfinance Management (IMM) at the Faculty of Economics and Business, Universitas Padjadjaran in Bandung, Indonesia. He also holds the positions of Co-Editor in Chief for the International Journal of Food and Beverage Manufacturing and Business Models (IJFBMBM), IGI Global, and Associate Editor for the Springer Cooperative Management Series.

Kurniawan Saefullah is Lecturer of the Faculty of Economics and Business, Universitas Padjadjaran, Bandung, Indonesia since 1998. He graduated from Universitas Padjadjaran and the International Islamic University, Malaysia. He has published several books and articles with his colleagues, including Introduction to Management and Introduction to Business. In 2009, he joined the International Project on Integrated Microfinance Management (IMM) for Poverty Reduction and Empowerment in Indonesia. As a Ph.D. Researcher at the Leiden Ethnosystems and Development Programme (LEAD), Faculty of Science of Leiden University, The Netherlands, he is finalising his Ph.D. Dissertation on 'Gintingan in Subang: An Indigenous Institution for

Sustainable Community-Based Development in the Sunda Region of West Java, Indonesia' under the supervision of Prof. Dr. L. J. Slikkerveer. The focus of his research is on the important role of the local cosmology of Tri Tangtu in the process of sustainable community development, while he also developed a model of integration of community-support institutions on the basis of the Indigenous Knowledge Systems (IKS) in combination with the IMM and ICMD approaches, such as financial, medical, educational, communication and social services through the indigenous institutions of the local communities in the Sunda Region of West Java.

Part I
Ethnoscience and the Paradigm Shift in Global Development Cooperation

Chapter 2
The Indigenous Knowledge Systems' Perspective on Sustainable Development

L. Jan Slikkerveer

> *We strongly recommend that the culture dimension of development be explicitly integrated in all the sustainable development goals.*
>
> Throsby (2013)

2.1 Cognitive Anthropology: Interest in the Emic Perspective on Culture

The foundations of *ethnoscience*, defined by Augé (1999: 118) as: "*...to reconstitute what serves as science for others, their practices of looking after themselves and their bodies, their botanical knowledge, but also their forms of classification, of making connections,* etc." were laid by the 19th century American anthropologist Franz Boaz, who introduced the concept of *cultural relativism*. Cultural relativism underscores people's differences and documents how they are the result of their own social, cultural, historical, and geographical conditions. It avoids the ideology of ethnocentrism being the normative application of one's own cultural standards to the assessment of other cultures (cf. Collins 1998).

The related relativist methodology is characterised by the emic (insider's) position which the researcher assumes in order to avoid his or her own cultural biases, while attempting to understand the knowledge, beliefs and practices of the participants themselves by using this emic view on the local context. In this methodology, ethnosemantics, ethnographic semantics, ethnographic ethnoscience, formal analysis, and componential analysis are used to acquire such an emic perspective on how other people tend to perceive, organise and use their culture in relation to their environment.

L. J. Slikkerveer (✉)
LEAD, Leiden University, Leiden, The Netherlands
e-mail: l.j.slikkerveer@gmail.com

L. J. Slikkerveer et al. (eds.), *Integrated Community-Managed Development*, Cooperative Management, https://doi.org/10.1007/978-3-030-05423-6_2

Ethnographic semantics soon evolved into *cognitive anthropology* which put its primary focus on the intellectual and rational perspectives of the people under study. Cognitive anthropology seeks to study patterns of shared knowledge and its transmission over time and space, using the *theories* and methods of the *cognitive sciences* such as *experimental psychology* and *evolutionary biology*, often in close collaboration with other scientists involved in the study and explanation of different cultures (cf. D'Andrade 1995).

In the 1960s, ethnoscience emerged from cognitive anthropology to refer to the scientific study of 'indigenous knowledge', based on the emic perspective, focusing initially on indigenous taxonomies, classifications and perceptions which local people use to categorise not only plants, animals, diseases, ecosystems, and landscapes, but also kinship relations, religions, cosmologies, deities and spirits. *Ethnoscience* was defined by Hardesty (1977) as: "*the study of systems of knowledge developed by a given culture to classify the objects, activities, and events of its universe.*"

Pioneering work by Conklin (1957), Goodenough (1957), Horton (1967) and others began to document that traditional peoples, such as for example the Philippine horticulturalists, often possess exceptionally detailed knowledge of local plants and animals and their natural environment, recognising in one case some 1600 plant species. In their efforts to conduct emic-oriented research, these scientists took a more multidisciplinary and collaborative position, not only with the participants, but also with their colleagues from among other disciplines.

As Ingold (2000: 406–7) rightly argues: "*ethnoscience is based on increased collaboration between social sciences and humanities (e.g., anthropology, sociology, psychology, and philosophy) with natural sciences such as biology, ecology, or medicine. At the same time, ethnoscience is increasingly transdisciplinary in its nature.*"

Soon thereafter, a growing number of specialisms of ethnoscience have evolved into particular sub-fields of ethnoscience, including ethno-history, ethno-ecology, ethno-biology, ethno-botany, ethno-linguistics, ethno-medicine and ethno-psychology, further substantiating the emerging ethnoscientific perspective on different human populations, their culture and their way of life (cf. Posey 1990, 2002). These studies, focusing on the indigenous peoples' own systems of knowledge, beliefs and practices, soon brought together a vast and interesting body of evidence-based information on how local people perceive their own history and their relations with their environment within the context of their cosmology.

However, as a result of the growing interest in ethnoscience from a number of scientists with a background in the *cognitive sciences*, most flourishing sub-fields remained largely oriented at the theoretical, phenomenological and philosophical level, rendering these studies in ethnoscience rather static. It is not surprising that these early studies in ethnoscience largely remained confined to the anthropological tradition of highlighting the unique, but often peculiar characteristics of remote non-Western cultures as compared to Western cultures, and as such also contributed to the previous misconceptions about indigenous knowledge as being 'isolated', 'geographically distant', 'something from the past', 'old-fashioned', 'static', or 'magic'. Although some anthropologists tend to oppose the definition of Atran (1991: 650) that: "*ethnoscience looks at culture with a scientific perspective*", the application of a specific empirical methodology of study and analysis of indigenous knowledge systems provides ethnoscience research with evidence-based results and conclusions.

In the meantime, however, a group of pioneering, more applied-oriented ethnoscientists started to study indigenous knowledge systems within the more dynamic context of socio-economic development and change, in which these local systems are identified as potentially valuable 'stepping stones' for development. A growing number of authors, including Warren et al. (1980, 1989, 1995), Richards (1985), Chambers et al. (1989), Warren (1991), Swift (1991), Slikkerveer (1999a, b), and Nazarea (1999) brought together numerous case studies in several sectors of the community which showed to be adaptable—and as such potentially sustainable—in programmes of development, based on indigenous knowledge. These case studies underscore that in response to local needs, the use and integration of indigenous knowledge in the form of 'bottom up' or 'grass roots' approaches prove to be more successful, largely because of the remarkable increase in indigenous peoples' participation in the local policy planning and implementation process. Through their applied-oriented work, these ethnoscientists have drastically changed earlier views that indigenous knowledge and practices were the domain of rather academically-oriented ethnographers, and an increasing number of scientists working in the context of development and change have recently identified, documented, and analysed numerous case studies in which systems of indigenous knowledge, practices, skills, technologies and decision-making processes have proved to be more ecologically and culturally adapted to the local settings. Moreover, these local systems proved often more sustainable than many imported—Western or 'scientific'—systems of knowledge and technology.

Thus, while initially, the ethnoscience studies remained rather theoretical in orientation in the documentation of local peoples' emic views, it was the work of Warren et al. (1980) that introduced the concept of IKS within the context of development. In their collection of case studies, they highlight the potential role which these Indigenous Knowledge Systems (IKS) can play in rural development, directly related to the discipline of *development studies.*

2.2 Neo-ethnoscience: Indigenous Knowledge Systems and Development

As mentioned above, in their publication *The Cultural Dimension of Development*, Warren et al. (1995) further extended the significance of applied ethnoscience into a broader, multidisciplinary field of study in relation with the processes of development and change in various sectors, also known as the 'new' ethnoscience or 'neo-ethnoscience', which transpired through their more dynamic definition of an indigenous knowledge system as: "*a system of knowledge, beliefs and practices which has evolved in a particular area or region, often outside universities and laboratories and transferred over many generations, and as such forming the base for the local-level decision-making process and which has been recorded in several sectors of the society, such as in animal and human health, environment, natural resources management, agriculture, fisheries, forestry, bio-cultural diversity conservation, and local economies.*"

Such an extension is further founded in a number of studies on the role of indigenous herbal medicine, veterinary medicine, traditional ecological knowledge, mixed cropping, forest gardens, indigenous pest management, wild food and non-food plants, situated in the socio-economic development process of many regions around the globe (cf. Leakey and Slikkerveer 1991; IK&D Monitor 1993–2001; Mathias-Mundy and McCorkle 1989; Innis 1997; Slikkerveer 1999a, b; Castro 2000). Eventually, the growing interest from among scientists, students and development experts in the field of 'neo-ethnoscience' also started to attract the attention of administrators and policy planners from development agencies and NGOs because of the functional role and positive contribution of indigenous knowledge systems to development programmes and projects in many Third World Countries.

As mentioned in the *Introduction*, the recognition and implementation of the IKS-based development paradigm of the '*cultural dimension of development*' has gained increasing support in international circles. For Asia and the Pacific, the Economic and Social Commission for Asia and the Pacific (ESCAP 2010) introduced the new concept of the 'Second Green Revolution' to encompass a combination between the IKS-based paradigm of 'development from the bottom' and the growing recognition and respect for indigenous agricultural knowledge systems—specifically in agricultural management and development—providing a promising solution to the food crisis so as to foster a 'Second Green Revolution', based on sustainable forms of agriculture.

Similarly, the International *World Culture in Development Forum* (WCF 2013) became the first in a series of international meetings in Bali which brings together key stakeholders to discuss vital issues in culture and development. In the Bali Forum, hosted by the President of Indonesia, Dr. Soesilo Bambang Yudhoyono and leaders from around 40 countries, as well as several Nobel Prize laureates, participated in the Conference, which resulted in a number of important strategic initiatives, as follows:

- to promote knowledge communities for intercultural, intergenerational and interfaith dialogue;
- to develop further ethical investment and business practices for cultural industries;
- to establish clearing houses for people-centred projects and practices, emphasising local knowledge systems; and
- to develop conceptual frameworks informing the *Post-2015 Development Agenda* of the United Nations (UN 2015).

The pioneering two-day *World Culture Forum* (2013) in Nusa Dua, Bali resulted in the "Bali Promise", which calls for a measurable and effective role and integration of culture, including indigenous knowledge systems in development at all levels in the *Post-2015 Agenda for Sustainable Development and the Sustainable Development Goals* (SDGs) of the United Nations (UN 2015).

In this way, and also in the environmental sciences, numerous kinds of specialised forms of Traditional Environmental Knowledge (TEK) were documented and described, showing not only the significance of an integrated system of knowledge, beliefs and practices for human adaptation and survival, but also the importance of the preservation of Traditional Environmental Knowledge (TEK) for development programmes, particularly in bio-cultural diversity conservation (cf. Inglis 1993; Berkes 1998; Gragson 1999). As the International Union of Conservation of Nature (IUCN 1986) underscored, the crucial role of Traditional Environmental Knowledge (TEK) for Conservation is manyfold: (1) it provides new insights in biological and ecological sciences; (2) it is relevant for area-specific resource management; (3) it supports indigenous people to continue their lifestyle in protected areas and education; (4) it facilitates the extension and validity of environmental assessment; and (5) it benefits development agencies to be more successful in involving local people in the planning of the environment, natural resources and production systems (cf. Inglis 1993).

The growing awareness of the richness and value of indigenous knowledge systems for development in the developing countries, which had been brought together by scientists from the above-mentioned sub-fields of 'neo ethnoscience' research, not only fostered a worldwide recognition of the great potential of indigenous knowledge systems for socio-economic development, but also drew attention to the way in which indigenous knowledge, beliefs and practices are expressed through the indigenous institutions (cf. Blunt and Warren 1996; Watson 2003; Slikkerveer 2012).

Meanwhile, in the practical setting of the international development debate, a revaluation of indigenous peoples and their knowledge systems had already emerged from experience as the early centralised, top-down policies of national and international programmes of development aid of the 1960s and 1970s largely failed to achieve their objectives. The planning had embarked on the false premise of developing the new nations to become modern, if not 'civilised' states, through an ill-conceived process of linear transfer of Western knowledge and technology to the 'empty vessels' of the developing countries, obviously leading up to the greatest failure of international development to this day, coined by Easterly (2006) as: '*the tragedy of aid*' (cf. Myrdal 1968).

Inspired by the new insights into the role of indigenous knowledge systems in development, the development experts and change agents began to recognise that these local knowledge systems would possess specific and unique qualities which not only provide the participants with a value of self-respect, but also render their role in policies most contributive to achieving the objectives of sustainable community development in many sectors of society. The old, largely failing development paradigm of 'development from the top' was soon replaced by a more effective policy of 'development from the bottom'. The radical shift in the development paradigm of the 1990s prompted Sillitoe (1998: 1) to comment on this major episode: "*A revolution is occurring in the pursuit of ethnography as the developed world changes its focus from top-down intervention to a grassroots participatory perspective.*"

Inspired by the practical achievements in applied ethnomedicine, the early expressions of the new approach in its rather successful implementation became manifest in the 1980s in medicine, where the integration of traditional medical knowledge and practice into modern primary health care eventually evolved into a new worldwide strategy of the World Health Organisation (1978, 2002, 2013) with the objective to improve health care for all people around the globe (cf. Warren 1982; Janzen 1982; Bannerman et al. 1983; Slikkerveer and Slikkerveer 1995; Djen Amar 2010; Leurs 2010; Ambaretnani 2012; Chirangi 2013; Aiglsperger 2014; Chan 2014).

At the same time, a similar re-orientation took place in the major sector of agriculture in the tropics. Embarking on the growing need for sustainable agriculture, international organisations and institutions, including the Food and Agriculture Organisation of the United Nations (FAO), the Institute of Development Studies (IDS), Compas, CIKARD and LEAD, started to highlight the significance of indigenous farmers' knowledge and practices in their efforts to attain sustainable agriculture (cf. Leakey and Slikkerveer 1991; Warren 1991; Reijntjes et al. 1994). The 'Farmer First' workshop held at the Institute of Development Studies (IDS) at the University of Sussex, UK in 1987 underscored the development of new approaches to farmer participation in agricultural research and extension (cf. Chambers et al. 1989). In a recent review, Scoones et al. (2008) observe the paradigm shift of the 1990s in the changing approaches to agricultural research and development since the 1960s. Following the era of a long history, central since the 1960s and known as 'Transfer of Technology', a period started in the 1970s and 1980s, called 'Farming Systems Research'.

Then, from the 1990s, the strategy of 'Farmer First/Farmer Participatory Research' developed, while from 2000 onwards, the 'People-centred Innovation and Learning' approach advanced. In the same review, the involvement and position of knowledge and related disciplines is showing an interesting pattern, departing from single-discipline driven (breeding) through inter-disciplinary (plus economics) and intra-disciplinary (including farmer experts) to extra/trans-disciplinary—holistic, multiple culturally-rooted approaches. The conceptualisation of 'sustainability' over the same period of time in the overview shows an evolution of the term from 'undefined' in the past through 'important' and 'explicit', up to the current significance of 'championed—and multi-dimensional, normative and political'.

2.3 From Endogenous Development to Neo-Endogenous Development

The recent interest in the IKS-based strategy of development '*from the bottom*' had succeeded the post-World War II period, where the early modernist model of rural development was based on classical interventions from outside, pertaining to development '*from the top*', known as '*exogenous development*'. In this model, rural areas were generally treated as technically, culturally and economically dependent on urban areas, while the main function of rural areas was to provide food and other resources for the expanding urban populations.

As this model did not really contribute to the sustainable development of local communities, largely because of a general lack of local peoples' participation, Lowe et al. (1998: 9–10) express their concern over exogenous approaches to development for being:

- dependent, as they are reliant on continued subsidies and the policy decisions of distant agencies or boardrooms;
- distorted, which boosts single sectors, selected settlements and certain types of business, but leaves others behind and neglects non-economic aspects of rural life;
- destructive, as they erase the cultural and environmental differences of rural areas; and
- dictated, as they are largely devised by external experts and planners.

Although Woods (2005) recognises that *exogenous development* was successful in certain aspects, such as in increasing employment rates, and improving technology, communication and infrastructure in rural areas, he also criticises this model since exogenous development is often dependent on external investment, causing not only the profits of development to be exported, but also creating a democratic discrepancy because of the non-participatory nature of the model.

With the renewed interest in the potential role of indigenous knowledge systems in development since the 1990s, and the subsequent shift in the international development paradigm, as mentioned above, rural development policies began to relocate their focus and embark on local structures, which soon became known as the more successful approach of *endogenous development*. Based on local peoples' own criteria of development, *endogenous development* takes into account not only the material, but also the social and spiritual well-being of peoples.

In the rural policies of Europe, for instance, endogenous development has shown a fundamental shift over the past decades from sectoral support policies (agriculture) to territorial development and spatial approaches (cf. Moseley 1997, 2000; Shortall and Shucksmith 2001; Organisation for Economic Co-operation and Development (OECD) (2006). As Gkartzios and Scott (2013) contend, such policy shift does recognise that territorial approaches and policies can integrate sectoral dimensions into public policy delivery—agriculture, housing, employment creation, transport—and offer: '*a holistic approach to balancing the economic, social and environmental processes which shape rural areas*'.

A leading organisation in the worldwide promotion of endogenous development is the international network implementing field programmes to develop, test and improve methodologies of endogenous development (ED), the *Comparing and Supporting Endogenous Development Network* (COMPAS). According to COMPAS (2016), the three pillars on which the endogenous development approach is based include: (1) *a Western bias*: The importance of participatory approaches and of integrating local knowledge into development interventions has become broadly recognised, but many of these approaches experience difficulties in overcoming an implicit Western bias; (2) *endogenous development and other participatory approaches*: The main difference between endogenous development and other participatory approaches is its emphasis on including spiritual aspects in the development process, in addition to the ecological, social and economic aspects; and (3) *key concepts*: These include: local control of the development process; taking cultural values seriously; appreciating worldviews; and finding a balance between local and external resources. The general aim of endogenous development is to empower local communities to take control of their own development process. While revitalising ancestral and local knowledge, endogenous development helps local people to select those external resources which best fit the local conditions.

Several field activities of COMPAS in endogenous development include support to livelihood development, organic and sustainable agriculture, ethno-veterinary services, management of sacred forests and groves, assessing and strengthening local health practices, and supporting traditional institutions, which have further substantiated the success of the endogenous development approach in cooperation with its partners in Africa, Asia and Latin America (cf. Haverkort and Hiemstra 2000; Haverkort et al. 2003; Millar et al. 2006; Balasubramanian and Nirmala Devi 2006; Haverkort and Reijntjes 2007).[1] As Moseley (1997) and Ray (2000) state, the endogenous approach to rural development is based on three principles: (1) a territorial and integrated focus; (2) the use of local resources; and (3) local contextualisation through active public participation. Walsh (1996) contends that the local area-based approach to rural development could realise new development objectives through multi-dimensionality, integration, co-ordination, subsidiarity and sustainability. According to Ray (1997: 345), the local approach to rural development has three primary characteristics:

- *it sets development activity within a territorial rather than sectoral framework, with the scale of territory being smaller than the nation-scale;*

[1]The Comparing and Supporting Endogenous Development (COMPAS) Programme is an international network implementing field programmes to develop, test and improve (ED) methodologies. Endogenous development is based on local peoples' own criteria of development, and takes into account the material, social and spiritual well-being of peoples. The COMPAS Programme is coordinated by the ETC Foundation in The Netherlands, which has discontinued its activities in 2016.

- *economic and other development activity is restructured in ways so as to maximise the retention of benefits within the local territory by valorising and exploiting local resources – both physical and human; and*
- *development is contextualised by focusing on the needs, capacities and perspectives of local people.*

Although the endogenous development strategy has been implemented rather successfully worldwide and continues to realise impressive achievements in sustainable development projects and programmes, clearly surpassing previous policies of exogenous development, several authors such as Barke and Newton (1997) have indicated the limitations of this approach. Similarly, Storey (1999) and Shucksmith (2000) document problems of participation and elitism, where local powerful elites tend to seize endogenous development initiatives for their own interest. Ward et al. (2005) also criticise the endogenous model implemented in European development policies by arguing that such socio-economic development of rural areas is not a practical proposition as it remains outside influences such as globalisation, external trade or governmental actions.

Indeed, while Dibden et al. (2009) point to the increased significance of neoliberal ideas, policies, and projects to the unfolding of social and spatial life in rural areas, Brunori and Rossi (2007) show that capital, consumers and regulatory bodies also play a significant role in shaping rural localities within the context of globalisation. As all rural areas experience a mix of exogenous and endogenous forces, where the local level must interact with the global level, Ward et al. (2005: 5) note that: "*the critical point is how to enhance the capacity of local areas to steer these wider processes, resources and actions to their benefit.*"

As it is clear that the endogenous development model could run the risk to pertain to an isolationistic strategy solely based on '*development from the bottom*', which in practice can hardly be found, the need for a more realistic hybrid model of local-global interaction has been expressed for various sectors of the society, including health, education, economics, environment, communication and wildlife (cf. Slikkerveer 1999a, 2012). In this context, Leakey (2012) also underscores the need for the development of such an interactive approach, currently operational in wildlife management in Africa: "*What we need is the best of both worlds in order to achieve truly sustainable community development.*"

Similarly, Ray (2001: 3–4) recognises the need for such a hybrid model in rural development which goes 'beyond endogenous and exogenous modes, by focusing on the dynamic interactions between local areas and their wider political and other institutional, trading and natural environments', and consequently introduces the term 'neo-endogenous development' to describe an approach to rural development which is locally rooted, but outward-looking and characterised by dynamic interactions between local areas and their wider environments.

Ray (2001: 8–9) contends that the term *neo-endogenous development* requires a recognition that development based on local resources and local participation can, in fact, be animated from three possible directions, separately or together:

First *it can be animated by actors within the local area.*
Second *it can be animated from above, as national governments and/or the EU respond to the logic of contemporary political-administrative ideology.*
Third *it can be animated from the intermediate level, particularly by non-governmental organisations which see in endogenous development the means by which to pursue their particular agendas. The manifestation of neo-endogenous development in any territory will be the result of various combinations of the development from above and the intermediate level sources interacting with the local level.*

The limitations of both the exogenous and endogenous approaches to achieve sustainable development in poverty reduction strategies underscore the need for a hybrid model of interacting exogenous and endogenous factors, which, against the background of rural development theory, has recently been developed for achieving poverty reduction at the community level through integrated models, implemented in Indonesia (cf. Ray 2006; Slikkerveer 2012).

In his perspective of the realisation of sustainable community development in sectors such as health care, education, environment and natural resources management, Slikkerveer (1999a, b) identifies the dynamic interaction between local and global systems of knowledge, beliefs and practices as a key concept, which can most effectively be implemented at the intermediate level between exogenous and endogenous approaches, as represented by the process of *neo-endogenous development.*

As will be elaborated below, here the concept of horizontal integration has been implemented to integrate the financial and non-financial aspects into an alternative strategy of poverty reduction, ranging from *Integrated Microfinance Management* towards *Integrated Sustainable Community Development*, in which community-based institutions provide a successful point of embarkation.

2.4 LEAD: A Multidisciplinary Input to Applied Neo-ethnoscience

Following the early developments related to the study and interpretation of indigenous knowledge systems since the 1970s, the *Leiden Ethnosystems and Development Programme* (LEAD) was established in 1987 together with the *Centre for Indigenous Knowledge and Agricultural and Rural Development* (CIKARD) in Ames, USA, and the *Centre for International Research and Advisory Networks* (CIRAN/NUFFIC) in The Hague as a global centre and co-founder of the *Global Network for Indigenous Knowledge and Development* supported by the *IKS Monitor Series.* The Global Network encompassed the regional centres ARCIK in Nigeria, REPPIKA in The Philippines, and the national centres such as BRARCIK in Brazil, BURCIK in Burkina Faso, GHARCIK in Ghana, NIRCIK in Nigeria, PHIRCSDIK in The Philippines, RIDSCA in Mexico, SARCIK in South Africa,

SLARIK in Sri Lanka, URURCIK in Uruguay and VERSIK in Venezuela. The extension of the cooperation between LEAD and a number of international universities working in the field of 'neo-ethnoscience' in the three above-mentioned target areas has also led the LEAD Programme to co-establish three IK-Centres with which further cooperation in education, research and joint projects has been realised, i.e. the *Indonesian Resource–Centre for Indigenous Knowledge* (INRIK) in Bandung, Indonesia, and the *Kenyan Resource-Centre for Indigenous Knowledge* (KENRIK) in Nairobi, Kenya.

Initially, the LEAD programme was established at the Institute of Cultural and Social Studies of Leiden University in The Netherlands, focusing on the new ethnoscience study of Indigenous Knowledge Systems, particularly in health and healing in the Mediterranean Region and South-East Asia, known as ethnomedicine, and in ethno-agriculture in South-East Asia and East Africa. In 1999, when the LEAD Programme joined the Faculty of Science of Leiden University as a special programme providing the complementary dimension of *ethnoscience* to *science*, it extended its activities into post-graduate research and training in the sub-fields of ethno-medicine, ethno-botany, ethno-pharmacy, ethno-communication, ethno-economics and ethno-mathematics. These sub-disciplines were recently integrated in three multidisciplinary clusters, i.e. *ethno-botany/pharmacy/medicine*, *ethno-economics/mathematics*, and *ethno-communication/history*, which, in combination with a focus on the three target regions of LEAD, has further substantiated its activities in international education and research in the field of neo-ethnoscience.

Underlying LEAD's mission to study, document and analyse Indigenous Knowledge Systems which are operational in sustainable development in many sectors of developing nations is the philosophy that humankind has to change its attitudes towards nature and the environment in order to achieve a more balanced relationship between humans, culture and nature as a prerequisite for the ultimate survival of humankind on Planet Earth in the future. The extended conceptualisation of indigenous knowledge includes not only the reference to local knowledge and practices, but also to perceptions, beliefs, cosmologies, values, wisdom, experience and last but not least institutions which have developed over many generations in a particular field of ethnographic study—or culture area—and which as such are unique to a specific culture. Moreover, the indigenous knowledge systems also include indigenous philosophies of nature and the environment, which, apart from their significance for fostering participatory development in collaboration with the local population, are important for the evolution towards a universal philosophy of nature at the global level. While most Western philosophies of nature are largely dedicated to the commercial exploitation of natural resources for the sake of surplus-building for current generations, in contrast, most indigenous philosophies of nature are dedicated to the use of their resources mainly for subsistence, leaving a larger part for future generations.

Since current research on complex systems of indigenous knowledge, beliefs and practices in different sectors requires a rather multidisciplinary approach of the 'new ethnoscience' to study, analyse and explain the emic dimensions of the indigenous knowledge systems, in which the subject matter sometimes appears to transcend rational knowledge when it comes to conceptualisations such as the above-mentioned 'beliefs', 'cosmologies' and 'philosophies', a specific research methodology had to be developed in order to fully understand and interpret the various material and immaterial dimensions of indigenous knowledge systems from an emic point of view. The study and analysis of indigenous knowledge within the context of development through an emic perspective had already drawn much attention for the design of an appropriate research methodology which would also include more empathic interest in local problems, such as poverty, expressed by the 'voices of the local people'. The quest for more participatory methods has resulted in several types of local appraisal techniques, such as 'Participatory Rural Appraisal' (PRA), 'Participatory Urban Appraisal' (PUA) and 'Participatory Action Research' (PAR).

However, since indigenous knowledge systems represent the participants' subjective, i.e. 'insider' view, which needs to be objectivised, made value-free and replicable in order to substantiate its evidence-based significance in the scientific discourse, a combination of complementary qualitative and quantitative techniques had to be developed. Starting with qualitative, in-depth research on case studies, the quantitative techniques would follow to measure the spread of phenomena, among larger—often comparative—populations, away from personal data towards the data among larger units such as groups or communities. Such 'objectivation' of research findings is all the more important since indigenous knowledge systems have largely been ignored and sometimes even ridiculed during the colonial era, in which 'primitive peoples' and their knowledge systems were often subordinated to 'superior' Western knowledge and technology from the colonial powers. Since, on the one hand, these systems have shown to be crucial to attain participatory sustainable development, while on the other hand, they tend to disappear rapidly in view of the current processes of modernisation and globalisation, neo-ethnoscience research has become most urgent to document these systems in a comparative, scientific methodological way for the future.

As will be further elaborated in Chap. 7, to this end, Slikkerveer (1999a) developed the 'Leiden Ethnosystems Approach', specifically designed with a view to studying indigenous knowledge systems within their dynamic context of processes of development and change, based on a combination of three concepts, i.e. the 'Historical Dimension' (HD), the 'Participant's View' (PV), and the 'Field of Ethnographic Study' (FES) (cf. Leakey and Slikkerveer 1991; Agung 2005; Leurs 2010; Djen Amar 2010; Ambaretnani 2012; Chirangi 2013; Aiglsperger 2014, Erwina 2019; Saefullah 2019).

An important factor in the outcome of recent development programmes has been an attempt to adopt the Participant's View of the indigenous target populations by the policy planners and project managers, as their attention has been drawn to the fact that one specific characteristic of the local people's participation refers to a form of local solidarity among participants/community members, which goes often beyond the economic or financial exchange of goods and services. Since community-based cooperation and solidarity is rooted in non-commercial, often institutionalised expressions of mutual aid and reciprocity, in which egalitarian rather than hierarchical principles are operational, the relative gap between the rich and the poor at the community level is often confined to the local socio-cultural environment.

Following the reorientation which ethnoscience has evoked towards indigenous peoples and their knowledge systems, it has also put indigenous peoples and their systems of knowledge and technology first in development. Given the holistic as opposed to the mainstream 'scientific' reductionist position in the assessment of objects and phenomena, the empirical value of the new ethnoscience perspective is represented in its multidisciplinary orientation towards practical problem-solving activities which tend to cross conventional boundaries of disciplines and sub-disciplines. Moreover, it is being well documented that complex Indigenous Knowledge Systems are often sustainable, participatory and particularly functional in small-scale level development programmes.

In the current globalisation process, the new perspective has introduced an important development-oriented picture of the interaction among local and global knowledge systems in various sectors of local communities, paving the way for a multidisciplinary integration and synergy of local and global knowledge and technology, which in a 'bottom-up' approach embarks on indigenous knowledge systems. Following the above-mentioned IKS-based paradigm of the cultural dimension of development, the LEAD Programme has also been able to implement its integrative approach in several sub-fields of the neo-ethnoscience, conveniently brought together in the three integrated R&D clusters as mentioned above: *ethno-botany/pharmacy/medicine*, *ethno-economics/mathematics*, and *ethno-communication/history*. In combination with the three target regions of South-East Asia, East Africa and the Mediterranean Region, LEAD has carried out a number of internationally-funded research programmes with its counterparts, such as the *Indigenous Agricultural Knowledge Systems for Agricultural Development in South-East Asia and East Africa* (INDAKS), the *Mediterranean Useful Wild Plants Programme* (MEDUSA), the *Over-the-Counter Medicines Provision and Consumption in South-European Countries* (OTC-SOCIOMED), and *Integrated Microfinance Management in Indonesia* (IMM).

As a follow-up to its participation in the remarkable two-day *World Culture Forum,* held in Nusa Dua, Bali, Indonesia in 2013, which pertained to the formulation of the '*Bali Promise*' calling for an effective role and integration of culture in development at all levels, the LEAD Programme is committed to contributing to the *Post-2015 Agenda for Sustainable Development and the Sustainable Development Goals* (SDGs) of the United Nations (UN 2015) by promoting the integration of local and global systems of knowledge and technology in order to implement *Integrated Microfinance Management* (IMM) and *Integrated Sustainable Community Development* (ISCD) for worldwide poverty reduction, particularly in Indonesia in the near future (cf. Agung 2013; Slikkerveer 2013).[2]

[2]The first 'World Culture Forum' (WCF) was held in Bali, Indonesia from 23 to 27 November, 2013, chaired by Dr. Susilo Bambang Yudhoyono and organised by the Government of Indonesia. Based on the general need for a more visible and effective integration and mainstreaming of culture into development policies and strategies at all levels, culture was identified as a driver in

2.5 The Challenge of Poverty Reduction for Human Development

As mentioned in the *Introduction*, the conventional description and definition of the concept of *poverty* in strictly monetary terms has failed to provide a realistic conceptualisation of poverty as it encapsulates more aspects: a general lack of access to proper services of health, education, justice, employment and freedom, rendering the condition to be a wider, more dramatic complex problem of the global society, which similarly needs a broader remedy than just the provision of a financial input of microcredit, microloans or microfinance.

Indeed, poverty as a human condition has recently been defined in various ways, as both national governments and international organisations take an interest to measure and compare the percentage of a population—or part of it—which falls within the definition selected to capture the concept of poverty. As an unacceptable condition of humankind, it has been the subject of various approaches designed to reduce or alleviate poverty, generally through the implementation of various socio-economic development-based policies and strategies.

In 2005, the United Nations published the *Handbook on Poverty Statistics*, followed by the World Bank, which in 2009 published the *Handbook on Poverty and Inequality* (cf. Haughton and Khandker 2009). Embarking on the classic definition of poverty by the World Bank (2000): '*poverty is pronounced deprivation in wellbeing*', questions have been raised about what is meant by 'well-being' and the reference point against which deprivation can best be measured. Different approaches to define poverty offer various conceptualisations, ranging from the strict availability of resources to a wider view of different dimensions related to the human condition.

More recently, the World Bank (2000) has not only broadened its thinking about the concept of poverty, its causal framework and its causal structure, but also widened its strategy of the 1990s of focusing on labour-intensive growth, social sector investments and transfers/safety-nets for those left out, to the recent attention for opportunity, empowerment and security of the poor. Another approach is to define 'well-being' as the command over commodities in general, so people are better off if they have a greater command over resources. As the main focus is on whether households or individuals have enough resources to meet their needs, poverty is then measured by comparing individuals' income or consumption with some defined threshold below which they are considered to be poor. This is the

sustainable development, as opposed to the past deficit model of culture as an obstacle. A general understanding was reached about the importance of culture and the processes for culture to be included in the Post-2015 Agenda for Sustainable Development and the Sustainable Development Goals (SDGs) of the United Nations. The outcome of the WCF is the "Bali Promise" in which all participants of the World Culture Forum call for the measurable and effective role and integration of culture in development at all levels in the Post-2015 Agenda Development Agenda.

most conventional view in which poverty is seen largely in monetary terms, being the starting point for most analyses of poverty.

A second approach to 'well-being'—and hence conversely to poverty—is to investigate whether people are able to obtain a *specific* type of consumption good: *Do they have enough food? Or shelter? Or health care? Or education?* In this way, the analyst goes beyond the more traditional monetary measures of poverty: *nutritional poverty* might be measured by examining whether children are stunted or wasted; and *educational poverty* might be measured by asking whether people are literate or how much formal schooling they have received.

The broadest approach to 'well-being' has been brought forward by Sen (1990, 1993), who argues that well-being comes from a capability to function in society, linking the concept of poverty to wider notions of social need and well-being. Sen's 'Capability Approach' (CA) provides an alternative economic framework for thinking about well-being, poverty and human development. In this way, poverty arises when people lack key capabilities and adequate income or education, or suffer from poor health, or insecurity, or low self-confidence, or a sense of powerlessness, or the absence of rights such as freedom of speech. Viewed in this way, poverty is a multi-dimensional phenomenon and less amenable to simple solutions: while higher average incomes will certainly help reduce poverty, these may need to be accompanied by measures to empower the poor, or insure them against risks, or to address specific weaknesses such as inadequate availability of schools or a corrupt health service. In this way, poverty arises when people lack key capabilities and have inadequate income or education, or poor health, or insecurity, or low self-confidence.[3]

More recently, the World Bank (2005) introduced a comprehensive definition for the people living in *extreme poverty* when they earn less than $1.25 per day and those living in *moderate poverty* when they earn between $1.25 and $.2.00 per day. The United Nations Development Programme (UNDP 2011) defines poverty as: "*a human condition characterised by the sustained or chronic deprivation of the resources, capabilities, choices, security and power necessary for the enjoyment of an adequate standard of living and other civil, cultural, economic, political and social rights.*" By taking such a wider approach, multi-dimensional poverty is made up of several factors which determine the poor people's experience of deprivation, including poor health, lack of education, inadequate living standards, lack of income, disempowerment, poor quality of work and threat from violence.

[3]The 'Capability Approach' was first voiced by the Indian economist and philosopher Amartya Sen in the 1980s, and is defined by its choice of focus upon the moral significance of individuals' capability of achieving the kind of lives they have reason to value. The approach differs from other perspectives on ethical evaluation, such as utilitarianism or resourcism, which tend to focus on subjective well-being. The individual's capability to live a good life is defined in terms of a set of value factors, such as good health or a balanced relationship with others. The Capability Approach has been used in the context of human development as a broader alternative to narrow economic metrics such as growth in GDP per capita. In this context, 'poverty' is understood as deprivation in the capability to live a good life, and 'development' is understood as capability expansion (http://www.iep.utm.edu/sen-cap/).

Also from the point of view of the poor people themselves, *i.e.* the participants, the need for a multi-dimensional approach to the conceptualisation of poverty is evident: not only do they perceive their poor condition as ill-being in terms of poor health and nutrition, insufficient education and housing, inadequate sanitation and drinking water, and a general lack of services, but they also suffer from another dimension of the concept: the stigma of poverty.

This 'missing' dimension of poverty refers to the relational perspective among individuals feeling shame, worthlessness and moral inferiority, causing the social stigma associated with poverty. In the same way, Gubrium et al. (2013) add to the *Multidimensional Poverty Index* the social, political and psychological—i.e. human—dimensions of poverty and stress the risk that anti-poverty policies and programmes inadvertently stigmatise their beneficiaries and aggravate their own shame. As indicated in the *Introduction*, the United Nations Development Programme (UNDP 2011) started to use such a composite index, known as the *Multidimensional Poverty Index* (MPI), based on: (a) the likelihood that a child is not surviving to the age of 60 years; (b) the functional illiteracy rate; (c) long-term unemployment; and (d) population living on less than 59% of the median national income.

Sometimes, poverty is also defined by the location of the people concerned. In Sub-Saharan Africa and Asia, there are the *village poor,* living in the numerous villages, while the *rural poor* are defined as living in small communities where a general situation of depression dominates in terms of drought, abandoned industry and unemployment. The *urban poor* are defined as people living in relative poverty in slums and shanty towns. Sachs (2005) distinguishes three degrees of poverty: (1) households living in *extreme poverty*; (2) households living in *moderate poverty*; and (3) households living in *relative poverty*.[4] In addition to population pressure, natural disasters and constraints of arable land, factors related to poverty include: "*adverse environmental factors, poor health, hard economic conditions, lack of basic infrastructures and services, poor access to education, strong social factors* etc." (cf. Kotler and Lee 2009).

Another negative phenomenon in this context is the '*vicious circle of poverty*', which seems to persist for generations among the poor and low-income households. As Kotler and Lee (2009) conclude: "*Every baby born into a poor family faces a higher-than-average chance of dying at birth or shortly thereafter due to inadequate health facilities and abominable living conditions.*" The impact of all these

[4]Households are living in 'extreme poverty' if: 'they cannot meet basic needs for survival'. The World Bank (2005) estimated that 1.4 billion people in 2005 were living in extreme poverty, i.e. on less than $1.25 per day. Households are living in 'moderate poverty' if: 'their basic needs are met, but just barely'. The World Bank (2005) estimated that 1.6 billion people in 2005 were living in moderate poverty, earning between $1.25 and $2.00 per day. Households are living in 'relative poverty' if: 'they have an income level below a given proportion of average national income'. This degree of poverty reflects the distribution of incomes in a country or region. Although estimates are lacking for the number of the 'relative poor', it can be expected that about 4 billion people around the world are living in relative poverty.

Table 2.1 Rank order of the top nine countries representing 84.24% of the world's poor living below the 'global poverty line' of $1.00 per day (2008)

Rank no.	Country	Percentage of the world's poor
1	India	41.01
2	China	22.12
3	Nigeria	8.03
4	Pakistan	3.86
5	Bangladesh	3.49
6	Brazil	1.82
7	Ethiopia	1.82
8	Indonesia	1.49
9	Mexico	1.43
10	Russia	0.99
Total		84.24

Source Nation Master (2008)

forces has further contributed to the increase in the number of the extreme poor in the past decade, specifically in developing countries.

Table 2.1 shows the rank order of the percentage of the world's poor who live in each of these 10 countries, where '*poor*' is defined as living below the 'global poverty line' of $1.00 per day (cf. Nation Master 2008), with almost two thirds (63%) living in India and China. The geographical spread of the 90% of people living in extreme poverty shows three regions on the globe: Sub-Saharan Africa, East Asia and South Asia. The global economic crisis of 2008 did not only hit the U.S. economy hard, but had a devastating effect on the monetary systems around the world.

In a recent study to seek an explanation for the fact that indigenous people suffer from poverty rates which are on average twice as high as for the remaining Latin Americans, Calvo-Gonzãlez (2016) found three additional factors which play a major role in the indigenous peoples' predicament of poverty:

First solid quantitative evidence of poverty and ethnicity is lacking in such poverty gaps as indicated in the recent report of the World Bank (2016): *Indigenous Latin America in the Twenty-First Century*.

Second the gap between the poverty rate of indigenous peoples and the rest of the population is not getting smaller, and in some countries it is stagnant or even widening. Such a poverty gap between the indigenous and non-indigenous populations is explained by the fact that indigenous peoples tend to live in rural areas, have lower education, etc.

Third apart from these demographic and economic factors, the fact that indigenous peoples seem still more likely to be poor remains a large 'unexplained gap', suggesting that indigenous peoples according to Calvo-Gonzãlez (2016): "*face specific challenges in benefiting from growth and getting out of poverty*." Figure 2.1 shows the unexplained part (8%) of the indigenous peoples in the poverty gap (19%) in Latin America.

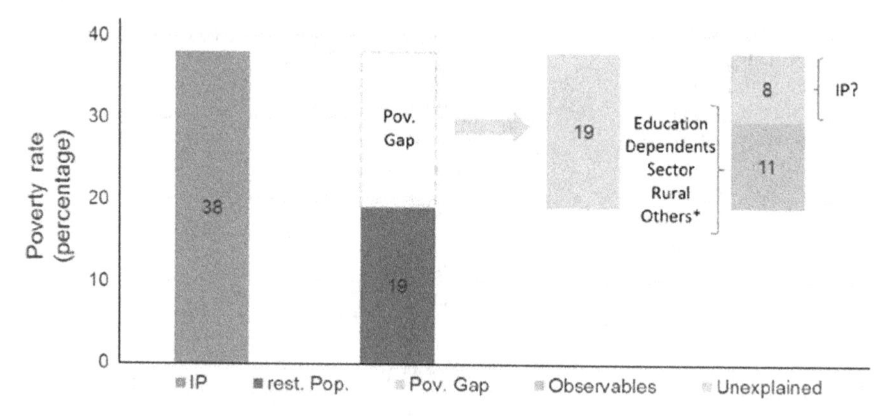

Fig. 2.1 The unexplained part (8%) of the indigenous peoples in the poverty gap (19%) in Latin America *Source* SEDLAC (World Bank and CEDLAS 2016)

2.6 IKS for Poverty Reduction and Sustainable Community Development

Following the misguided 'wisdom' among planners, development experts, change agents and government extension workers implementing their top-down policies up to the 1980s, that indigenous knowledge had 'little relevance' to the local decision-making process because modern Western science would manage and solve all development-related problems from an objective point of view, both neo-ethnoscience research and experience from the field show a different picture of the failure of many projects of local development planning, often going hand in hand with the suffering of physical, moral, emotional, and economic conditions by the local population. The growing number of scientists who support the view that indigenous knowledge is the key concept for sustainable development—referred to by Agrawal (2011) as 'neo-indigenismo'—agree that the integration of indigenous knowledge into development planning is the main requirement to achieve successful strategies (cf. Warren et al. 1995; Slikkerveer 1999a, b).

Meanwhile, with the reassessment in neo-ethnoscience of indigenous knowledge systems within the context of development, soon a growing number of organisations also started to acknowledge the usefulness of indigenous knowledge in sustainable socio-economic development, promoted by international organisations such as the *World Council of Indigenous Peoples* (WIPO), the *Canadian International Development Corporation* (CIDA), the *World Bank* (WB) and the *United Nations* (UN) (cf. Emery 1999).

Almost parallel to the sequence in the evolution of the major sub-fields of neo-ethnoscience after the 1980s, the wider recognition and use of the positive contributions from the integration of indigenous knowledge with global knowledge for development started in the field of medicine, where about 80% of the population in developing countries is using traditional medicine as their primary source for health and healing. Closely related developments had occurred in the sub-fields of ethno-botany and ethno-pharmacy, which refer to the interdisciplinary science investigating the perception and use of pharmaceuticals within a given human society for development. The first steps preceding the development of these sub-fields had already been taken since the formative period of time for modern medicine in the 16th century, where the study and functionalisation of various forms of indigenous, often herbal, medicines from non-Western areas were found to be useful institutions, not only for local health care development, but even more so for the production of modern medicines (cf. Lee and Balick 2001; Slikkerveer and Slikkerveer 1995; Bodeker et al. 1997; Leurs 2010; Aiglsperger 2014).

After the formal inclusion of Traditional Medicine (TM) and Complementary and Alternative Medicine (CAM) in Primary Health Care (PHC) since the 1980s by the World Health Organisation (WHO 1978, 2002, 2013; Bannerman et al. 1983), the practical significance of Traditional Ecological Knowledge (TEK) became widely accepted and integrated in development programmes, particularly in natural resource management, conservation of bio-cultural diversity, environmental assessment and planning, and the prevention of land degradation (cf. Inglis 1993; Berkes 1993; Posey 1999; Slikkerveer 1999a, b; Agung 2005).

Thereupon, the breakthrough in the integration of indigenous knowledge systems in agricultural development took place in many developing nations in Sub-Saharan Africa, Asia and Latin America, prompting the Food and Agricultural Organisation of the United Nations to officially include *Indigenous Agricultural Knowledge* (IAK) as a key factor to achieve sustainable agriculture, as it has demonstrated its potential to play an important role in conserving resources, finding food alternatives and reducing the use of chemicals in agricultural production (cf. Food and Agricultural Organisation 2010). Closely related to this sub-field is agroforestry, generally defined as focused on species-rich, low-input agricultural techniques including a diverse array of indigenous tree crops, rather than on high-input monocultures with only a small set of staple food crops. Tree domestication in agroforestry is aiming at the promotion of the cultivation of indigenous trees which have economic potential as new cash crops (cf. Michon and De Foresta 1993; Leakey and Tchoundjeu 2001).

The closely affiliated sub-field of ethno-economics evolved from the work of a group of scientists rethinking the contribution of economic anthropology to the study of 'primitive societies', in which the use of conventional economic models of people using their scarce resources to reach specific ends had been perceived as having global value, albeit in diverse variations among different peoples around the globe.

The revolutionary reorientation in ethnoscience of the 1980s as the study of the emic view of participants within the dynamic context of development in order to assess indigenous knowledge systems as the primary object of a new form of ethnoscience research—also known as 'neo-ethnoscience'—has rendered the existing economic-anthropological models no longer practicable, particularly as a result of the recognition that the planet's finite natural resources are imposing a limit to economic growth (cf. Schumacher 1973; Daly 2007). In a broader perspective, the same considerations seem to bear on the field of 'applied anthropology', championed by authors such as Sillitoe (1998) and Purcell (1998). As Halani (2004) comments on a recent publication of the papers presented at the ASA Conference 2000 by Sillitoe et al. (2003), it is: "*the attempt by anthropologists to contribute to the debate and carve their own niche in the process.*"

Cavalcanti (2002: 39) rightly indicates the implications of the adoption of the concept of sustainable development, that the economic problems which all communities have to solve can no more be analysed merely in terms of efficient allocation: "*It is now of the greatest relevance to contemplate the concomitant requirement of sustainability, which means finding solutions both efficient and compatible with the limits imposed by nature.*" Referring to the definition of ethno-economics, previously coined by Posey (1983), Cavalcanti (2002: 40) extends this sub-field to encompass a holistic domain: "*not only of the economic perspectives of traditional and indigenous peoples, but also of the latter's perceptions of a higher order of reality in which the economy is integrated with nature, social organisation, culture and the supernatural world, as just another element of this larger whole.*" Interesting contributions to this sub-field of ethno-economics include Crawford (1982), Posey (1983, 1999), Umoh-Akpan (2000), Sayle (2000), Posey and Balick (2006), and Saefullah (2019).

An overarching sub-field of neo-ethnoscience refers to ethno-communication as the study of the interaction between indigenous and modern systems of information and communication within the context of development in various sectors, which has evolved from the ethnography of communication referred to as the analysis of communication within the social and cultural practices and beliefs of the members of a particular culture or community (cf. Hymes 1964). While the ethnography of communication – unlike ethnography per se – used to focus on various forms of communication and their function within a particular culture, back in the 1960s and 1970s, the concept of 'development communication', had been put forward by authors such as Lerner (1958), Schramm (1964) and Rogers (1976), who proposed a new communication model based on social justice and equality instead of economic growth (cf. Moemeka 1994).

As in other neo-ethnoscience sub-fields, ethno-communication refers to what Eilers and Oepen (1991) define as: "*the description and study of communication means, structures and processes in a cultural unit*", further substantiating the integration of the new development paradigm in communication science in various sectors of the community (cf. Jamieson 1991; Servaes 1991; Djen Amar 2010; Erwina 2019).

Although these various sub-fields of neo-ethnoscience are primarily focused on different forms of indigenous knowledge as a contribution to attain sustainable community development in the vital sectors of the society, such as human and animal health, food security, education, and natural resource management, so far, their impact on poverty reduction seems only indirect. Notwithstanding the fact that the condition of poverty is much more complex than just a matter of low income and a scarcity of financial resources, the process of overall sustainable community development, particularly in general terms of socio-economic growth, creation of employment, increase in wages, improved health and education, and security of food and nutrition would in general also provide a better environment for the poor segment of the population.

As it has become clear that poverty is not just a matter of lack of finances, the conventional 'remedies' of international organisations, government agencies, MFIs and NGOs providing subsidised funds through microcredit and, more recently, commercial microfinance and banks, are only covering part of the problem on a short-term basis, eventually leaving the poor at the community level empty-handed, if not worse off.

During the past centuries, the position of the poor people in rural communities has largely been safeguarded and protected by the indigenous systems of knowledge, beliefs and practices sustaining the livelihood of the entire community for many generations. During the colonial period, however, most of these local systems have been marginalised—or even completely swept away—by the penetration from outside Western systems of commercialisation, resource exploitation and political dominance. Despite the beginning of development aid after World War II, followed by development cooperation since the 1960s, a large segment of people in the rural communities has remained marginalised or poor, largely dependent on the still operational indigenous knowledge systems and indigenous institutions of informal cooperation, self-help and mutual aid which have continued to take care of poor community members.

While, today, the recognition, functionalisation and integration of indigenous knowledge systems in programmes and projects of development cooperation have eventually been widely accepted in so far as they have survived and not been threatened by extinction because of the influx of the processes of modernisation and globalisation, it is essential to assess to what extent the indigenous knowledge systems have directly contributed to poverty reduction.

Referring to the above-mentioned achievements documented in the various sub-fields of neo-ethnoscience, indigenous knowledge systems have indeed contributed to the process of sustainable community development in many sectors, such as health, education, agriculture, natural resources management and communication. A comprehensive collection of case studies is provided by Warren et al. (1995), which documents the direct contribution of activities implementing the integration of indigenous knowledge systems in development projects and programmes in sectors such as human and animal health, agriculture, environment, forestry, education and communication. Underscoring that all communities have developed their own body of knowledge over many generations, Gorjestani (2000) broadens this view, contending that indigenous knowledge (IK) constitutes not only a significant resource which could contribute to the increased efficiency and sustainability of the development process, but that: "*Harnessing of IK empowers local communities and could help improve aid effectiveness in poverty reduction*".

Kimbwarata (2010) documents the interesting discussion of 'Harnessing and Using Local Knowledge for Development' organised in 2009 by the *Kenya Knowledge Network for Policy, Research and Development* (KNET).[5]

In this context, Øyen et al. (2002) present an overview of 'best practices' in poverty reduction, in which the role of indigenous knowledge in poverty reduction is also examined. As the authors try to treat the concept of 'best practices' as a scientific tool to understand the many failures of poverty reduction programmes, they identify 'participation' as a crucial factor in the intervention process, albeit that these processes tend to last not long enough to warrant participation as a 'best practice'. Although they list a number of best practices databases and linked knowledge resources presented in several web sites, there remains the difficulty of what exactly is meant by 'best practices' and the lack of the related general consensus. In the list of indigenous knowledge databases, the *CIRAN Indigenous Knowledge Database* is mentioned, encompassing 27 best practices in this area, largely referring to cost-effective and sustainable survival strategies.[6]

2.7 The Critical Role of Indigenous Institutions in Sustainable Development

Throughout history, people have not only relied on their own activities and that of their relatives to maintain their health and well-being, but have also created community-based associations and institutions to secure and increase their individual and family resources for their livelihood and survival, and that of the entire community. Although there is no universal agreement, institutions are defined by North (1990) as: "*constraints that human beings impose on themselves*" implicating that institutions prohibit, permit or require particular economic, political, legal or

[5]KNET is a partnership project funded by the African Capacity Building Foundation (ACBF). The aim of the network is to facilitate sharing of knowledge, expertise, resource materials, best practices, lessons, and experiences. Its members include the Institute of Policy Analysis and Research (IPAR), Kenya Institute for Public Policy Research and Analysis (KIPPRA), Kenya Private Sector Alliance (KEPSA), Kenyan Economic Association (KEA), Association of Professional Societies in East Africa (APSEA), Association of Local Government Authorities of Kenya (ALGAK) and Kenya's Ministry of Higher Education, Science and Technology (MoHEST). IPAR is currently hosting the network.

[6]The Centre for International Research and Advisory Networks (CIRAN) at Nuffic was established in cooperation with the Leiden Ethnosystems and Development Programme (LEAD) at Leiden University and the Center for Indigenous Knowledge and Agricultural Development (CIKARD) at Iowa State University, supported by 35 local and regional IK centres worldwide. The Indigenous Knowledge and Development Monitor (IK&D 1993–2001) promotes the dissemination of local knowledge as a service to the international development community and all scientists who share a professional interest in indigenous knowledge systems and practices. The journal was published until the end of 2001, after which most of the IK&D Centres had expanded, now being involved in further bringing together a growing body of indigenous knowledge and practices within the context of development in the many sub-fields of ethnoscience.

social actions. In the Western countries, the formation of early trade unions and workers' organisations for the protection of their well-being during the industrialisation of the 19th century led to increased political influence of the 20th century, pressing government agencies to take over such tasks and implement communal social service and income regulations, which eventually pertained to the welfare state. As Livermore and Midgley (2009: 181) document, the institutional approach was based on progressive liberalism and social democratic ideology, contending that: "*social needs should be addressed through a range of statutory interventions, including fiscal measures, legal regulations, and the provision of a comprehensive system of social services.*"

Although the institutional approach has recently been undermined by the neo-liberal ideology, several authors argue that the institutional ideas should continue to provide a basis for the provision of social policies. While Etzioni (1993) is convinced: "*that institutional ideas should be united with communitarian beliefs and that the state should actively encourage community participation in social policy formulation and implementation*", Gilbert and Gilbert (1989), Sherraden (1991) and Midgley (1995, 1999) contend that the institutional ideas underlying the state responsibility for all members of the society underscore the significance of the role of the state for the social welfare of the society.

The comparable evolution of the growing recognition of the potential of institutions in development programmes by international development organisations and agencies emerged on the basis of two factors:

Firstly the historical development of government institutions into the welfare state of the 20th century in the industrialised countries, which transpired also as a promising 'replica model' through the policies towards socio-economic development in the developing countries, and

Secondly the centralist, 'top-down' approach in development cooperation, which had been introduced after the independence of many new states, facilitating the implementation of often forced socio-economic development programmes from outside the community.

As regards the situation in developing countries, a number of scientific studies started to express a growing concern regarding the rather limited success—and sometimes even right-off failure—of international development programmes in their efforts to impose modern, Western-oriented norms, rules and practices on local systems of knowledge and use of resources, such as the work carried out by De Boef et al. (1993), Bruce and Migot-Adholla (1994) and Uphoff (1996). These and other studies eventually called for a drastic reorientation towards development cooperation, in which bilateral projects were gradually replaced by multilateral programmes under the auspices of international development organisations and their agencies, including the United Nations and the World Bank.

In addition, socio-economic development became rather risk-group oriented, focused on promoting the improvement of remote communities outside the mainstream areas, embodied by development policies of rural development and community development, largely through the intermediary of newly-established 'external' institutions introduced from outside such as NGOs, and later on, MFIs. Despite these international efforts, the practical experience, which had been accumulated in the field by some development experts and change agents documented the notion that those development projects which linked up with the locally existent indigenous institutions were generally enjoying increased participation, and by consequence, resulted often in more positive impacts on the development of the local communities.

Several studies document that in the developing countries, for many centuries, the indigenous institutions have provided the local socio-cultural arrangements for the guidance, organisation and regulation of the activities of all community members, embedded in the local historical and cultural setting (cf. Chamlee 1993; Appiah-Opoku and Mulamoottil 1997). With regard to development, Williamson (2000) introduces a broader definition of institutions which also includes the organisational, procedural and regulatory frameworks, allowing for the assessment of institutional quality in relation to development outcomes.

As mentioned in the *Introduction*, the interest in the role of indigenous institutions and organisations in development had emerged in the course of the 1990s from the work of scholars concerned about the rather limited success of international development programmes in imposing their modern, Western-based norms, rules and practices on local systems of knowledge and use of resources. Later research has further strengthened the central position of indigenous institutions in the development process (cf. Blunt and Warren 1996; Metha et al. 1999; Watson 2003). As the convergence of this indigenous institutional approach with the interest in indigenous ecological knowledge and practices has generated new challenges in various sectors, Watson (2003) contends that: "*if the indigenous NRM institutions can be identified and harnessed, it is thought that this strengthened NRM will be achieved, that it will have the 'added value' of drawing on local expertise, and that the projects will be participatory*".

In his literature review of institutions and development, Jütting (2003) refers to traditional or indigenous institutions as informal, often dating back many centuries, and embedded in social norms, customs and traditions. Jütting (2003) provides an analytical framework to measure the role of indigenous social institutions in development, which in the view of Katseli (2003): "*are of major importance for low-income countries, but still receive only marginal attention.*"

Indigenous institutions have shown in several case studies to be crucial for the understanding of the management of natural resources, the development of markets and conflict management. Similar research on the positive role of indigenous institutions in development include case studies in Africa by Brautigam (1998), Platteau (2000) and Mazzucato and Niemeyer (2002), all strongly recommending policy makers to include these informal institutions in their development plans. Linquist and Adolph (1996) document the positive experience in development with the integration of indigenous institutions in the development planning of the Gabra in Northern Kenya.

In his article on *Indigenous Economic Institutions and Ecological Knowledge in Ghana*, Appiah-Opoku (1999: 217) discusses the nature and operation of indigenous economic institutions in conjunction with their ecological knowledge, norms, beliefs, and practices which pertain to sustainable utilisation of natural resources and environmental management, underscoring that: "*In this sense, a meaningful discussion of indigenous ecological knowledge and practices should not be separated from the institutions which give rise to such knowledge and practices.*"

Although indigenous institutions are generally perceived as inextricably linked to the related indigenous knowledge systems being a key concept for sustainable development, so far they are still under-represented in IKS-based policies of sustainable development, let alone poverty reduction. As mentioned in the *Introduction*, the shift in the development paradigm evoked not only a growing interest in indigenous knowledge, but also soon led to follow-up research on the role of the closely related indigenous institutions and organisations in 'bottom-up' development policies and programmes. In contrast to formal institutions, the indigenous institutions have been developed over many generations at the community level, based often on unwritten social values and norms of behaviour and conventions which prohibit or permit individuals to undertake activities within their socio-cultural and natural environment.

The significant role of indigenous institutions in the economic sector in Ghana is well documented by Chamlee (1993), illustrating that indigenous financial arrangements provide a practical alternative to the formal banking system. Despite the fact that in general the potential of these alternative institutional arrangements hardly tends to be considered as promising, Chamlee (1993) lists several reasons why these indigenous arrangements should not be immediately dismissed. First, the bulk of investment activity is financed through indigenous institutions and not through the formal banking institutions. Secondly, larger loans are not always needed as some small development programmes such as the microloans in Bangladesh indicate that small loans can make a substantial difference. Thirdly, in cases where these indigenous institutions are disrupted, it is often the result of state regulations and restrictions on trade.

In the area of the development of politics and governance, Beal and Ngonyama (2009) embark in their study of the indigenous institution of chieftaincy in South Africa on the notion that such customary forms of governance have remained salient, as they are deeply rooted in local traditions. In their assessment of the role of leadership and inclusive elite settlements, they conclude that: "*the forging of development coalitions shows the importance of institutional arrangements for consolidating and scaling-out of local success stories into broad-based development strategies.*" These and related studies on the role of different categories of indigenous institutions in sustainable development, which potentially could indirectly pertain to a significant contribution to poverty reduction at a global scale, include in addition to the above-mentioned economic, political and legal institutions, also indigenous medical institutions, indigenous educational institutions and indigenous communication institutions.

Moreover, as regards the specific input into the alleviation of poverty at the community level, a specifically important category includes the indigenous institutions of mutual aid, solidarity, self-help and cooperation which over many generations have not only sustained the life and livelihood of all community members of a particular culture, but also supported the survival of the poor and low-income households. By consequence, the role of these indigenous socio-economic institutions is particularly important in the joint efforts of local development to achieve poverty reduction at the community level. Indeed, despite the marginalisation of their position as the result of outside interventions since the 1960s, the indigenous institutions have maintained their socio-cultural role and continued to be operational at the village level, taking care of the poor and low-income families. In this way, the largely ignored indigenous institutions constitute a key factor in poverty reduction which merits more attention from the international development circles.

In 2005, a major initiative was taken by the *Asia Indigenous Peoples Pact* (AIPP), with the objective to provide a venue for indigenous peoples in Asia to come to a common understanding about the concepts and issues, and to identify different aspects and needs concerning indigenous development. A series of Indigenous Development Conferences were organised between 2005 to 2009 in Tulongan, Mindanao, Philippines (2005), Toraja, Indonesia (2006), Pokhara, Nepal (2007) and Sabah, Malaysia (2008). These conferences were attended by representatives from 13 Asian countries, i.e. Bangladesh, Burma, Cambodia, India, Indonesia, Laos, Malaysia, Myanmar, Nepal, Philippines, Timor Leste, Thailand, and Vietnam. As the *Asia Indigenous Peoples Pact* (AIPP) indicates, the overall objectives of the Indigenous Development Conferences were:

- to restore the integrity and cohesiveness of indigenous communities in the region;
- to empower and affirm self-determination of communities in terms of the type of development based on indigenous concepts;
- to provide a venue for indigenous peoples in Asia to come to a common understanding about the concept and issues to identify different aspects and needs on indigenous development; and
- to come up with strategies to revitalise the different aspects of indigenous development.

By defining the concept of ‘indigenous development’ as ‘the growth or progress of an indigenous community in their originality or within the context of their ethnic identity in a holistic way’, the *Asia Indigenous Peoples Pact* (AIPP 2015) based its conception of ‘indigenous integrity’ on 10 aspects of indigenous systems which are interrelated, indivisible, and interdependent. These aspects are based on cultural, social, spiritual, political/institutional, juridical, economic, technological, health-related and educational ways of learning, including natural resources management.

Recently, the *Asia Indigenous Peoples Pact* (AIPP 2015) further documented that it had achieved significant progress in the expansion of its programmes at all local, national, regional and global levels, and strengthened the solidarity and cooperation among indigenous peoples, rendering its activities to further strengthening of the Indigenous Peoples' Movements in the Asian Region.

The adoption of the United Nations *Declaration on the Rights of Indigenous Peoples* in 2007 has further provided affirmative support to the indigenous peoples' perspective in developing these elements as well as to the framework of normative rights for indigenous peoples to pursue such a development model, and to inform the governments of their duties and obligations to change the models of development which so far have been detrimental (cf. United Nations 1948).

References

Agrawal, A. (2011). Dismantling the divide between indigenous and scientific knowledge. *Paper for the Conference on Indigenous Knowledge*. Tampa, FL: University of South Florida.

Agung, A. A. G. (2005). *Bali endangered paradise? Tri Hita Karana and the conservation of the Island's biocultural diversity*. Ph.D. Dissertation. Leiden Ethnosystems and Development Programme. LEAD Studies No. 1. Leiden University. xxv + ill., pp. 463.

Agung, A. A. G. (2013, November 24–27). *The key role of the balinese cosmology of Tri Hita Karana for sustainable conservation and development in Indonesia*. Paper presented at the World Culture Forum, Nusa Dua, Bali, Indonesia.

Aiglsperger, J. (2014). *'Yiatrosofia yia ton Anthropo': Indigenous knowledge and utilisation of MAC plants in Pirgos and Pretoria, Rural Crete: A community perspective on the plural medical system in Greece*. Ph.D. Dissertation. Leiden Ethnosystems and Development Programme. LEAD Studies No. 9. Leiden University. xxv + ill., pp. 336.

Ambaretnani, P. (2012). *Paraji and Bidan in Rancaekek: Integrated medicine for advanced partnerships among traditional birth attendants and community midwives in the Sunda Area of West Java, Indonesia*. Ph.D. Dissertation. Leiden Ethnosystems and Development Programme. LEAD Studies No. 7. Leiden University. xx + ill., 265 pp.

Appiah-Opoku, S. (1999). Indigenous economic institutions and ecological knowledge: A Ghanaian case study. *Environmentalist, 19*(3), 217–227.

Appiah-Opoku, S., & Mulamoottil, G. (1997). Profile: Indigenous institutions and environmental assessment: The case of Ghana. *Environmental Management, 21*, 159–171.

Asia Indigenous Peoples Pact (AIPP). (2015). *Annual report 2015: Strengthening indigenous peoples' movements*. Thailand, Chiang Mai: AIPP. http://www.aippnet.org.

Atran, S. (1991). Social science information/Sur Les Sciences Sociales. *Ethnoscience Today, 30* (4), 595–662.

Augé, M. (1999). *The war of dreams: Exercises in ethno-fiction*. London, Sterling, Va.: Pluto Press.

Balasubramanian, A. V., & Nirmala Devi, V. D. (Eds.). (2006, July 3–5). *Traditional knowledge systems of India and Sri Lanka*. Papers presented at the COMPAS Asian Regional Workshop on Traditional Knowledge Systems and their Current Relevance and Applications, Bangalore.

Bannerman, R. H., Burton, J., & Wen-Chieh, C. (1983). *Traditional medicine and health care coverage: A reader for health administrators and practitioners*. Geneva: WHO.

Barke, M., & Newton, M. (1997). The EU LEADER initiative and endogenous rural development: The application of the programme in two rural areas of Andalucia. *Journal of Rural Studies, 13* (3), 319–341.

Beal, J., & Ngonyama, M. (2009). Indigenous institutions, traditional leaders and elite coalitions for development: The case of Greater Durban, South Africa. *Crisis States Working Papers Series*, No. 2. London: DESTIN Development Studies Institute.

Berkes, F. (1993). Traditional ecological knowledge in perspective. In J. T. Inglis (Ed.), *Traditional ecological knowledge: Concept and cases*. Ottawa, Canada: IDRC.

Berkes, F. (1998). Exploring the basic ecological unit: Ecosystem-like concepts in traditional societies. *Ecosystems, 1*(4), 409–415.

Bodeker, G. m., Bhat, K. K. S., Burley, J., & Vantomme, P. (Eds.). (1997). *Medicinal plants for forest conservation and health care*. Non-Wood Forest Products. No. 11. Rome: FAO.

Blunt, P., & Warren, D. M. (1996). *Indigenous organisations and development: A training manual for non-government organisations*. IT Studies on Indigenous Knowledge and Development. London: Intermediate Technology Publications Ltd.

Brautigam, D. (1998). Substituting for the state institutions and industrial development in Eastern Nigeria. *IRIS Reprint*. No. 81. University of Maryland, MD.

Bruce, J. W., & Migot-Adholla, S. E. (1994). *Searching for land tenure security in Africa*. Dubuque, IA: Kendall Hunt Publishing Company.

Brunori, G., & Rossi, A. (2007). Differentiating countryside: Social representations and governance patterns in rural areas with high social density: The case of Chianti, Italy. *Journal of Rural Studies, 23*(2), 183–205.

Calvo-Gonzãlez, O. (2016). *Why are indigenous peoples more likely to be poor?* Washington, D. C.: The World Bank.

Castro, A. P. (2000). *Facing Kirinyaga: A social history of forest commons in Southern Mount Kenya*. IT Studies on Indigenous Knowledge and Development. London: Intermediate Technology Publications Ltd.

Cavalcanti, C. (2002). Economic thinking, traditional ecological knowledge and ethnoeconomics. *Current Sociology, 50*(1), 39–55.

Chambers, R., Pacey, A., & Thrupp, L. A. (1989). *Farmer first: Farmer innovation and agricultural research*. London: Intermediate Technology Publications Ltd.

Chamlee, E. (1993). Indigenous African institutions and economic development. *Cato Journal, 13* (1) (Spring/Summer).

Chan, M. (2014). Supporting the integration and modernisation of traditional medicine. *Science, 346*(6216 Suppl), S2.

Chirangi, M. M. (2013). *'Afya Jumuishi': Towards interprofessional collaboration between traditional & modern medical practitioners in the Mara Region of Tanzania*. Ph.D. Dissertation. Leiden Ethnosystems and Development Programme. LEAD Studies No. 8. Leiden University. xxvi + ill., 235 pp.

Collins, H. (1998). *What's wrong with relativism? Physics world*. Bristol, UK: IOP Publishing.

Comparing and Supporting Endogenous Development Network (COMPAS). (2016). *Vision and mission*.

Conklin, H. C. (1957). *Hanunoo agriculture: A report on an integral system of shifting cultivation in the Philippines*. Northford, CT: Elliot's Books.

Crawford, I. M. (1982). *Traditional aboriginal plant resources in the Kalumburu area: Aspects in ethno-economics*. Records of the Western Australian Museum Supplement. No. 15.

D'Andrade, R. G. (1995). *The development of cognitive anthropology* (286p). Cambridge: Cambridge University Press.

Daly, H. E. (2007). *Ecological economics and sustainable development. Selected essays of Herman Daly. Advances in ecological economics*. Edward Elgar Publishing.

de Boef, W., Manor, K. A., Wellard, K., & Bebbington, A. (Eds.). (1993). *Cultivating knowledge. Genetic diversity, farmer experimentation and crop research*. London: Intermediate Technology Publications Ltd.

Dibden, J., Cocklin, C., & Potter, C. (2009). Productivist and multifunctional trajectories in the European Union and Australia. *Journal of Rural Studies, 25*(3), 299–308.

Djen Amar, S. C. (2010). *Gunem Catur in the Sunda Region of West Java: Indigenous communication on the MAC plant knowledge and practice within the Arisan in Lembang*. Ph.D. Dissertation. Leiden Ethnosystems and Development Programme. LEAD Studies No. 6. xx + ill., 218 pp.

Easterly, W. (2006). *The white man's burden: Why the West's efforts to aid the rest have done so much ill and so little good.* New York: The Penguin Press.

Economic and Social Survey of Asia and the Pacific (ESCAP). (2010). *Sustaining recovery and dynamism for inclusive development.* Bangkok, Thailand: United Nations. http://www.unescap.org/resources/economic-and-social-survey-asia-and-pacific-2010.

Eilers, F., & Oepen, M. (1991). Communication and development: Mainstream and off-stream perspectives—A German view. In L. Casmir (Ed.), *Communication in development.* Norwood N.J.: Ablex Publishing Corporation.

Emery, A. R. (1999). *Guidelines on traditional knowledge in environmental assessments.* Canada: KIVUNATURE.

Erwina, W. (2019). *Iber Kasehatan in Sukamiskin: Utilisation of the Plural Health Information & Communication System in the Sunda Region of West Java, Indonesia.* Ph.D. dissertation. Leiden Ethnosystems and Development Programme. LEAD Studies No. 10. Leiden University.

Etzioni, A. (1993). *The spirit of community: Rights, responsibilities and the communitarian agenda.* New York: Crown.

Food and Agriculture Organisation (FAO). (2010). *The state of food insecurity in the world addressing food insecurity in protracted crises.* Rome: FAO.

Gilbert, N., & Gilbert, B. (1989). *The enabling state: Modern welfare capitalism in America.* New York: Oxford University Press.

Gkartzios, M., & Scott, M. (2013). Attitudes to housing and planning policy in rural localities: Disparities between long-term and mobile rural populations in Ireland. *Journal of Land Use Policy, 31,* 347–357.

Goodenough, W. H. (1957). Cultural anthropology and linguistics. In P. L. Garvin (Ed.), *Report on the 7th annual round table meeting in linguistics and language study* (pp. 109–173). Washington, D.C.: Georgetown University.

Gorjestani, N. (2000). *Indigenous knowledge for development: Opportunities and challenges.* Washington, D.C.: The World Bank.

Gragson, T. L. (1999). *Ethnoecology: Knowledge, resources, and rights.* Athens, Georgia: University of Georgia Press.

Gubrium, E. K., Pellissery, S., & Lødemel, I. (2013). *The shame of it: Global perspectives on anti-poverty policies.* Bristol: Policy Press, University of Bristol.

Halani, L. (2004). Book review. Sillitoe, P., Bicker, A., & Pottier, J. (2002). Participating in development: Approaches to indigenous knowledge systems. *Anthropology Matters Journal, 6* (1).

Hardesty, D. L. (1977). *Ecological anthropology.* Hoboken, New Jersey: Wiley.

Haughton, J., & Khandker, S. R. (2009). *Handbook on poverty and inequality.* Washington, D.C.: The World Bank.

Haverkort, B., & Hiemstra, W. (Eds.). (2000). *Food for thought: Ancient visions and new experiments of rural people.* London: Zed Books.

Haverkort, B., & Reijntjes, C. (Eds.). (2007). *Moving worldviews: Reshaping sciences, policies and practices for endogenous sustainable development.* Compas Series on Worldviews and Sciences, No. 4. Leusden: ETC/COMPAS.

Haverkort, B., van 't Hooft, K., & Hiemstra, W. (Eds.). (2003). *Ancient roots, new shoots: Endogenous development in practice.* Leusden, Netherlands: ETC/Compas. London: Zed Books.

Horton, R. W. G. (1967). African traditional thought and western science. *Africa: Journal of the International African Institute, 37*(1): 50–71.

Hymes, D. (1964). Introduction: Toward ethnographies of communication. *American Anthropologist, 66*(6).

Indigenous Knowledge & Development (IK&D) Monitor. (1993–2001). *Centre for International Research and Advisory Networks (CIRAN).* The Haque: Nuffic.

Inglis, J. T. (1993). *Traditional ecological knowledge: Concepts and cases.* Ottawa, Canada: International Development Research Centre (IDRC).

Ingold, T. (2000). *The perception of the environment: Essays on livelihood, dwelling and skill.* London, UK: Routledge.

Innis, D. (1997). *Intercropping and the scientific basis of traditional agriculture.* IT Studies on Indigenous Knowledge and Development. London: Intermediate Technology Publications Ltd.

International Union of Conservation of Nature (IUCN). (1986). *Traditional conservation and development.* Occasional Newsletter of the Ecology Commission's Working Group on Traditional Ecological Knowledge (TEK), No. 4. Gland.

Jamieson, N. (1991). Communication and the new paradigm of development. In L. Casmir (Ed.), *Communication in development.* Norwood N.J.: Ablex Publishing Corporation.

Janzen, J. M. (1982). *The quest for therapy: Medical pluralism in Lower Zaire. Comparative studies of health systems and medical care.* Berkeley, CA: University of California Press.

Jütting, J. (2003). *Institutions and development: A critical review.* Working Paper No. 210. OECD Development Centre.

Katseli, L. (2003). Preface. In J. Jütting (Ed.), *Institutions and development: A critical review.* Working Paper No. 210. OECD Development Centre.

Kimbwarata J. (2010). Want sustainable development? Try indigenous knowledge. *Baobab, 57.*

Kotler, P., & Lee, N. R. (2009). *Up and out of poverty: The social marketing solution.* Prentice Hall, New Jersey: Pearson Education, Inc.

Leakey, R. E. (2012, June 29). Ryan Shaffer: Evolution, humanism, and conservation: The humanist interview with Dr. Richard Leakey. *The Humanist.*

Leakey, R. E., & Slikkerveer, L. J. (Eds.). (1991). *Origins and development of agriculture in East Africa: The ethnosystems approach to the study of early food production in Kenya.* Studies in Technology and Social Change, No. 19 (302p). Ames: Iowa State University Research Foundation.

Leakey, R. R. B., & Tchoundjeu, Z. (2001). Diversification of tree crops: domestication of companion crops for poverty reduction and environmental services. *Experimental Agriculture, 37,* 279–296.

Lee, R., & Balick, M. J. (2001). Ethnomedicine: Ancient wisdom for contemporary healing. *Alternative Therapies in Health and Medicine, 7*(3), 28–30.

Lerner, W. (1958). *The passing of the traditional society.* Glencoe, Ill.: The Free Press.

Leurs, L. N. (2010). *Medicinal, aromatic and cosmetic (MAC) plants for community health and bio-cultural diversity conservation in Bali, Indonesia.* Ph.D. dissertation. Leiden Ethnosystems and Development Programme. LEAD Studies No. 5. Leiden University. xx + ill., 343 pp.

Linquist, B. J., & Adolph, D. (1996). The drum speaks—are we listening? Experiences in development with a traditional Gabra institution—The Yaa Galbo. In: Blunt, P., & Warren, D. M. (1996). *Indigenous organisations and development: A training manual for non-government organisations.* IT Studies on Indigenous Knowledge and Development. London: Intermediate Technology Publications Ltd.

Livermore, M., & Midgley, J. (2009). *Handbook of Social policy.* Thousand Oaks, CA: SAGE Publications Inc.

Lowe, P., Ray, C., Ward, N., Wood, D., & Woodward, R. (1998). *Participation in rural development: A review of european experience.* Research Report, Centre for Rural Economy. University of Newcastle Upon Tyne. http://www.ncl.ac.uk/cre/publish/pdfs/rr98.1a.pdf. Last accessed February 25, 2011.

Mathias-Mundy, E., & McCorkle, M. C. (1989). *Ethnoveterinary medicine: An annotated bibliography.* Bibliographies in Technology and Social Change. No. 6. Technology and Social Change Programme. Ames, USA: Iowa State University.

Mazzucato, V., & Niemeyer, D. (2002). Population growth and the environment in Africa: Local informal institutions. The missing link. *Economic Geography, 78,* 2.

Metha, L., Leach, M., Newell, P., Scoones, I., Sivaramakrishnan, K., & Way, S. A. (1999). Exploring understandings of institutions and uncertainty: New directions in natural resource management. *IDS Discussion Paper, No. 372.*

Michon, G., & De Foresta, H. (1993). *Indigenous agroforests in Indonesia: Complex agroforestry systems for future development*. Bogor, Indonesia: Ostom-Biotrop.

Midgley, J. (1995). *Social development: The developmental perspective in social welfare*. Thousand Oaks, CA: Sage.

Midgley, J. (1999). Growth, redistribution and welfare: Toward social investment. *Social Service Review, 77*(1), 3–21.

Millar, D., Kendie, S. K., Apusigah, A. A., & Haverkort, B. (2006). *African knowledges and sciences: Understanding and supporting the ways of knowing in Sub-Saharan Africa*. Barneveld: BDU.

Moemeka, A. A. (1994). *Communicating for development: A new pan-disciplinary perspectie*. Albany: State University of New York Press.

Moseley, M. (1997). New directions in rural community development. *Built Environment, 23,* 201–209.

Moseley, M. (2000). Innovation and rural development: Some lessons from Britain and Western Europe. *Planning Practice and Research, 15,* 95–115.

Myrdal, G. (1968). *Asian drama: An inquiry into the poverty of nations*. New York: Twentieth Century Fund.

Nation Master. (2008). *Population below poverty line: Countries compared*. http://www.nationmaster.com/country-info/stats/Economy/Population-below-poverty-line.

Nazarea, V. (1999). *Ethnoecology: Situated knowledge/located lives*. Tucson: The University of Arizona Press.

North, D. C. (1990). *Institutions, institutional change and economic performance*. Cambridge: Cambridge University Press.

Organisation for Economic Co-operation and Development (OECD). (2006). *The new rural paradigm—Policies and governance*. Paris: Pallier & Prévost.

Øyen, E., et al. (2002). *Best practices in poverty reduction: An analytical framework*. London: Zed Books.

Platteau, J. P. (2000). *Institutions, social norms and economic development*. Amsterdam: Harwood Publishers.

Posey, D. A. (1983). Indigenous ecological knowledge and development of the Amazon. In E. F. Moran (Ed.), *The dilemma of Amazonian development* (pp. 225–250). Boulder: Westview Press.

Posey, D. A. (1990). Introduction to ethnobiology: Its implications and applications. In D. A. Posey et al. (Eds.), *Ethnobiology: Implications and applications: Proceedings of the First International Congress of Ethnobiology*. Belém: Museu Goeldi.

Posey, D. A. (1999). *Cultural and spiritual values of biodiversity: A complementary contribution to the global biodiversity assessment*. London/Nairobi: Intermediate Technology Publications Ltd./United Nations Environment Programme (UNEP).

Posey, D. A. (2002). *Kayapo ethnoecology and culture*. New York: Routledge.

Posey, D. A., & Balick, M. J. (Eds.). (2006). *Human impacts on Amazonia: The role of traditional ecological knowledge in conservation and development*. New York: Columbia University Press.

Purcell, T. W. (1998). Indigenous knowledge and applied anthropology: Question of definition and direction. *Human Organization, 57*(3).

Ray, C. (1997). Towards a theory of the dialectic of local rural development within the European Union. *Sociologia Ruralis, 37*(3), 345–362.

Ray, C. (2000). The EU LEADER programme: Rural development laboratory. *Sociologia Ruralis, 40*(2), 163–171.

Ray, C. (2001). *Culture economics*. Newcastle, UK: Centre for Rural Economy, Newcastle University.

Ray, C. (2006). Neo-endogenous rural development in the EU. In P. J. Cloke, T. Marsden, & P. Mooney (Eds.), *Handbook of rural studies* (pp. 278–291). London: Sage.

Reijntjes, C., Haverkort, B., & Waters-Bayer, A. (1994). *Farming for the future: An introduction to low-external-input and sustainable agriculture*. Leusden/London: ILEIA/Macmillan Press.

Richards, P. (1985). *Indigenous agricultural revolution: Ecology and food production in West Africa*. London: Hutchinson.

Rogers, E. M. (1976). *Communication and development: Critical perspectives*. Beverly Hills: Sage Publications.

Sachs, J. (2005). *The end of poverty: Economic possibilities for our time*. New York: Penguin Press.

Saefullah, K. (2019). *Gintingan in Subang: An Indigenous Institution for Sustainable Community-Based Development in the Sunda Region of West Java, Indonesia*. Ph.D. Dissertation. Leiden Ethnosystems and Development Programme. LEAD Studies No. 11. Leiden University.

Sayle, M. (2000). Ethno-economics in Japan, *Asian Perspectives, 24*(4).

Schramm, W. (1964). *Mass media and national development*. Stanford University Press.

Scoones, I., Thompson, J., & Chambers, R. (2008). *Farmer first revisited: Innovation for agricultural research and development workshop summary*. Brighton, UK: Institute of Development Studies, STEPS Centre, University of Sussex.

Sen, A. K. (1990). Development as capability expansion. In K. Griffin & J. Knight (Eds.), *Human development and the international development strategy for the 1990s* (pp. 41–58). London: Macmillan.

Sen, A. K. (1993). Capability and well-being. In M. C. Nussbaum & A. K. Sen (Eds.), *The quality of life* (pp. 30–53). Oxford: Clarendon Press.

Servaes, J. (1991). Towards a new perspective for communication and development. In L. Casmir (Ed.), *Communication in development*. Norwood N.J.: Ablex Publishing Corporation.

Sherraden, M. (1991). *Assets and the poor: A new American welfare policy*. Armonk, NY: M. E. Sharpe.

Shortall, S., & Shucksmith, M. (2001). Rural development in practice: Issues arising in Scotland and Northern Ireland. *Community Development Journal, 36*(2), 122–134.

Shucksmith, M. (2000). Endogenous development, social capital and social inclusion: Perspectives from LEADER in the UK. *Sociologica Ruralis, 40*(2), 208–218.

Shumacher, E. F. (1973). *Small is beautiful: A study of economics as if people mattered*. London: Blond & Briggs.

Sillitoe, P. (1998). The development of indigenous knowledge: A new applied anthropology. *Current Anthropology, 39*(2).

Sillitoe, P., Bicker, A., & Pottier, J. (2003). *Participating in development: Approaches to indigenous knowledge*. London: Psychology Press.

Slikkerveer, L. J. (1999a). Ethnoscience, TEK and its application to conservation. In D.A. Posey (Ed.), *Cultural and spiritual values of biodiversity: A complementary contribution to the global biodiversity assessment* (pp. 169–177). London/Nairobi, Intermediate Technology Publications Ltd./United Nations Environment Programme (UNEP).

Slikkerveer, L. J. (1999b). Traditional ecological knowledge (TEK): Practical implications from the African experience. In K. Adimihardja & M. Clement (Eds.), *Indigenous knowledge systems research & development studies* (Vol. 3). Bandung: UPT.

Slikkerveer, L. J. (Ed.). (2012). *Handbook for lecturers and tutors of the new master course on integrated microfinance management for poverty reduction and sustainable development in Indonesia (IMM)*. Leiden/Bandung: LEAD/UL/UNPAD/MAICH/ GEMA PKM.

Slikkerveer, L. J. (2013). *Traditional ecological knowledge (TEK): The cultural dimension of sustainable environmental development.* Paper presented at the World Culture Forum, 24–27 November, 2013, Nusa Dua, Bali, Indonesia.

Slikkerveer, L. J., & Slikkerveer, M. K. L. (1995). Taman Obat Keluarga (TOGA): Indonesian medicine for self-reliance. In D. M. Warren, L. J. Slikkerveer, & D. Brokensha (Eds.), *The cultural dimension of development: Indigenous knowledge systems*. IT Studies on Indigenous Knowledge and Development. London: Intermediate Technology Publications Ltd.

Storey, D. (1999). Issues of integration, participation and empowerment in rural development: The case of LEADER in the Republic of Ireland. *Journal of Rural Studies, 13*(3), 307–315.

Swift, A. (1991). Local customary institutions as the basis for natural resource management among the Boran Pastoralists in Northern Kenya. In M. Leach & R. Mearns (Eds.), *Environmental change: Development challenges*. Sussex: IDS Bulletin.

Throsby, C. D. (2013). *The Bali promise calls for culture integration with development*. Nusa dua, Bali, Indonesia: World Culture Forum.

Umoh-Akpan, A. (2000). Ethnoeconomics in sustainable agricultural development: The Nigerian case. *Journal of Sustainable Agriculture, 17*.

United Nations (UN). (1948). *Universal declaration of human rights*. New York: United Nations.

United Nations (UN). (2015). *Post-2015 agenda for sustainable development and sustainable development goals (MDGs)*. New York: United Nations.

United Nations Development Programme (UNDP). (2011). *Human development report 2011. Sustainability and equity: A better future for all*. New York: UNDP.

Uphoff, N. (1996). *Local institutional development: An analytical sourcebook with cases*. West Hartford: Kumarian Press.

Walsh, J. (1996). Local development theory and practice: Recent experience in Ireland. In J. Alden & P. Boland (Eds.), *Regional development strategies: A European perspective*. London: Regional Studies Association.

Ward, N., Atterton, J., Kim, T., Lowe, P., Phillipson, J. & Thompson, N. (2005). Universities, the knowledge economy and neo-endogenous rural development. Centre for Rural Economy. *Discussion Paper Series,* No. 1.

Warren, D. M. (1982). The Techiman-Bono ethnomedical system. In Yoder, S. P. (Ed.), *African health and healing systems: Proceedings of a Symposium*. Los Angeles, California: Crossroads Press.

Warren, D. M. (1991). Using indigenous knowledge in agricultural development. *World Bank Discussion Paper 127*. Washington, D.C.: World Bank.

Warren, D. M., Slikkerveer, L. J., & Brokensha, D. (Eds.). (1995). *The cultural dimension of development. Indigenous knowledge systems*. IT Studies on Indigenous Knowledge and Development. London: Intermediate Technology Publications Ltd.

Warren, D. M., Slikkerveer, L. J., & Titilola, S. O. (1989). *Indigenous knowledge systems: Implications for agriculture and international development*. Ames: Iowa State University.

Warren, D. M., Werner, O., & Brokensha, D. (1980). *Indigenous knowledge systems and development*. New York: Latham Publishers.

Watson, E. E. (2003). Examining the potential of indigenous institutions for development: A perspective from Borana. *Development and Change, 34*(2), 287–310.

Williamson, O. E. (2000). The new institutional economics: Taking stock, looking ahead. *Journal of Economic Literature, 38*(3), 595–613.

Woods, M. (2005). *Rural geography*. London: Sage.

World Bank (WB). (2000). *World development report 2000/2001. Attacking poverty*. Washington, D.C.: World Bank.

World Bank (WB). (2005). *World development report 2005: A better investment climate for everyone*. Washington, D.C.: World Bank.

World Bank (WB). (2016). *Indigenous Latin America in the twenty-first century: The first decade*. Washington, D.C.: World Bank.

World Culture Forum (WCF). (2013). *World culture forum: The power of culture in sustainable development*. Paris: UNESCO.

World Health Organisation (WHO). (1978). *Primary health care: Report on the International Conference on Primary Health Care.* Alma Ata, USSR; Geneva, Switzerland: WHO.
World Health Organisation (WHO). (2002). *Traditional medicine strategy 2002–2005.* Geneva, Switzerland: WHO.
World Health Organisation (WHO). (2013). *Traditional medicine strategy 2014–2023.* Geneva, Switzerland: WHO.

L. Jan Slikkerveer is Professor of Applied Ethnoscience and Director of the Leiden Ethnosystems and Development Programme, Faculty of Science, Leiden University, Leiden, The Netherlands. He received his Ph.D. on his fieldwork in the Horn of Africa from Leiden University in 1983 and an Honorary Degree from the Faculty of Medicine of Universitas Padjadjaran in Bandung, Indonesia in 2005. He further extended advanced training and research in the newly-developing field of applied ethnoscience in several sub-disciplines, including ethno-economics, ethno-medicine, ethno-biology/botany, ethno-pharmacy, ethno-communication, ethno-agriculture and ethno-mathematics in combination with a focus on three target regions of South-East Asia, East-Africa and the Mediterranean Region. He is the supervisor of 25 Ph.D. students at Leiden University and has published more than 100 books and 300 articles on the subject. While he has also received a number of substantial subsidies from the European Union in Brussels for international development programmes in South-East Asia, East-Africa and the Mediterranean Region, he also conceptualised both the IMM & ICMD approaches in Indonesia, and is Senior Lecturer in the International Master Course on Integrated Microfinance Management (IMM) at the Faculty of Economics and Business of Universitas Padjadjaran in Bandung, Indonesia.

Chapter 3
The Failure of Financial-Economic Policies to Reduce Global Poverty

Kurniawan Saefullah

As we have accepted that poverty is not merely a condition of people's lack of material assets, our search for its reduction should design holistic integration models which also encompass immaterial components.

L. Jan Slikkerveer (2007)

3.1 Poverty, Inequality and the Dream of a 'World Free of Poverty'

Poverty is one of the major problems in international development today. It is so widespread around the globe that at the beginning of this century the United Nations (2000) had designated the eradication of poverty as the first and foremost objective to be achieved in 2015 by the *Millennium Development Goals* (MDGs). Although in general, global poverty rates have indicated a noteworthy decline in all countries, the countries in Sub-Saharan Africa have consistently lagged behind. While the overall numbers of poverty rates have declined, however, the inequalities have actually widened. Most of the 35 member countries of the Organisation for Economic Cooperation and Development (OECD) have experienced the widening of their income disparities during the last 30 years. The Gini coefficient, which measures the inequalities, is indicating that 20% of the highest income groups have a tendency to widen the gap, with 20% of the lowest income groups (cf. World Bank 2013).

The general impression is that the rich groups of the population are becoming richer, while the poorest groups become poorer. Income inequality does not only lead to a decreasing happiness index and a lower overall life expectancy of some groups in society, but it also causes their lesser access to health care, education, and

K. Saefullah (✉)
FEB, Universitas Padjadjaran, Bandung, Indonesia
e-mail: kurniawan.saefullah@unpad.ac.id

L. J. Slikkerveer et al. (eds.), *Integrated Community-Managed Development*, Cooperative Management, https://doi.org/10.1007/978-3-030-05423-6_3

other public services. In the long run, the widening gap can even reduce the productivity of a particular country, while it may also cause social problems. As Dabla-Norris et al. (2015) mention: "*Irrespective of ideology, culture, and religion, people care about inequality. Inequality can be a signal of lack of income mobility and opportunity - a reflection of persistent disadvantage for particular segments of the society*". The authors also indicate that the increase of inequality can generate economic instability as was the case in 2008 during the global financial-economic crisis and the related obstacles to global growth and employment.

The growing trends in income inequality are not only taking place in the OECD countries, but also in the developing countries as documented by Dabla-Norris (2015). A similar analysis is also presented by Yap (2013) in which he addresses the increased income inequality in South-East Asia. Similarly, Rama et al. (2015) document the rising income inequality in South Asia.

The application of a different measure of income disparity—known as the *Mean Log Deviation* (MLD) index—indicates that the inequalities in most of the South-Asian Countries are increasing.[1] With the exception of the Maldives and Buthan, the other countries in the region, including Bangladesh, Sri Lanka, India, Pakistan, Nepal, and Afghanistan, are currently experiencing a similar increase of inequality. As mentioned above, the inequalities in these countries have also affected the access of the poor and lower income groups to health, education, sanitation, water consumption, etc. (cf. Rama et al. 2015).

In another way, inequality can also indicate a sign of poverty. Although the poverty rates show a decrease, the increased inequality explains a different trend as it shows that poverty is not yet resolved from the perspective of disparities in access to resources and services. When the inequality increases with a high *Gini coefficient*, some groups of the population with the highest income are enjoying more access to resources and services. At the same time, however, some other groups with the lowest income are suffering from less access. The overall economic situation shows that the growth of the economy is generally limited to certain groups in the community. In this context, Dabla-Norris et al. (2015) argue that policy makers should pay more attention to the poor groups instead of to the rich groups of the society, since the absence of progress among the poor will lead to an overall decline in growth and sustainability. By consequence, any programme which is established to improve the position of the poor should provide them with special support by empowering them with skills and increased access to financial and non-financial services with the objective to alleviate poverty.

[1]The 'Mean Log Deviation' (MLD) index is mainly used in statistics and econometrics as a measure of degree of income inequality. For instance, the MLD is zero when everyone has the same income, and shows larger positive values when incomes become more unequal, especially at the higher levels. The concept of economic inequality—also known as 'income inequality', 'wealth inequality', or the 'wealth gap'—refers to the difference in various measures of economic well-being among individuals in a group, among groups in a population, or across countries.

According to Robinson (2001), such financial support could be implemented through microfinance, which in the opinion of some authors would serve as a 'poverty reduction tool'. Robinson (2001) suggests dividing the poor population into three groups of people and two types of service approaches, which should be provided to each of the groups. The two types of financial service approaches include subsidised poverty alleviation programmes and commercialised financial services. The first approach to combating poverty is through the provision of subsidised poverty alleviation programmes. The financial services are meant for those groups of people who are living under the poverty line or for those groups of the extremely poor or the poorest of the poor. Prahalad (2004) identifies these groups as living on the 'bottom of the pyramid' (BOP). The objective of the financial support programmes is to ensure that these groups of poor people could have adequate access to the basic needs, such as food and shelter, as well as health services (Fig. 3.1).[2]

The second approach is the provision of commercial financial services, and is meant to support two categories of groups according to their income level: the economically active poor and the poor with a lower-middle income. For the economically active poor, the interest-bearing savings accounts for small savers and commercial microloans could be used in order to enhance their welfare. The services could stimulate these two groups in order to sustain their lives by carrying out economic activities which could generate income. Their income could enable them not only to have access to some basic needs, such as food, clothes, health and shelter, but also to reserve some savings for a sustainable future. In addition, for those groups which are living at the lower-middle income level, the standard commercial bank loans and a full range of savings services are part of the financial services approach, which could be made available to extend their economic activities with capital support in order to generate additional income for a decent life.

In her publication for the World Bank, entitled: *Our Dream: A World Free of Poverty,* Granzow (2000) acknowledges, that reducing poverty is a complex and difficult challenge with many dimensions. Apart from the usual definition of poverty as the lack of human and material assets and vulnerability to ill-health, drought, job loss, economic decline, violence, and societal conflict, the author contends that: "*it often means a deep condition of disempowerment, even humiliation.*" Despite the limitation, that the projects presented were selected by a team of World Bank staff which only included those activities which had shown poverty impact, the conclusion by Granzow (2000) that effective public action can make a difference to

[2]The concept of the 'Bottom of the Pyramid' (BOP) is a socio-economic concept, elaborated by Prahalad (2004), which allows a form of grouping of a large segment—in excess of about four billion—of the world's poorest citizens which constitutes an invisible and unserved market, blocked by challenging barriers which prevent them from realising their human potential for their own benefit, those of their families, and that of society's at large. Technically, a member of the BOP is part of the largest but poorest groups of the world's population, who live on less than $2.50 a day and who are excluded from the modernity of the globalised, civilised societies, including consumption and choice as well as access to organised financial services.

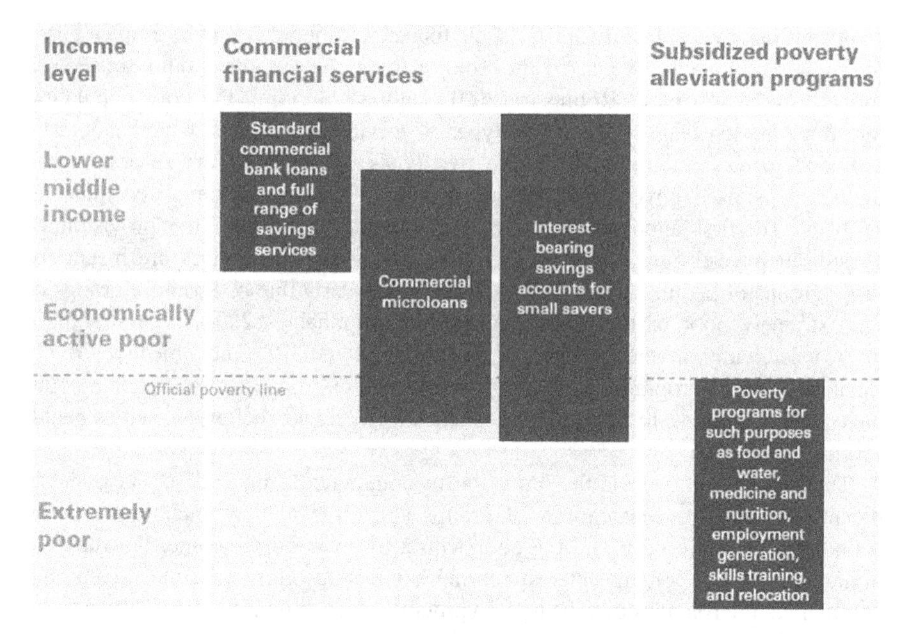

Fig. 3.1 Financial services in the poverty alleviation toolbox *Source* Robinson (2001)

poverty in all its complexities links up well with the above-mentioned plea for a particular focus on the involvement of local peoples' decision-making systems and related active participation in all development policies and programmes.

3.2 The *Emic-Etic* Dichotomy and the Process of Development

Development is one of the most important subjects in many countries, attracting the attention of not only scientists from different academic disciplines, but also from policy planners and practitioners in the field. It encompasses a variety of themes, ranging from different economic theories to anthropological descriptions of the transition of people living in communities and countries as a whole. While Hill (2014) discusses a development model for South-East Asian countries, based on the existence of heterogeneities among these countries, Slikkerveer and Dechering (1995) suggest treating those various states differently. In fact, the development model which applies to a particular country should be differentiated from other countries in similar situations. In order to fully understand the local cultures, both the development approach as well as the related models should be based on an *emic* (insider's) view, rather than an *etic* (outsider's) view.

In contrast to the *etic* perspective on local communities, which are not able to fully understand and explain local structures, the *emic* view provides a deeper understanding of the culture of the target group and their decision-making processes, most relevant to the introduction and adoption of changes which would fit within the local context. Since particular cultures illustrate how *emic views* can differ from one community to another, the related development approaches should similarly be different. As Clammer (2005) states, the debate on development theories has recently brought the concept of culture back to the central stage of development, especially as a result of obvious failures of conventional approaches to economic growth and social transformation. In Chap. 7, the specifically designed emic research methodology of the *Leiden Ethnosystems Approach* will further elaborate on the need for such emic-oriented understanding of indigenous peoples and their culture as a stepping stone for sustainable development.

The concept of culture itself, however, has also undergone a critical re-assessment. Recently, several cultural studies have challenged the self-defined specialisation of anthropology in the social-scientific examination of culture. The subsequent reaction on the inclusion of culture in development has been put forward by Loeffelman (2010), who states: "*Understanding the local cultural and gender dimension of any community is critical to the success of any development project. How can a development project succeed if the clients or population are not included in the organisation and creation? How else will development practicioners, typically in the West, know what needs to be done in the local communities unless those people are directly involved.*" Loeffelman (2010) concludes that: "*In some development organisations, when women in the global East and South are the recipients of aid, they are either left completely out of the development process or considered one homogenous group that has the same life experiences, needs, and goals for themselves and their families.*"

The adoption of the *emic* view on development has had important consequences for the general discourse on development. The concept of development, the objective of development, and the indicators of development should no longer be approached from the outsider's point of view, but from the insider's point of view of the local people in order to achieve sustainable development.

As mentioned in the *Introduction*, scientists working in the newly-developing field of neo-ethnoscience since the 1990s have further elaborated on the practical aspects of ethnoscience in relation to socio-economic development and change, with the result that the "cultural dimension of development" can no further be ignored or left outside self-respecting, effective development programmes and projects worldwide (cf. UNESCO 1994; Warren, Slikkerveer and Brokensha 1995).

A similar approach has also been implemented by the *Compas Group of Sustainable Development and Bio-Cultural Diversity* (COMPAS 2007) with the introduction of the *endogenous development* approach, where the development process is regarded as *developing from within* the local communities. Key factors in this approach of integrating culture into development include the understanding of the indigenous knowledge systems, the peoples' participation in development programmes and the involvement of local institutions in the community. As mentioned above, the attention on the cultural aspects in development has also pertained to the concept of *community development* since the late 1960s. As indicated above, community development basically refers to a process of development which is based on the community itself in terms of the local people, their systems and their cultures. Originally, community development was implemented as an approach in which local people were invited to participate in the process of development, but the first implementation during the 1960s and 1970s eventually declined because of disappointing results. The recent approach of endogenous development, promoted by COMPAS, involves a continuous process of adaptation and innovation, starting from within the local community, and seeking to achieve improvement of local peoples' *well-being* by also taking their indigenous worldview or cosmology into account. In these cosmologies, in the universe of humankind, the closely interrelated human world, natural world and spiritual world are observed so as to remain in balance. In such a configuration, the effectiveness of development cannot merely be measured only by material progress, but it also has to assess the spiritual gains (cf. Hiemstra 2008).

The *emic* perspective on the concept of *balance* is important as it relates to the harmonic relationships between the three perceived worlds of the indigenous cosmologies in which humans take a central position. In Indonesia, as a multicultural country, there are several hundreds of ethno-cultural groups spread over the Archipelago, each with its own particular form of indigenous worldview. These sub-cultures are largely rooted in their cosmologies, in all of which the concept of balance between the human, the natural and the spiritual worlds play an important role in their daily life. Interesting examples of such generations-old indigenous institutions are documented by the study of Agung (2005) on *Tri Hita Karana* ('three foundations of well-being') of the Balinese cosmovision, and the study of Saefullah (2019) on *Tri tangtu* ('three realms of life') of the Sundanese cosmovision in West Java, in which the harmonious balance between the three worlds represents the status of well-being of the local population. These studies also found that the indigenous knowledge and wisdom as part of the cosmologies is often harboured and preserved among the poor and low-income groups of the rural communities, rendering the protection and improvement of these often vulnerable members most urgent in development programmes and projects (cf. Sumardjo 2010; Siregar 2010).

The role of the local culture and the peoples' participation are two major key concepts brought up in today's discourse on the 'cultural dimension of development' (cf. Warren et al. 1995). These concepts are directly related to the conceptualisation of the community and to the way in which sustainable development should be implemented at the community level.

3.3 The Road to Reach Sustainable Community Development

The presently fashionable concept of 'sustainability', transpiring through the Sustainable Development Goals (SDGs), as well as through almost all development-oriented objectives, had already been introduced during the 1960s by authors such as Carson (1962), Boulding (1966), Meadows et al. (1972) and Schumacher (1973), along with the global environmental movement, where the practice of exploiting natural resources was cautioned to avoid harm to nature and reserve some of the resources for future generations. In 1987, the United Nations *World Commission on Environment and Development* (WCED 1987) released the report *Our Common Future*, also known as the "Brundtland Report", which defines sustainable development as: "*development that meets the needs of the present without compromising the ability of future generations to meet their own needs.*" This definition, which is based on two priority concepts of the essential needs of the world's poor and the idea of limitations imposed by technology and social organisation on the environment's ability to meet present and future needs, was later elaborated and extended by several authors.

Slikkerveer (1999) argues, however, that the concept of "sustainability" is not so much a novelty of the development rhetoric which expanded in the course of the 1980s, but can be traced back to various indigenous cosmologies and philosophies of nature and the environment throughout the developing world, guiding local peoples' knowledge, beliefs and practices in their balanced relations with the universe over many generations. As Agung (2005) documents for Bali, Indonesia, these communities have sought to maintain their traditional practice of generations-long "subsistence" in their indigenous management systems for mere sufficient use of resources by the community in harmony with their environment, as opposed to the modern policy of short-term, commercial "surplus-building", irrespective of the condition of resources and the environment, which could threaten the resource base, and as such eventually human life on Earth.

The study of Agung (2005) in Bali contributes to the theory that the transfer over many generations of these local methods and practices of the sustainable use of resources and the environment is embedded and protected—as in the case of sacred groves and forests—in the local cosmologies still prevailing today in the indigenous institutions among many local peoples and communities in different regions of the world. Sustainable development has been described by Passet (1979) and Dréo (2007) in three spheres: dimensions, domains or pillars, *i.e.* the environment, the economy and society.

The United Nations Conference on Sustainable Development (UNCSD 2012) in Rio de Janeiro was the Third International Conference on sustainable development which aimed at reconciling the economic and environmental goals of the global community, and in which information, integration, and participation were identified as the key building blocks to help countries achieve sustainable development. The conceptualisation of 'sustainability' within the context of development is well illustrated in the agricultural sector, where it basically refers to the capacity to maintain both productivity and the resource base. The Technical Advisory Committee of the Consultative Group on International Agricultural Research (TAC/CGIAR 1988) defines sustainable agriculture as: "*the successful management of*

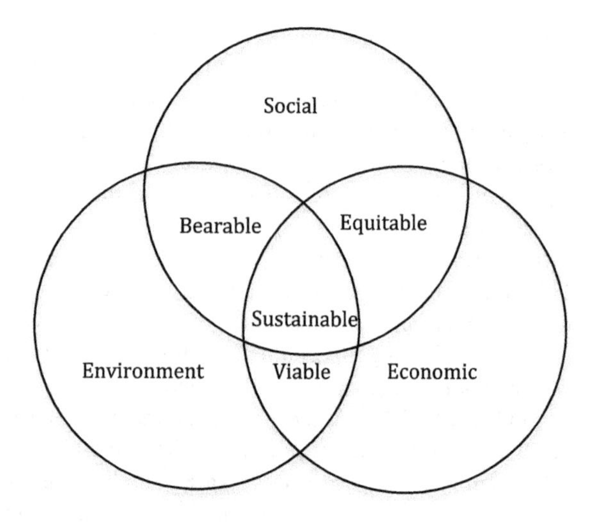

Fig. 3.2 Schematic representation of sustainable development at the confluence of three constituent parts *Source* Dréo (2007)

resources for agriculture to satisfy changing human needs while maintaining or enhancing the quality of the environment and conserving natural resources" (Fig. 3.2).

A broader definition of Gips (1986), later elaborated by Reijntjes et al. (1992), seeks to formulate an operational description built on five requirements of development which have to be met in order to merit the label of 'sustainable', as follows:

- *ecologically sound:* the quality of natural resources is maintained and the vitality of the ecosystem is enhanced;
- *economically viable:* farmers can produce enough for self-sufficiency and/or income, and gain sufficient returns to warrant the labour and costs involved;
- *socially just:* resources and power are distributed equally so that the basic needs of all members of the community are met and their rights assured;
- *humane:* all forms of life (plant, animals and human) are respected. The fundamental dignity of all human beings is recognised and relationships and institutions are based on human values of trust, honesty, self-respect, cooperation and compassion; and
- *adaptable:* rural communities are capable of adjusting to the constantly changing conditions of the community and the environment, such as population growth, development policies, markets etc.

Warren et al. (1995), in their book *The Cultural Dimension of Development*, added the crucial factor of *cultural appropriateness* to the above-mentioned five requirements of sustainable development. As will be further highlighted below, the operational set of criteria of 'sustainability' links up well with the focus on the crucial role of indigenous knowledge and institutions in the current approach of this Volume towards the design and implementation of alternative, human-cantered approaches which are based on integrated strategies of *Integrated Microfinance*

Management (IMM) and *Integrated Community-Managed Development* (ICMD) for worldwide poverty reduction and sustainable development.

While several authors have contributed to the discussion in relation to the conventional model of community development, Toledo (2001) proposes a holistic model of community development in which all factors and dimensions are clearly identified and integrated. The author introduces the concept of *sustainable community development*, which he has adapted from the previous concept of sustainable rural development. Toledo (2001) contends that there are six dimensions of sustainable community development which need to be taken into account. They are territorial, ecological, cultural, social, economic, and political. In order to realise a process of sustainable community development, all of these six dimensions need to be integrated. In addition, the control over resources should be complete and integrated with these six dimensions. In this way, it is not possible to maintain the culture of the society, while the destruction of the natural resources persists. By consequence, the protection of both culture and nature and the maintenance and improvement of the quality of life of the members of a community will become difficult to realise, if an adequate political organisation does not exist. Toledo (2001) concludes that the six dimensions of sustainable community development can only be fulfilled if the members of the indigenous communities fully participate in the process of development.

The holistic perspective of the development model elaborated by Toledo (2001) had also previously been highlighted by Matin et al. (1999) and Khan (1996), who contend that development efforts can only be implemented successfully if a comprehensive and holistic idea exists about development itself.

Furthermore, Khan (1996) also states that: "*most programmes developed for the poor in the Third World failed because they are designed by professionals who belong to the upper classes and are not fully conversant with the sociology, economics, and culture of the low-income communities or the causes of conditions in low-income settlements*". Today, there are about six thousand different cultures around the globe with more than 2500 different languages, spread over the five continents.

Although globalisation has certainly made some positive impacts on peoples and cultures, such as increased access from one country to another, it however also challenges those communities where globalisation tends to gradually replace the cultural diversity by a homogenous global culture.

The inclusion of a particular culture into a development plan is not only meant to preserve the distinctive characteristics of that culture, but also to ensure that the way of life of the people around the globe can be maintained in a sustainable way (Fig. 3.3).

The understanding of the position of indigenous communities, and in particular their poor and low-income members, is a first prerequisite to include them into development programmes in order to achieve sustainability of the entire community. It includes a keen interest in their indigenous cosmovision and its local components of knowledge, beliefs and practices, as well as their indigenous institutions, which guide and determine local peoples' behaviour at the community

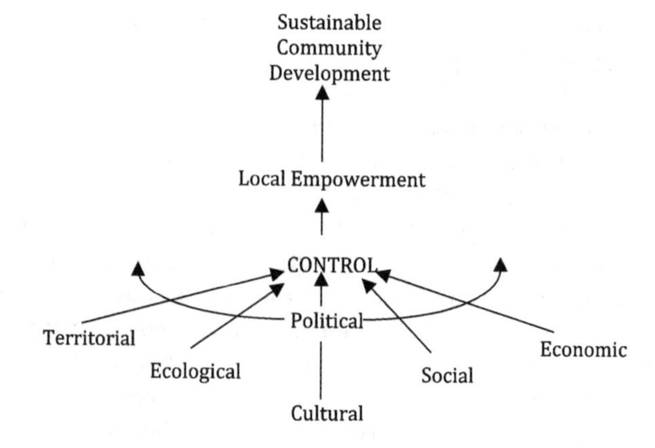

Fig. 3.3 Principles and processes of sustainable community development at the village community level *Source* Toledo (2001)

level. As regards the practice of development activities, such understanding also helps to identify how the community members recognise and solve their own problems in various sectors, such as in health, education, agriculture and natural resources management, but also in the socio-cultural problems of poverty and related limited access to public services. The holistic perspective on development has also been incorporated in the LEAD approach to integrated models of IMM and ICMD, where it provides a major principle in the related analytical frameworks.

3.4 The Drift from Microcredit to Microfinance: Commercialisation of Solidarity

As mentioned in the *Introduction*, over the past decades, financial organisations, development experts and planners have launched a whole array of poverty reduction approaches and development strategies, varying from humanitarian development aid, welfare and debt relief, through the financial-economic approaches of capital investment and liberalisation, to the human development approaches of capacity building and investment in the services of education, health and employment for the improvement of the standard of living of the poor.

The new financial concept which had emerged in the course of the 1970s in Asia with a view to assisting the poor stemmed from the pioneering work of Mohammad Yunus and the Grameen Bank which he had established in Bangladesh. The concept of small-scale finance known as 'microcredit' embarked on the narrow monetary definition of poverty—the general lack of financial funds—and introduced the 'first stage' in providing small and group-based finances to the poor (cf. Yunus and Weber 2007). Microcredit refers to the provision of credit services to poor people who do not have any access to traditional banking services due to a lack of collateral. Initially, money lenders, pawnshops and consumption credit providers were the usual instances providing this form of small or 'micro' credit. In this way, microcredit was originally established as a social movement of solidarity with a view to enabling lower income people to gain access to production opportunities which could increase their income. Following its initial success, the 'microcredit model' soon started to attract the interest of many international organisations and NGOs, rendering microcredit an international strategy in the endeavour to reduce poverty by the end of the 1980s.

In the meantime, the approach of Yunus (1999) to deliver microcredit to the poor had grown fast, and in 1982 the number of members of his organisation had increased to about 28,000. Later onwards, the success of the Grameen Bank encouraged other countries to initiate similar approaches of micro-loans through credit institutions with similar responses to the financial needs. Not long thereafter, directors of large financial companies and institutions, such as the philanthropist George Soros and the co-founder of eBay Pierre Omidyar, started to capitalise in microcredit, soon followed by global commercial banks, including the Citigroup Inc. and the Deutsche Bank AG, which invested hundreds of millions of dollars into microcredit (cf. Karnani 2007).

The various financial demands of the members of the community soon evoked a wider response in the form of extended financial services. While some groups needed small credit for production, such as for working capital or buying fertilisers etc., other groups demanded credit for consumption, such as for health and education. The difference in demand for small loans introduced the microcredit approach into wider services beyond the original target groups of the poor, such as the middle class groups, and soon, the development of microcredit led to the introduction of the extended financial services of the early 1990s, including not only small group loans, but also insurance, savings and deposits, which were managed by a growing number of microfinance institutions (MFIs) and their managers.

This process of the 'second stage' in the provision of extended financial services to the poor and middle-class groups became known as 'microfinance', referring to a complex process of distributing a whole range of financial services, with the main objective to make poor and low-income groups of the community 'bankable'. Since the microcredit-providing institutions had to change their social mission in order to run their institutions as independent and sustainable companies, the original humanitarian non-profit orientation of the provision of subsidised loans was replaced by a commercial ideology of providing financial products with no more interest-free or low-interest loans. At the same time, many NGOs used a more profitable opportunity to make the transition to become MFIs.

The shift of the early 1990s from the original social motives of NGOs to the commercial interests of MFIs is also reflected in the subsequent statements of Muhammad Yunus, made during subsequent periods of time. As Engler (2009) documents, Yunus initially believed that: "*Not every business should be bound to serve the single objective of profit maximisation, but rather pursue specific social goals*". However, in another interview by Yanagidaira in 2009, as recorded later by Valadez and Buskirk (2011), Yunus (1999) also acknowledged that "*(a) company can make (a) profit.*", and: "*that profit stays within (the) company*". In the last statement, Yunus admits that the profit was used to pay for the expenses of the MFIs, and that the investors can claim back their original money. These considerations underscore the dramatic turn of the initially small-scale solidarity-based philosophy behind microcredit into the neo-liberal ideology of large-scale commercialisation of capital investment in international development cooperation.

In this context, the term ‘transformation’ or ‘commercialisation’ of the MFIs refers to a change in legal status from an unregulated non-profit or non-governmental organisation (NGO) into a regulated, for-profit institution. In their study on the *Mission Drift in Microfinance Institutions*, Ghosh and van Tassel (2008) give an interesting explanation for the way in which the change in the portfolio of a poverty-minimising MFI might be linked to the phenomenon of increasing commercialisation through the advent of large profit-oriented donors, underscoring the general concern that: “*an emphasis on profitability implies a de-emphasis on poverty reduction and related development goals.*”

As a result, the transition of the early 1990s from *microcredit* to *microfinance* has further increased the distance of the microfinance approach to the poorest of the poor, for whom the Grameen Bank initially was set to provide small-scale financial services. In joining similar conclusions, that microfinance is not an appropriate ‘poverty reduction tool’, Al-Mamun et al. (2013) notice that poor people are deprived of their basic needs in life, such as food, shelter, clothing, clean water, health care, education and employment opportunities, and they conclude that microfinance alone cannot alleviate poverty from the grass roots level of the society or the poorest of the poor.

The results of the empirical study by Al-Mamun et al. (2013) in Bangladesh, using the three categories of the poor, as suggested by Robinson (2001), indicate that only economically active people are able to gain access to the financial services of the MFIs, while the extremely poor people could not make use of microfinance as the solution for their economic problems. It is not surprising that earlier, Fernando (2004) claimed that instead of eliminating poverty, microfinance is in fact perpetuating it. Fernando (2004) argues that: “*The users of microfinance are generally those who are living within poverty lines, and those who are among the poorest in the society remain neglected and invisible by the microfinance. The requirement set by microfinance cannot be fulfilled by the poorest or extremely poor groups in the society*”. As will be further elaborated in the next paragraph, microfinance has become ‘exclusive’, i.e. hardly accessible to poor people, particularly to the extreme poor, since they are not ‘bankable’, do not possess any collateral, and have no generating income activities, so they are predicted as unable to pay back any possible “borrowing money” from the MFIs.

As a possible solution, Al-Mamun et al. (2013) suggest that the government should provide basic infrastructure facilities, such as adequate roads and transport, schools, hospitals, continuous power supply, etc. in order to ensure that people would get the benefits of microfinance. In this context, it has become evident that microfinance itself cannot alleviate poverty, and as such cannot be labeled as a ‘poverty reduction tool’. Although a basic contribution could be expected from an effort to make microfinance inclusive in terms of its clients, and not exclusive to the poor and the extremely poor people, the overall financial-economic approach to poverty reduction should drastically be redirected towards a more human-centered, holistic poverty reduction approach in order to attain a more effective process of sustainable community development in the near future.

While microfinance subsequently proliferated in countries with limited bank infrastructures, such as most countries in Africa, Asia, Latin America, and Eastern Europe, it was found that in some of these countries, less than one-fifth of the population has a bank account, particularly the group of people living at the 'bottom of the pyramid' (cf. Ayayi and Sene 2010). Similarly, Walt (2012) poses in *Time Magazine* the key question and provides the answer: "*Does Microfinancing really work? A New Book Says No.*"

These results further render the use of the term 'poverty reduction tool' for commercial microfinance as not only inadequate, but also misleading, since the characteristics of people in the community reflect various types of problems which are bound to different cultures and traditions, as indicated in the previous paragraph. Thus, microfinance should at least be integrated with other public and community services in order to provide a contribution to the sustainable community development approach as mentioned above.

3.5 The Fundamental Failure of Microfinance to Reduce Poverty

As indicated above, the shift of the early 1990s from *microcredit* to *microfinance* introduced a new era of extension of the commercialisation of the MFIs into conventional banking institutions, often at the cost of their poor and low-income clients, eventually rendering the financial-economic approaches to reducing poverty worldwide a mere disillusion. As the widely documented financial approaches have even weakened the position of the poor, alternative approaches have been sought since the beginning of this century. In addition to the financial-economic approach, promoted by the World Bank in its annual *World Development Reports* since 1978, a more human approach has been promoted by the United Nations and its agencies —specifically UNESCO and UNDP—as transpiring through their *Human Development Reports* since 1990, focused on renewed orientations of community development approaches on a global scale. It has been estimated that around 10,000 MFIs were active in 2006, serving over 113 million clients. Since then, MFIs have been growing and although no global comprehensive figures are available because of the difficulty of the assessment of the large number of informal organisations, the available data all point to strong double-digit growth in the sector. As Hsu Ming-Yee (2007) documents, the growth of a sample of 200 MFIs in the years between 2003 and 2005 shows that: "*the total assets and loan portfolios have doubled within two years, their number of active borrowers has grown by over*

50%, they have become more profitable, they have more and more recourse to debt and commercial funding, and they increasingly integrate into the formal financial sector". These data provide ample evidence of an industry in full expansion.

Following the continuing debate on poverty and microfinance, as indicated in the *Introduction*, the advocates of the 'new wave' microfinance model include authors such as Pitt and Khandker (1998) and Littlefield et al. (2003). In addition, a large number of international organisations tend to support MFIs in providing financial services to low-income clients, such as Americans for Community Co-operation in Other Nations (ACCION 2007), the Foundation for International Community Assistance (FINCA 2016) and the Vietnam Bank for Social Policies (VBSP 2016). Such a largely positive view is also transpiring through some recent statements from international organisations, such as the International Labour Organisation (ILO 2015) and the (Microcredit Summit Campaign 2015). Most of these statements and reports support the suggestion that microfinance could be a useful poverty reduction tool. So, a number of studies have focused on the success stories of microfinance in development, such as Obaidullah (2008). Hulme and Mosley (1996) indicate that microfinance contributes a positive impact to the income of the borrowers by an increase of 30% compared to that of those who do not take credit from microfinance. Some other research indicates that the positive impact of microfinance is not only affecting the economic aspects of peoples' life, but also the improvement in their nutrition status and the decline in gender inequalities (cf. Pitt and Khandker 1998).

A more nuanced standpoint in the current debate is reflected in the recent conclusion of the contribution of Gaiha and Kulkarni (2013), who state that: "*while the magic of microfinance has eroded with financial sustainability overriding social goals, there are ample grounds for optimism about resolving this trade-off.*" Meanwhile, however, it did not take long until many of the generally favourable impact studies of microfinance evoked growing substantial criticism, and several authors started to reveal the failure and incapacity of the "new wave" microfinance to improve the situation of the poor, including Morduch (1998), Copestake et al. (2005), Karnani (2007), Roodman and Morduch (2009), Bateman (2010), Roodman (2011b), Slikkerveer (2012), Dogra and Gorbachev (2016) and Bateman and Chang (2012). Not surprisingly, the positive microfinance evaluations from studies conducted by the MFIs themselves under the Assessing the Impact of Microenterprise Services (AIMS 2002) project led to much criticism from several academicians and experts, mainly blaming the authors of 'marking their own papers', reflecting an obvious 'publication bias' in their findings to highlight the positive results, while ignoring the negative outcomes in their own interest. Karnani (2007), in his article *Microfinance Misses its Mark*, argues that: "*Despite the hoopla over microfinance, it doesn't cure poverty. But stable jobs do. If societies are serious about helping the poorest of the poor, they should stop investing in microfinance and start supporting large, labour-intensive industries. At the same time, governments must hold up their end of the deal, for market-based solutions will never be enough.*"

Another criticism on microfinance contests the widely used claim that microfinance is capable of building social capital and solidarity among members of poor communities. As Leys (2001) notices, the introduction of commercialisation of community development and poverty reduction through microfinance results in the opposite: a decrease in social solidarity, communication and interaction among community members, threatening the mission of local voluntary associations and institutions. In India, recent research is documenting that the new microfinance model has driven some poor households into a debt trap, in some cases even leading to suicide (cf. Biswas 2010). Such a growing wave of defaults and suicides among numerous debt-ridden farmers in India has recently also been reported by *Business Insider* (2012) and Roy and Biswas (2014).

Bateman (2010) documents his criticism of the glory of microfinance by showing that instead of helping poor people to become independent and economically active, microfinance has been commercialised by investors who are expecting huge returns through capital investment in MFIs. Extremely poor people are regarded by these financiers as 'unbankable' and thus not regarded as profitable. As a result, the commercialisation of microfinance tends to limit the microfinance services only to economically active people.

The positive impact assessment of the SAAD Microfinance Project by SMERU (2005), as described above, has also been criticised by Bateman (2010), who raised an important question about impact assessment: "*does impact assessment produce a genuine reflection of what microfinance can achieve economically and socially?*" David Ellerman (2007), a former World Bank staff member, answered '*no*' to the question, while arguing that some microfinance impact assessment methodologies are fundamentally wrong. His claim has been supported by Duvendack et al. (2011) who carried out an impact analysis of the role of microfinance in poverty alleviation. Duvendack et al. (2011) evaluate various impact assessment methods of microfinance, including *Randomised Control Trials* (RCT), *pipeline designs, natural experiments, and general purpose surveys*. Their research surprisingly indicates that by using the RCTs, no convincing impact has been found of microfinance on the well-being of poor people. Similarly, many studies on the positive impact of microfinance on women's empowerment have been found to be methodologically weak with inadequate representative data.

Hulme and Mosley (1996) found the same limitations in the provision of microfinance services to the extremely poor people, and argue that microfinance is only effective in improving the status of the middle and upper segments of the poor. Clients below the poverty line or the extremely poor were even worse off after borrowing than before. Fernando (2004) claims that instead of eliminating poverty, microfinance is in fact perpetuating it. Fernando (2004) contends that: "*The users of microfinance are generally those who are living within poverty lines, and those who are among the poorest in the society remain neglected and invisible by the microfinance. The requirement set by microfinance cannot be fulfilled by the poorest or extremely poor groups in the society*".

In addition to the criticism, the existence of '*moral hazard*' in the microfinance impact analysis has also been questioned.[3] The largely positive figure of the Ekonomi Program Kredit Mikro Project (SAAD) as reported by SMERU (2005), for instance, has been criticised on the basis of the possible existence of moral hazard in the assessment, considering that the assessment of the SAAD Microfinance Project by the SMERU Research Institute was funded by the World Bank, which also financed the SAAD Project. Also, it is not a surprise that the Report does not explain the conditions of the group of extremely poor people which is accommodated by the project. In this context, Bateman (2010:35) rightly contends that: "*you do not bite the hand that feeds you.*"

From these and other related rather negative outcomes of several independent microfinance impact studies, it becomes apparent that the 'new wave' microfinance model not only fails in reducing poverty, but also by consequence—as Bateman (2010:202) rightly notes—it even represents an anti-development approach which: "*largely works against the establishment of sustainable economic and social development trajectories*".

The lasting albeit largely misconceived popularity of the microfinance movement within international development agencies, national governments and some public figures can largely be attributed to the way in which microfinance was initially presented by Yunus (1999) as a subsidised model for the creation of micro-income and employment for the poor through the informal sector, enabling them to escape from their disadvantaged situation. The neoliberal revolution, however, soon started to convince donors to capitalise MFIs and replace subsidised loans by investments with commercial market interest rates, which further worsened the position of the poor. In this way, the initially positive intentions of Mohammad Yunus and the Grameen Bank to help the poor with microcredit as a means to provide small subsidised group loans with low interest rates to the poor have been overtaken by the commercial interest of MFIs, NGOs and Village Banks, supported by a neoliberal ideology which seeks to promote a private elite of 'micro entrepreneurs' rather than a public democratic and social movement to assist the poor around the globe.

The World Bank publication *Moving out of Poverty* (2009), a study conducted across fifteen countries in Africa, East Asia, South Asia and Latin America involving more than sixty thousand interviews with the poor, concludes that: "*Microcredit can help the poor subsist from day to day, but in order to lift them out of poverty, larger loans are needed so that the poor can expand their productive activities and thereby increase their assets*". Such results render the argument that microfinance could lift up the poorest of the poor as rather unconvincing.

---

[3]In general, 'moral hazard' is defined as a situation in which one party gets involved in a risky event, knowing that it is protected against the risk and the other party will incur the cost. It arises when both parties have incomplete information about each other. For instance, in a financial market, there is a risk that the borrower might engage in activities which are undesirable from the lender's point of view because they make him/her less likely to pay back a loan (cf. The Economic Times http://economictimes.indiatimes.com/).

Rahmat et al. (2006) provide an answer to the question: "*How could large money lift the poor out of poverty?*" They found that if a loan is given in a double amount or more, the tendency of decreasing performance could occur.

In their study of Islamic Microfinance Institutions (MFIs) in Indonesia, Seibel and Agung (2005) conclude that while Islamic Banking tends to raise their assets, the performance of the Islamic MFIs is diminishing. Not only does the Islamic Rural Bank show stagnation, but so do the Islamic Cooperative Microfinance Institutions (BMTs), which have dwindled significantly. Out of 3000 institutions operating in the 1990s, only 20% had maintained sustainability in 2005 (Saefullah 2010).

Similarly, criticism is growing concerning current research on microfinance impact assessment, usually operationalised by *Randomized Control Trials (RCT).* Bateman (2010) mentions: "*displacement or spillover effects and client exit/failure effects*", referring to loss of jobs and income among the non-clients of micro-enterprises. An interesting example of displacement effects is given by the Grameen Bank anti-poverty programme, known as *GrameenPhone*. While it encouraged Bangladeshi women to borrow money in order to run small telephone enterprises, initially successful by selling air time for mobile telephones, after a certain period of time their income dramatically decreased as the result of an oversaturated market, implying difficulties in sustainability.

Another problem is *client exit/failure*, referring to the business failure caused by client exit from business, which in fact worsens the position of the poor in comparison to their previous situation, *i.e.* before they started to run a business with borrowed money, as their repayment still has to continue.

These and similar dramatic cases underscore the conclusion of Karnani (2007) that microfinance misses its mark: microfinance does not cure poverty, but in some instances microfinance makes life at the bottom of the pyramid even worse. The situation prompted Roodman (2011a) to state that: "*On current evidence, the best estimate of the average impact of microcredit on the poverty of clients is zero*".

As regards the general disillusion which transpires through the prevailing case studies of microfinance, Slikkerveer (2015) points to the fundamental delusion in the current financial-economic approaches to poverty reduction by drawing attention to the growing distance they engender between, on the one hand, microfinance as a multi-million dollar industry focused on investing share-holders' capital for profit in middle-class enterprises and business companies, and on the other hand, the poor and extremely poor, who are in fact marginalised and excluded from benefitting from these kinds of services as the "non-bankable" segment of the population. In his view, the failures stem from the basic incompatibility between the neo-liberal ideology and the humanitarian solidarity movement and could only be bridged by the transformation towards a solidarity economy, based on approaches to increase peoples' quality of life mainly through humanitarian not-for-profit policies.

3.6 The Need for Community-Centered Integration and Management of Services

The exclusion of the poorest of the poor or extreme poor from financial-economic approaches including microfinance has been among the major criticisms on the role of microfinance in poverty reduction, largely as the result of the absence of integration of economic sustainability and social development in such predominantly financial programmes. Even the recently introduced concept of 'inclusive microfinance', offering people access to a range of financial services, does not contribute to poverty reduction, as it basically embodies a mere extension of the profit-making activities of MFIs and commercial banks. An interesting factor, which has been identified as a 'missing link' on the side of policy makers and planners, is a general lack of interest and understanding of the target groups and their culture, largely represented by the indigenous peoples in developing countries (cf. Warren et al. 1995; Woodley et al. 2006).

Another major factor refers to the consideration of cultural diversity. Highland people tend to have a different culture than people living in the lowlands, and Northern cultures are different to cultures in the South. The differentiation, however, is not only based on the particular geographical area, but also on the adaptation and interaction with the environment, the management of resources, livelihood, social structure and the specific cultural identity. By consequence, the consideration and understanding of the diverse cultural aspects of a particular community are major prerequisites for a successful introduction of a microfinance programme at the community level, which should not embark on the conventional notion that 'the poor' are a homogenous group in any community worldwide.

Some studies have recently been conducted with the aim to show the impact of cultural understanding in managing MFIs. In a study of microcredit in Mali, Deubel (2006) shows how culture can be integrated in developing microfinance. Rana (2008) also confirms the need to develop a cultural orientation on the management of microfinance in Bangladesh, where the evaluation systems used by MFIs should include specific cultural aspects of the people of Bangladesh, which are mainly influenced by religion and the colonial legacy. Rana (2008) argues that if the management of microfinance is based on the Western management control system, it will be less effective. In other words, the integration of the cultural factors of the local population in such financial development programmes is a basic prerequisite for MFIs to sustain.

A microfinance programme which is accommodating the local culture also links up with the ideas of Bateman (2010:166) who refers to an alternative approach towards 'conventional microfinance'. The author documents his notion on the basis of several examples, such as Japan's fast recovery after World War II, the Basque experience in Spain, the township and village enterprise in China, India's Kerala model of development, Venezuela's experiment with the people-economy, and the development model of Vietnam. Bateman (2010) argues that integrating microfinance with the local culture is much more 'inclusive' in comparison to the conventional notion of 'inclusive microfinance'. The author adds that implementing a local financial system which is operated by a bottom-up participatory approach also tends to provide a positive impact on the local population. Bateman (2010) indicates that in this way, unemployment has decreased from 15% in 1999 to just below 8% in 2008, followed by the decrease in poverty from more than 50% in 1998 to just over 33% in 2007. Moreover, extreme poverty—which is neglected by conventional finance programmes—has declined from 20% in 1998 to only 10% in 2007. As regards the provision of services to the local people, and in particular to the poor and low-income families, the research on microfinance amply documents that there is a general lack of participation of the poor, largely because of the commercial undertones of these imported financial-economic programmes and their incapacity to link up with the peoples' local knowledge, beliefs, practices and their institutions.

The case of Indonesia, further elaborated in Part III of this Volume, shows that the poor people in the rural communities have virtually been excluded from such government–driven financial services, but when the outside experts, organisations and agencies make an effort to embark on—or at least include—the indigenous culture in their so far sporadic projects, the local people are more inclined to participate.

Such an increase in participation is particularly important in advanced education and training programmes, where not only financial services are included, but also health, education, communication and social services are regarded as part of the poverty-reduction package. As is further described in Chap. 11, the understanding of the culture of poor people, including the extreme poor who have been marginalised, can substantially contribute to the alleviation of poverty. This ability of development experts and integrated microfinance managers is part of their newly-developed higher education and training programmes, pioneered by the *Faculty of Economics and Business* of *Universitas Padjadjaran* in Bandung in collaboration with the *LEAD Programme* of *Leiden University* in The Netherlands with the implementation of the new Master Course on *Integrated Microfinance Management* (IMM) since 2012 (cf. Slikkerveer 2007; 2012). The Integrated Microfinance Management (IMM) approach seeks to avoid the *etic* perspective of the 'outsider', but instead implements the *emic* view of the local people as they ultimately decide, on the basis of their indigenous knowledge systems, how poverty reduction and development programmes should be introduced and implemented in a sustainable way (cf. Warren et al. 1995; Loeffelman 2010).

Another shortcoming of conventional microfinance programmes refers to their determination of the requirements and objectives set for the poor people, based on the ideas of the outsiders, the microfinance providers, instead of the voice of the poor people themselves. Right from the start, the local people have to participate in the decision-making process concerning the objectives of the development programmes, based on their own knowledge, beliefs and livelihood.

As mentioned in the previous paragraph, Slikkerveer (2015) draws much attention to the failures of the microfinance approach to reduce poverty, stemming from the basic incompatibility between the neo-liberal ideology and the humanitarian solidarity movement, rendering *solidarity* crucial in helping the poor to escape from their predicament. His view is also supported by Bateman (2010:38) who contends that: "*Poor people as a group lack cash, assets, education, market know-how, and connections with the rich and powerful. When poor people associate only with each other, they bring only their own meager resources to the table. Poor people understand these constraints and affirm that there is a limit to how much one hungry man can feed another. The challenge is to extend these positive local traditions of mutual help so that they reach across social lines to involve those who can bring in new resources, ideas, and skills.*"

A closely related cultural consideration of poor people in relation to development is that *mutual help* and *support networks* are most prone to breakdown among these vulnerable groups in the community. The poor need the respect and dignity that flows from their own way of identifying and managing their resources and creating their own livelihood (cf. Getubig et al. 2000). The poor people do indeed understand that the best way to secure their dignity and respect is actually collective, using their most important asset, i.e. their numbers.

It is clear that there is a strong need for the implementation of the approach of Integrated Community-Managed Development (ICMD), not only to engender the full participation of the indigenous people in the development programmes from the very beginning onwards, but also to focus on the integration of their indigenous knowledge, beliefs and management practices with global knowledge systems, implemented by the local peoples' own institutions.

The Indonesian experience provides the hope that, in this way, a contribution can be further extended to poverty reduction by integrating the indigenous knowledge and institutions of the local population in the transformation towards a sustainable economy for the benefit of all members of the entire community.

References

Agung, A. A. G. (2005). *Bali Endangered Paradise? Tri Hita Karana and the Conservation of the Island's Biocultural Diversity.* Ph.D. Dissertation. Leiden Ethnosystems and Development Programme. LEAD Studies No. 1. Leiden University. xxv + ill., p. 463.

Al Mamun, C. A., Hasan, N. & Rana, A. (2013). Microcredit and poverty alleviation: The case of Bangladesh. *World Journal of Social Sciences, 3*(1), 102–108. PDF available at https://www.researchgate.net/publication/306323524.

Americans for Community Co-operation in Other Nations (ACCION). (2007). *ACCION International's Strategic Plan 2008–2011.* Boston, MA: ACCION. http://www.accion.org/about-us.

Assessing the Impact of Microenterprise Services (AIMS). (2002). USAID—Assessing the Impact of Microenterprise Services. Washington, D.C.: USAID. https://www.microfinancegateway.org.

Ayayi, A. G., Sene, M. (2010). What drives microfinance institution's financial sustainability? *The Journal of Developing Areas, 44* (1).

Bateman, M., & Chang, H. J. (2012). Microfinance and the illusion of development: From hubris to nemesis in thirty years. *World Economic Review, 1,* 13–36.

Bateman, M. (2010). *Why doesn't microfinance work? the destructive rise of local neo-liberalism.* London: Zed Books.

Biswas, S. (2010). India's micro-finance suicide epidemic. BBC News. 16 Dec 2010. http://www.bbc.com/news/world-south-asia-11997571.

Boulding, K. (1966). The economics of the coming spaceship earth. In H. Jarret (Ed.), *Environmental quality in a growing economy*. Baltimore, MD: Johns Hopkins University Press.

Business Insider. (2012). Hunderds of suicides in india linked to microfinance organisations. http://www.businessinsider.com/hundreds-of-suicides-in-india-linked-to-microfinance-organizations-2012-2?international=true&r=US&IR=T.

Carson, R. (1962). *Silent spring*. New York: Penguin Books.

Clammer, J. (2005). Culture, development, and social theory: on cultural studies and the place of culture in development. *The Asia Pacific Journal of Anthropology, 6*(2), 100–119.

COMPAS. (2007). *Learning endogenous development: build on biocultural diversity*. Rugby: Intermediate Technology Publications Ltd. (Trading as Practical Action Publishing).

Copestake, J., Greeley, M., Johnson, S., Kabeer, N., & Simanowitz, A. (2005). *Money with a mission. Microfinance and poverty reduction*. London: Intermediate Technology Publications Ltd.

Dabla-Norris, E., Kalpana, K., Suphaphiphat, N., Rika, F., & Tsounta, E. (2015). *Causes and consequences of income inequality: A global perspective*. Washington, D.C.: International Monetary Funds.

Deubel, F, T. (2006). Banking on culture: microcredit as incentive for cultural conservation in mali. In Ringmar, E & Fernando, J. L. *Perils and prospects. Routledge series on economic development*. London: Routlege-Taylor & Francis Group.

Dogra, K., & Gorbachev, O. (2016). Consumption volatility, liquidity constraints and household welfare. *The Economic Journal, 126*(597), 2012–2037.

Dréo, J. (2007). Illustration of sustainable development. https://en.wikipedia.org/wiki/File:Sustainable_development.svg.

Duvendack, M., Palmer-Jones, R., Copestake, J. G., Hooper, L., Loke, Y. & Rao, N. (2011). What is the evidence of the impact of microfinance on the well-being of poor people? *Systematic Review*. London: EPPI-Centre, Social Science Research Unit, Institute of Education, University of London.

Ellerman, D. (2007). Microfinance: Some conceptual and methodological problems. In T. Dichter & M. Harper (Eds.), *What is wrong with microfinance?* London: Intermediate Technology Publications Ltd.

Engler, M. (2009). From microcredit to a world without profit? *Dissent, 56,* 81–87.

Fernando, N. A. (2004). Microfinance outreach to the poorest: A realistic objective? *ADB Finance for the Poor, 5*(1).

Foundation for International Community Assistance (FINCA). (2016). *Mission and Vision*. http://www.finca.org/.

Gaiha, R., & Kulkarni, V. S. (2013). Credit, microfinance and empowerment. Paper Contribution to the Expert Group Meeting: Policies and Strategies to Promote Empowerment of People in Achieving Poverty Eradication, Social Integration and Full Employment and Decent Work for All, 10–11 September 2013. New York: United Nations.

Getubig, M., Gibbons, D., & Remenyi, J. (2000). Financing a revolution. In J. Remenyi & B. Quinones Jr. (Eds.), *Microfinance and poverty alleviation: Case studies from Asia and the Pacific*. London and New York: Pinter.

Gips, T. (1986). What is sustainable agriculture? In P. Allen & D. van Dusen (Eds.), *Global perspectives on agroecology and sustainable agricultural systems*. Proceedings of the 6th International Scientific Conference of the International Federation of Organic Agriculture Movements (IFOAM). Santa Cruz: University of California.

Ghosh, S., & van Tassel, E. (2008). A model of mission drift in microfinance institutions. *Economic papers*. Florida Atlantic University. http://econpapers.repec.org/paper/falwpaper/08003.htm.

Granzow, S. (2000). *Our dream: A world free of poverty*. Washington, D.C.: World Bank. https://openknowledge.worldbank.org/handle/10986/2411.
Hiemstra, W. (2008). Preface: Towards endogenous development. In D. Milar, A. A. Apusigah, & C. Boonzaijer (Eds.), *Endogenous development in Africa: Towards a systematisation of experiences*. Leusden: ETC/COMPAS.
Hill, H. (2014). Is there a southeast Asian development model? discussion paper series. No. 26. Freiburg, Germany: Freiburg University.
Hsu M.-Y. (2007). *The International funding of microfinance institutions: An overview*. Commissioned by LuxFLAG. ADA–Appui au Développement Autonome (pp. 21–25). Luxembourg: Allée Scheffer.
Hulme, D., & Mosley, P. (1996). *Finance against poverty*. London: Routledge.
International Labour Organisation (ILO). (2015). *World employment and social outlook—trends 2015*. Genève: ILO. http://www.ilo.org/global/about-the-ilo.
Karnani, K. (2007). Microfinance misses its mark. Stanford social innovation review: Informing and inspiring leaders of social change. Summer 2007.
Khan, A. H. (1996). *Orangi pilot project—reminiscences and reflections*. Karachi: Oxford University Press.
Leys, C. (2001). *Market driven politics*. London: Verso.
Littlefield, E., Morduch, J., & Hashemi, S. (2003). Is microfinance an effective strategy to reach the millenium development goal? *Focus Note, 24*(2003), 1–11.
Loeffelman, C. (2010). *Gender and development through western eyes: An analysis of microfinance as the west's solution to third world women, poverty, and neoliberalism* (Thesis). George Washington University, Washington, D.C.
Matin, I., Hulme, D., & Rutherford, S. (1999). *Financial services for the poor and poorest: Deepening understanding to improve provision*. Manchester, UK: Institute of Development Policy Management.
Meadows, D. H., Meadows, D. L., Randers, J., & Behrens, W. H., III. (1972). *The limits to growth*. New York: Universe Books.
Microcredit Summit Campaign. (2015). State of the campaign Report 2015. https://stateofthecampaign.org.
Morduch, J. (1998). The microfinance schism. *World Development*, *28*(4), 617–629 (April 2000).
Obaidullah, M. (2008). *The role of microfinance in poverty alleviation: Lessons from experiences in selected IDB members countries* (1st ed.). Islamic Research & Training Institute of the Islamic Development Bank (IRTI-IDB). http://www.isdb.org.
Passet, R. (1979). *L'economique et le vivant (Economics and the living)* (p. 287). Paris: Petite Bibliotheque Payot.
Pitt, M. M., & Khandker, S. R. (1998). The impact of group-based credit programs on poor households in Bangladesh: Does the gender of participantrs matter? *The Journal of Political Economy, 106*(5), 958–996.
Prahalad, C. K. (2004). *Fortune at the bottom of the pyramid: Eradicating poverty through profits*. Prentice Hall: Upper Saddle River: NJ.
Rahmat, T., Megananda., & Maulana, A. (2006). *The Impact of microfinance on micro and small enterprise's performance and the improvement of their business opportunity*. Working Papers in Economics and Development Studies (WoPEDS). 2006-01. Bandung: Department of Economics, Padjadjaran University.
Rama, M., Beteille, T., Yu, L., Mitra, P. K., & John, L. N. (2015). *Addressing inequality in South Asia*. Washington, D.C.: The World Bank Group.
Rana, B. M. (2008). Cultural oriented management systems of microfinance institutions in Bangladesh. http://kamome.lib.ynu.ac.jp/dspace/bitstream/10131/3797/1/5-Rana.pdf.
Reijntjes, C., Haverkort, B., & Waters-Bayer, A. (1992). *Farming for the future: An introduction to low external-input and sustainable agriculture*. London: Macmillan Education Limited.
Robinson, M. (2001). *Microfinance revolution: Sustainable finance for the poor*. Washington, D.C.: The World Bank.

Roodman, D. & Morduch, J. (2009). The impact of microcredit on the poor in Bangladesh: Revisiting the evidence. Working Paper No 174. Washington, D.C.: Center for Global Development.

Roodman, D. (2011a). *Due diligence: The impertinent inquiry into microfinance*. Washington, D.C.: Center for Global Development.

Roodman, D. (2011b). David Roodman's microfinance open book blog. http://blogs.cgdev.org/open-book/.

Roy, M. K., & Biswas, T. K. (2014). Micro-credit and poverty reduction in Bangladesh. *ASA University Review*, *8*(1).

Saefullah, K. (2010). Cultural aspects on the islamic microfinance: An early observation on the case of Islamic microfinance institutions in Bandung, Indonesia. Paper's presented in the second workshop on Islamic finance: What Islamic finance does (not) change. March 17th 2010. EM Strasbourg Business School and European Research Group.

Saefullah, K. (2019). *Gintingan in Subang: An Indigenous Institution for Sustainable Community-Based Development in the Sunda Region of West Java, Indonesia.* Ph.D. Dissertation. Leiden Ethnosystems and Development Programme. LEAD Studies No. 11. Leiden University.

Schumacher, E. F. (1973). *Small is beautiful: Economics as if people mattered*. New York: Harper & Row.

Seibel, H. D., & Agung, W. D. (2005). *Islamic microfinance in Indonesia*. Economic Development and Employment Division. Financial Systems Development Sector Project Eschborn: Deutsche Gesellschaft fur Technische Zusammenarbeit (GTZ).

Siregar, A. M. (2010). *The role of microfinance in alleviating HIV/AIDS-related poverty: A case study in Bandung*. West Java, Indonesia, Term Paper IMM, Bandung: Universitas Padjadjaran.

Slikkerveer, L. J., & Dechering, W. H. J. C. (1995). LEAD: The leiden ethnosystems and development programme. In D. M. Warren, L. J. Slikkerveer, & D. W. Brokensha (Eds.), *The cultural dimension of development: Indigenous knowledge systems* (pp. 435–440). London: Intermediate Technology Publications Ltd.

Slikkerveer, L. J. (1999). Ethnoscience, 'TEK', and its application to conservation. In D. A. Posey (Ed.), *Cultural and spiritual values of biodiversity: A complementary contribution to the global biodiversity assessment* (pp. 167–260). London: Intermediate Technology Publications Ltd.

Slikkerveer, L. J. (2007). *Integrated microfinance management and health communication in Indonesia*. Cleveringa Lecture Jakarta. Trisakti School of Management and Sekar Manggis Foundation. Leiden: LEAD, Leiden University.

Slikkerveer, L. J. (Ed.). (2012). *Handbook for lecturers and tutors of the new master course on integrated microfinance management for poverty reduction and sustainable development in Indonesia (IMM)*. Leiden/Bandung: LEAD-UL/FEB-UNPAD.

Slikkerveer, L. J. (2015). Integrated microfinance management for poverty reduction and sustainable development in Indonesia. In *International Workshop in Integrated Microfinance Management*. Bandung: Universitas Padjadjaran.

SMERU. (2005). *Ekonomi Program Kredit Mikro Project (SAAD)*. Jakarta: SMERU Research Institute.

Sumardjo, J. (2010). *Estetika Paradoks*. Bandung: Sunan Ambu Press.

Technical Advisory Committee of the Consultative Group on International Agricultural Research (TAC/CGIAR). (1988). *Report of the Technical Advisory Committee forty-sixth meeting*, 13–21 June 1988. Rome: CGIAR Technical Advisory Committee Secretariat.

Toledo, V. M. (2001). Biocultural diversity and local power in Mexico: Challenging Globalisation. In L. Maffi (Ed.), *On biocultural diversity: Linking language, knowledge and the environment*. Washington, D.C.: Smithson Institute Press.

United Nations. (2000). *Millennium Development Goals Report.* New York: United Nations.

United Nations Conference on Sustainable Development (UNCSD). (2012). *Third International Conference on Sustainable Development*. Rio de Janeiro, Brazil: UNCSD. https://sustainabledevelopment.un.org/rio20.

United Nations Educational, Scientific and Cultural Organisation (UNESCO). (1994). *World decade of culture and development 1994–2004*. Paris: UNESCO.
Valadez, M. R., & Buskirk, B. (2011). From microcredit to microfinance: A business perspective. *Journal of Finance and Accountancy*. Academic and Business Research Institute.
Vietnam Bank for Social Policies (VBSP). (2016). *Mission*. Hanoi, Vietnam. http://www.vbsp.org.vn/evbsp.
Walt, V. (2012). Does microfinancing really work? A new book says No. *Time Magazine*. 6 January 2012.
Warren, D. M., Slikkerveer, L. J., & Brokensha, D. (Eds.). (1995). *The cultural dimension of development: Indigenous knowledge systems. IT studies on indigenous knowledge and development*. London: Intermediate Technology Publications Ltd.
Woodley, E., Crowley, E., Pryck, J. D. de & Carmen, A. (2006). Cultural Indicators of Indigenous Peoples' Food and Agro-Ecological Systems. SARD Initiatives. Paper on the 2nd Global Consultation on the Right to Food and Food Security for Indigenous Peoples. Nicaragua. 7–9 September 2006.
World Bank. (2009). *Moving out of poverty*. Washington, D.C.: The World Bank.
World Bank. (2013). *Poverty and Equity Data*. Washington, D.C.: The World Bank. http://povertydata.worldbank.org/poverty/home.
World Commission on Environment and Development (WCED). (1987). *Our common future—The brundtland report*. New York: United Nations.
Yap, J. T. (2013). *Addressing Inequality in East Asia through Regional Economic Integration*. ERIA RIN Statement No. 3. Jakarta, Indonesia: ERIA Research Institute Network.
Yunus, M. (1999). *Banker to the poor: Micro-lending and the battle against world poverty*. New York: Public Affairs.
Yunus, M., & Weber, K. (2007). *Creating a world without poverty: Social business and future capitalism*. New York: Public Affairs.

Kurniawan Saefullah is Lecturer of the Faculty of Economics and Business, Universitas Padjadjaran, Bandung, Indonesia since 1998. He graduated from Universitas Padjadjaran and the International Islamic University, Malaysia. He has published several books and articles with his colleagues, including Introduction to Management and Introduction to Business. In 2009, he joined the International Project on Integrated Microfinance Management (IMM) for Poverty Reduction and Empowerment in Indonesia. As a Ph.D. Researcher at the Leiden Ethnosystems and Development Programme (LEAD), Faculty of Science of Leiden University, The Netherlands, he is finalising his Ph.D. Dissertation on 'Gintingan in Subang: An Indigenous Institution for Sustainable Community-Based Development in the Sunda Region of West Java, Indonesia' under the supervision of Prof. Dr. L. J. Slikkerveer. The focus of his research is on the important role of the local cosmology of Tri Tangtu in the process of sustainable community development, while he also developed a model of integration of community-support institutions on the basis of the Indigenous Knowledge Systems (IKS) in combination with the IMM and ICMD approaches, such as financial, medical, educational, communication and social services through the indigenous institutions of the local communities in the Sunda Region of West Java.

Chapter 4
Towards a Model of Integrated Community-Managed Development

L. Jan Slikkerveer

One of the important effects of globalisation is decentralisation which means recognition of different backgrounds and ethnic values. It assures the reorientation of traditional values and norms, discarding the Western hierarchical concept of Transfer of Technology (TOT), replacing it with more participatory policies of a 'bottom-up' approach where participation and sustainable development form the order of how things are to be done.

Anak Agung Gde Agung (2007)

4.1 Multidimensional Conceptualisations of Global Poverty

One of the most prevalent and pervasive development problems on the globe today relates directly to poverty. As an unacceptable condition of humankind, it has been the subject of an advancing process of conceptualisations, definitions, analyses and levels, which have extended its meaning over the past decades from a single condition of a mere lack of finances to a multidimensional complex of socio-cultural, economic, and political factors influencing human deprivation in well-being. From a psychological perspective, poverty can be caused by two factors: those related to the individual's role and those related to the social-cultural role. Depression, alcoholism and anti-social personality disorder are some causes of poverty at the individual level, where these cases commonly occur in urban areas (cf. Murali and Oyebode 2004).

Economic disparity, income difference, social class and prejudicial stereotypes are among the causes of poverty related to the social system (cf. Turner and Lehning 2006). The anthropological view sees poverty as the result of the growing

L. J. Slikkerveer (✉)
LEAD, Leiden University, Leiden, The Netherlands
e-mail: l.j.slikkerveer@gmail.com

L. J. Slikkerveer et al. (eds.), *Integrated Community-Managed Development*, Cooperative Management, https://doi.org/10.1007/978-3-030-05423-6_4

imbalance between global and local systems, where the processes of international economic development and globalisation tend to exclude indigenous cultures, causing poverty and deprivation to rise among local peoples and communities, particularly in developing countries.

As mentioned in the *Introduction*, a realistic conceptualisation of poverty is not only important for a credible assessment of the actual position and numbers of the poor living on the planet, but also for the design of appropriate strategies to reduce this intolerable condition on a global scale. Poverty has also been defined in various ways in order to enable a reliable measurement and comparison of the percentage of different categories of poor people living within and among populations. Three basic approaches have emerged over the past decades in an effort to conceptualise poverty, respectively, as a material condition, as a multidimensional condition and as a relational condition of the poor, which have been influenced by a number of global trends, including globalisation, financial crises, climate change and political instability. While poverty has initially been defined in monetary terms as a human condition where people lack money to meet their basic needs, the definition has become more realistic over the past decades to encapsulate the persistent condition of a general lack of access to adequate services of health, education, justice, employment and freedom, rendering poverty to a wider, more socio-cultural complex problem in the society.

The challenge of addressing poverty from a multidisciplinary perspective grew out of diverse views and opinions from scientists of different disciplines. From a historical point of view, poverty can be understood in the public consciousness over the past centuries. Following the colonial period of time of dominance and marginalisation of the rural people in non-Western areas, mass immigration and industrialisation have contributed to both rural and urban poverty in the early 20th century, whereas in the 1930s the stock market crash and depression have further increased poverty among the population in both Western and Non-Western countries.

A wider approach to assess poverty in conjunction with well-being has been introduced by Sen (1990), arguing that well-being is directly related to the capability to function in the society, linking the concept of poverty to wider notions of social need and well-being. According to this approach, poverty arises when people lack such capabilities to obtain sufficient income, education, and health, pertaining to insecurity, low self-confidence, a sense of powerlessness, or the absence of rights, such as freedom of speech. According to this view, poverty is a multidimensional phenomenon and by consequence less amenable to simple solutions. Although in general a higher income could contribute to the alleviation of poverty, it would need to be preceded by measures to empower the poor, insure them against risks and address specific weaknesses, such as inadequate schools, or limited health services. Poverty tends to arise when people are lacking these key capabilities.

Later onwards, Anand and Sen (1997) further developed the multidimensional perspective of the concept of poverty in relation with human development, in which poverty is assessed from a human development point of view. Their wider multidimensional perspective focuses not just on poverty of income, but on poverty within the context of development as a denial of choices and opportunities for living a tolerable life. By acknowledging that development refers to a multidisciplinary process, by consequence, the related factor of poverty similarly requires a multi-disciplinary approach. Recently, Austin et al. (2005) have further elaborated the multidimensional perspective on poverty in a framework for social services agencies to move their services towards a more extended family and neighborhood approach. The related global *Multidimensional Poverty Index* (MPI) was later proposed by Alkire and Santos (2010), and since then implemented by the UNDP's *Human Development Reports* since 2010.

A similar multidisciplinary approach to focus on poverty reduction in Indonesia through the provision of not only financial, but also medical, educational, communication and cultural services to the poor has also been initiated in the advanced training of community-based managers at Universitas Padjadjaran in Bandung with the introduction in 2012 of the new master Course on Integrated Microfinance Managers (IMM).

Among the leading international organisations working in the field of poverty reduction through a focus on *human* development is the United Nations Development Programme (UNDP). It has been implementing the above-mentioned multidimensional perspective on the concept of poverty in relation with human development, as substantiated by its annual *Human Development Reports* since 1990. In contrast, the other major international organisation involved in poverty reduction, the World Bank, supports *economic* development, as documented in its annual *World Development Reports* published since 1997.

What followed was the increased priority which poverty reduction has recently received on a global level, not only by the World Bank (2016a) in its dual approach to reach the end of chronic extreme poverty by 2030 and the promotion of shared prosperity, but also by the United Nations (2015) in its latest Post-2015 Agenda for Sustainable Development and the Sustainable Development Goals (SDGs).

Recently, the World Bank (2017) introduced an important aspect into the debate on poverty reduction, being the need for an improved methodology to collect and compare reliable data on poverty and the related factor of inequality worldwide, providing a more solid framework for policy making. The recommendations as to how the monitoring of the progress up to 2030 should be conducted are presented under three headings: raw materials (data), analysis, and presentation.

Also, the promotion of shared prosperity has recently attracted attention since high income inequality is found to constrain economic systems and international collaboration, and as such the primary goal of ending poverty by 2030. As the Report on *Taking on Inequality* by the World Bank (2016a: 9) shows, the more equal countries appear to have healthier people and be more economically efficient than highly unequal countries. Moreover, those countries which focus their policies on reducing inequality are likely to experience more sustained economic growth than those which don't: "*Less inequality can benefit the vast majority of the world's population.*"

As mentioned before, one of the most commonly used ways to measure the incidence of poverty has been to assess the material condition of people based on the 'dollar-a-day' criterion elaborated by the World Bank (2005a). In this way, people are presumed to live in *extreme poverty* when they earn less than $1.25 per day, while the population living in *moderate poverty* earn between $1.25 and $2.00 per day. According to this indicator, the number of people who are living under the poverty line was estimated in 2011 at about 1.4 billion (21.7%) and 2.6 billion (40.2%), observed in all parts of the world, including the developed nations, further substantiating the urgent need for the reduction of extreme poverty as a global challenge. These most recently available poverty estimates are shown in Table 4.1, which, however, are not considering the recent global food crisis and increased cost of energy, which would add about 100 million more people living below the poverty line (cf. Shah 2011). After the recalculation by the World Bank (2012) of the poverty line of $1.25 to $1.90 per day in order to introduce a new international poverty standard, however, the overall level of global poverty has remained basically unchanged.

The World Bank recently published two books on the subject of poverty, which deserve attention: *Introduction to Poverty Analysis* (2005b) and *Handbook on Poverty and Inequality* (Haughton and Khandker 2009) which elaborate on the recent widening of the concept of poverty. Embarking on the previous definition of poverty by the World Bank (2000) that: "*poverty is pronounced deprivation in well-being*", new questions have been raised of what actually is meant by 'well-being' and of what the reference point is against which to measure deprivation.

The conventional approach is to define well-being (and hence poverty) as the command over commodities in general, meaning that people are better off if they have a greater command over resources. As the main focus is on whether households or individuals have enough resources to meet their needs, poverty is then measured by comparing individuals' income or consumption with some defined threshold below which they are considered to be poor. However, a broader approach to well-being is to establish whether people are able to obtain a specific type of consumption good, such as food, shelter, health care or education, further extending the traditional monetary measures of poverty. Since additional factors should be considered, such as the inflation rate of the country, the differentiation

Table 4.1 The World Bank's latest estimate of poverty at different poverty levels (2011)

Poverty line (US $ per day)	Population in poverty (in billions of people)	Population above that level of poverty (in billions of people)	Percentage in poverty
1.00	0.88	5.58	13.6
1.25	1.40	5.06	21.7
1.45	1.72	4.74	26.6
2.00	2.60	3.86	40.2
2.50	3.14	3.32	48.6
10.00	5.15	1.31	79.7

Source Shah (2011)
URL: http://www.globalissues.org/article/4/poverty-around-theworld/
World Bank's Poverty Estimates Revised

among household members, the prevailing health conditions in terms of diseases, access to education, etc. the World Bank (2016a) further extends its approach by linking *poverty* with *shared prosperity*, which also takes the disadvantages of inequality into account within the context of the widening gap between the rich and the poor. Shared prosperity is measured as the growth in the income or consumption of the bottom 40% of the population in a country.

For the World Bank, poverty is now conceptualised as a situation of pronounced deprivation in well-being, comprising several dimensions of low income and the inability to acquire the basic goods and services necessary for survival, low levels of health and education, poor access to clean water and sanitation, inadequate physical security, lack of voice, and insufficient capacity and opportunity to better one's life. As the result of its experience over the past decade, the World Bank (2016b) has added another dimension to the definition of poverty with the concept of 'inequality'. Such inequality among people has continued in opportunities, gender disparities, and deprivations in many sectors of the society, which need to be brought into balance, implying that prosperity must be shared meaningfully within developed and developing countries. In the latest publication of the World Bank (2016c), an estimated 767 million people are living under the new international poverty line of $1.90 a day, meaning that almost 11 people in 100, or 10.7%, were poor. Given the low standard of living implied by the $1.90-a-day threshold, poverty continues to remain unacceptably high around the globe.

The *United Nations Development Programme* (UNDP 1997), one of the world's leading bodies in the field of poverty reduction and international development, had introduced the *Human Poverty Index* (HPI), as a composite index which combined national estimates of deprivations in health, education and standards of living in a single number to complement the *Human Development Index* (HDI) in order to reflect the extent of deprivation. As mentioned before, the conventional measure of poverty only considers income, and people living on less than $1.25 a day are regarded as extremely poor. People, however, can also be deprived of schooling, proper nourishment, safe drinking water, etc. rendering them poor in a broader perspective. The *Multidimensional Poverty Index* (MPI) encompasses such a broader, weighted average of 10 indicators, which allows for a realistic assessment of people who can be considered in multidimensional poverty if they are deprived in at least a third of these indicators, with each indicator having a defined deprivation level. Longitudinal measurement of MPI changes in the actual situation of poverty became manifest in developing countries as a group, where human poverty affects more than a quarter of the population. Sub-Saharan Africa and South Asia are sharing the population living in extreme poverty with an income of less than $1.90 per day and human poverty at about 40% (cf. World Bank 2014).

The *Human Poverty Index* (HPI) was later supplanted by Alkire & Foster (2009) to the *Multidimensional Poverty Index* (MPI) which was introduced in the *Human Development Report* of the United Nations (2010), entitled: *The Real Wealth of Nations, Pathways to Human Development*. The MPI overcomes the overlapping deprivations at the household level by using three dimensions of human development: health, education, and living standards. These dimensions consist of ten indicators which were weighted equally in the MPI. The ten indicators include nutrition, child mortality, years of schooling, children's school enrollment, cooking fuel, sanitation, water, electricity, floor and assets.

The United Nations basically refers the concept of poverty to the inability of people of having choices and opportunities, a violation of human dignity characterised by lack of a basic capacity to participate effectively in society and of access to services. The above-mentioned *Multidimensional Poverty Index* (MPI) responds to the complexity of the concept and identifies deprivations across the same three dimensions as the *Human Development Index* (HDI), reflecting a long and healthy life, access to knowledge and a decent standard of living. The MPI shows the number of people who are multi-dimensionally poor, i.e. suffering deprivations in 33% or more of weighted indicators and the number of deprivations with which poor households are typically confronted.

The MPI can be deconstructed by region, ethnicity and other groupings as well as by dimension, rendering it a useful tool for national and regional policymakers to pay particular attention to specific target groups in the society.

According to the United Nations *Human Development Report* (2015), around 1.5 billion people live in multidimensional poverty, estimated by using the MPI measure for 101 countries. At least about one third of the indicators reflect severe divestiture in access to health care and education, and a low standard of living. In addition, about 800 million people are potentially vulnerable to fall into poverty. The five countries with the largest populations in multidimensional poverty include Ethiopia, Nigeria, Bangladesh, Pakistan and China. However, the countries with the highest proportions of their population living in severe poverty, i.e. deprived in more than half the dimensions, are Niger, South Sudan, Chad, Ethiopia, Burkina Faso and Somalia, at more than 60%, and Guinea-Bissau and Mali, at more than half.

4.2 The Challenge of the New Century: Global Poverty Reduction

Since the problem of poverty has been acknowledged to pervade human life into worldwide miserable conditions, efforts to solve this predicament on a global scale have recently moved further upwards on the list of global priorities. It is clear that among the leading international organisations and agencies concerned about the worldwide problematic position of the poor, such as the World Bank and the United Nations, a general consensus has eventually been reached about the complexity and multiplicity of the conceptualisation of poverty. Similarly, these organisations agree, that the current state of global poverty is still unacceptably high, and as such in urgent need of a comprehensive and effective strategy of poverty reduction at the global level.

The absence of relevant considerations concerning poverty in the past has also created difficulties to envisage effective policies and solutions towards poverty alleviation. Scientists and practitioners agree that approaching poverty requires a broadened view, involving different categories of factors ranging from economic to social, medical, educational and cultural circumstances in which poor people are living, and by consequence, the related efforts to reduce poverty demand a similarly holistic approach.

In general, the different approaches designed to lift people permanently out of poverty include humanitarian, financial, economic and social measures. Since the concept of poverty in itself is already complicated, as indicated above, it is not surprising that likewise, the various remedies have shown to be rather problematic as well.

Humanitarian aid is largely material and logistic assistance is given to people in need, usually provided as short-term help until the long-term aid by the government and other institutions take over in the form of measures of poverty-efficient allocation of aid. The allocation of aid among countries generally reflects multiple objectives. It may be used to rebuild post-conflict societies, or to meet humanitarian emergencies. However, the core objective is most commonly poverty reduction. Humanitarian aid is closely related to the *Universal Declaration of Human Rights* (UDHR) adopted by the United Nations General Assembly on 10 December 1948 (cf. Pogge 2012). The Declaration represents the first global expression of what many people believe to be the rights to which all human beings are inherently entitled, including: "*promoting and encouraging respect for human rights and for fundamental freedoms for all without distinction as to race, sex, language, or religion, and member states pledge to undertake 'joint and separate action' to protect these rights.*" (cf. UN-UDHR 1948).

Back in the 1990s, the financial policies towards poverty alleviation were initially focused on programmes of structural adjustment and financial liberalisation with a view to improving economic growth in developing countries. The results of financial sector reform, however, have been disappointing (cf. World Bank 1989; Cull 1997; Williamson and Maher 1998). The subsequent financial crisis had a severe impact on the position of the poor, and poverty levels showed an increase in most developing countries. In their study on the relationship between financial development and poverty reduction in developing countries, Jalilian and Kirkpatrick (2001) contend that, despite the limitations of their data, financial development can contribute to poverty reduction. The authors support the position of several agencies and NGOs, including the World Bank, that improved access of the poor to financial services strengthens the productive assets of the poor, and as such would reduce poverty.

However, as indicated in the *Introduction* and further elaborated in Chap. 2, following the transition by the mid-1990s from socio-economic policies of poverty reduction to a new financial strategy of *microcredit*, introduced by Mohammad Yunus and the Grameen Bank in Bangladesh, a shift took place in the late 1990s to commercial *microfinance*. However, the 'new wave' of *microfinance* led eventually to growing disillusions as the neoliberal approach in microfinance began to reveal the failure and factual incapacity of microfinance to improve the situation of the poor and low-income families. Soon, an international debate emerged between the defenders of the *financial systems* approach and those who support the *poverty lending* approach.

While the first approach advocates the notion that the capacity of MFIs to achieve self-sustainability can only be generated from income from lending money to clients—including the new target group of the 'bankable poor'—and reducing operational costs of the institutes, the *poverty lending* approach, however, emphasises the importance of providing credit with subsidised interest rates to help overcome poverty as the poor cannot afford high interest rates (cf. Robinson 2001; Hermes and Lensink 2011; Slikkerveer 2012). As described above, an increasing number of independent studies are documenting that the microfinance approach does not really reach the poor and low-income families, and in only a few cases does it improve the situation of the middle-income groups, largely because of the commercial interests of MFIs, self-sustaining NGOs and banks, rendering microfinance being a 'poverty reduction tool' extremely doubtful.

Following the successful U.S. Programme of the Marshall Aid after World War II, aimed at helping the European countries in their efforts to reconstruct the war-torn circumstances, largely focusing on 'investment in capital' to strengthen the relations between relief, rehabilitation and development, the development aid to the new independent nations of the Third World attempted to implement a similar socio-economic development approach (cf. Myrdal 1968).[1] During the successive phases of 'development aid' and 'development cooperation' with developing countries, the community development movement expanded rapidly, but in the course of the 1970s, it declined largely because of the disappointing results of the 'top-down' 'Transfer of Technology' (TOT) process which failed to encourage self-help efforts and community participation for socio-economic development. As mentioned before, soon thereafter, a new development assistance approach of *Integrated Rural Development* (IRD) was launched, which, after its promising take-off to direct its efforts towards improving the productivity and welfare of the rural poor in the poorest countries, gradually declined as the result of its incapability to meet its objectives of increased agricultural production and human well-being. As Cohen (1987: 11) concludes: "*In the end, the strategy of integrated rural development suffered the same fate as community development: rejection.*"

[1]The Marshall Plan (officially the European Recovery Programme, ERP) was an American initiative to aid Western Europe, in which the United States gave over $12 billion (approximately $120 billion in current dollar value as of June 2016) in economic support to help rebuild Western European economies after the end of World War II. The plan was in operation for four years beginning April 8, 1948. The goals of the United States were to rebuild war-devastated regions, remove trade barriers, modernise industry, make Europe prosperous again, and prevent the spread of communism. The Marshall Plan required a lessening of interstate barriers, a dropping of many regulations, and encouragement towards an increase in productivity, labour union membership, as well as the adoption of modern business procedures.

By the end of the 20th century, experience in international economic growth, especially in the developing countries, had shown that, despite the widening gap between the rich and the poor, the process of the reduction of poverty could indeed provide a contribution to human development, which, in turn could eventually also contribute to the improvement of the position of the poor on the long term. So far, the overall consensus underscores that poverty reduction and sustainable development are inextricably linked and mutually dependent.

As Schaffer (2001) contends, new ways of thinking started to reflect through the international discourse on poverty and development, exemplified by new approaches towards poverty reduction by the World Bank and the United Nations. As described above, the concept of poverty had already been widened by the shift from the physiological model of deprivation to a social model, encompassing issues of vulnerability, inequality and human rights. In addition, the interpretation of the causes of poverty was also broadened to include a wider range of new variables related to social, political, cultural, and environmental factors. The interest in the causal context of poverty was further deepened with a focus on the fluctuations and movements of the poor in and out of poverty. Special international development-based poverty reduction programmes have been designed and implemented as concerted efforts for longer periods of time in different ways of economic liberalisation, self-determination, returning property rights to the poor—especially land and resources—providing various financial, health, education and social services to all—and the fight against corruption and political instability on an international scale in order to achieve socio-economic development.

Several international organisations and their agencies started to focus their attention on the global aspects of poverty reduction in conjunction with socio-economic development, and soon, poverty reduction became acknowledged to embody the main development problem of the new century. Wiggins and Higgins (2007) introduced a strategy of pro-poor growth and development based on the theorem of the 1990s that economic growth rates in developing countries would have to increase in order to close the gap with the developed countries. However, since poverty reduction still refers to a rather complicated and yet indeterminate process in which economic growth is one among many factors, uncertainty remains about its role. Moreover, the authors agree that no blueprints for growth and poverty reduction exist, and that each country would need detailed and specific analysis.[2]

The humanitarian approach to poverty reduction was further underscored by UNESCO, a United Nations agency which seeks to contribute to peace and security around the globe by encouraging global collaboration between countries through

---

[2]The idea of the 1990s, that economic growth could play a role in the reduction of poverty, has led to a renewed interest in pro-poor growth, in which two concerns dominate the discussion: rates of growth in developing countries have to increase in order to narrow the gap between the developing and the developed countries, and poverty has to be reduced on a worldwide scale. According to Wiggins and Higgins (2007), economic growth is usually necessary for poverty reduction, but it is far from sufficient, and poverty reduction through growth depends on access to markets.

education, science, and culture in order to substantiate universal respect for justice, the rule of law and human rights together with fundamental freedom, affirmed in the *Universal Declaration of Human Rights* (UDHR) of the United Nations (1948). The constitution of UNESCO declares that peace must be recognised upon the intellectual and moral solidarity of humanity. In addition, in 2005, the 33rd General Conference of UNESCO in Paris announced the *Universal Declaration on Bioethics and Human Rights* (UDBHR). The reason behind the declaration is based on the lack of bioethical guidelines, particularly in developing countries. Although the Declaration attempts to guide global ethical considerations in the field of bioethics and development, a more general concern to individuals and communities, particularly in developing countries, is well indicated. The main implication of the UDBHR for poverty reduction and development is that policy makers and government agencies should also incorporate ethical factors and humanitarian considerations in the policy planning process, which relate to human rights, both individually and socially, towards the interests of local peoples and communities.

In this respect, UNESCO's Declaration also supports the integration of the *emic view* in development, as promoted by Warren et al. (1995). The recommendation that sustainable development should be planned and executed on the basis of local people's perspective and participation later found wide support from other scientists, including Morris et al. (1999), Woodley et al. (2006) and Deubel (2008). The broader implications of the *Universal Declaration on Bioethics and Human Rights* (UDBHR 2005) encompass, on one hand, the support to the global recognition towards ethical considerations in development, while on the other hand, it also acknowledges peoples' own perspective of how development should be planned and implemented in a sustainable way.

4.3 World Bank and United Nations: *Economic and Human Development*

In addition to the above-mentioned advancement of ethical guidelines for sustainable development by UNESCO, two major organisations had taken up the challenge of poverty reduction for sustainable development by the end of the former century, i.e. the World Bank and the United Nations and its major agencies.

Although a consensus has been reached about the multidimensional aspects of poverty, the strategies are still different. While the World Bank continues to approach poverty reduction from an economic development perspective, the United Nations does so from a human development point of view. Such differences are also expressed in the respective annual reports: the *World Development Reports* of the World Bank and the *Human Development Reports* of the United Nations.

Since 1978, the World Bank had started to publish its annual *World Development Reports*, providing a comprehensive and timely overview on the economic dimension of development, in which each year a specific aspect of development is highlighted as a reflection of the progress and experience in international world development. In this way, the *World Development Reports* have not only provided a view of the evolution in the way of thinking on socio-economic development, but also the progress in policy recommendations on relevant topics ranging from agriculture, the role of the state, transition economies, and labor to infrastructure, health, the environment and poverty reduction.

After the first publication of the *World Development Report* (WB 1978) on some major economic development issues confronting the developing countries, the prospects for progress in accelerating growth were already highlighted within the context of the related policy issues operational at the time. A few years later, the *World Development Report* (WB 1980) identified two major challenges facing the world at the time: to continue the social and economic progress of the past 30 years in an international climate which looked less helpful, and to tackle the plight of the 800 million people then living in absolute poverty. A few years later, the *World Development Report* (WB 1985) focused on the contribution which international capital was making to economic development, especially how the institutional and policy environment affected the volume and composition of financial flows to developing countries.

Then, the *World Development Report* (WB 1990) focused entirely on the position of the poor: a broad definition of poverty was adopted to include not only income, but also literacy, nutrition, and health. In addition, two elements have been put forward as being important to strive for sustainable progress on poverty reduction: the promotion of efficient use of the poor's most abundant asset—labour—and the provision of basic social services to the poor, e.g. primary health care, family planning, nutrition, and primary education.

Following subsequent *World Development Reports* thereafter, focusing on relevant topics for development, such as the role of the environment, the infrastructure, the labour market, and health, the *World Development Report* (WB 1998/1999) entitled *Knowledge for Development* further deepened the way of thinking about poverty reduction and economic development by acknowledging that not capital, but *knowledge* is the key to sustained economic growth and improvement in human well-being. Such important recognition did not only pay due attention to the work of many scientists working on the crucial role of the exchange and transfer of different systems of knowledge in international development, but also enabled the World Bank to acquire more appropriate tools for sustainable policy planning and implementation, particularly with regard to developing countries. The Report starts with an interesting discussion of the importance of knowledge for development, also touching on the international debate on the role of indigenous knowledge in development. Although the Report rightly draws the attention on the central issue of the enduring knowledge gap in terms of inequality between developing and developed nations which affects the poor in a disproportional way, and identifies some critical steps which developing countries should take in order to narrow the knowledge gaps, such as the acquisition of knowledge through research and development and building on indigenous knowledge, the potential and functionality of indigenous knowledge for poverty reduction are not fully reflected in the conclusions of the Report.

The *World Development Report* (WB 1999/2000), entitled *Entering the 21st Century,* signalled the new challenges posed by the transforming economic, political and social development landscape at the turn of the century. Two forces of change are dealt with: the integration of the world economy and the increasing demand for self-government, which both affect responses to key issues including poverty reduction. The *World Development Report* (WB 2000/2001), entitled *Attacking Poverty,* has further set the tone for the high priority of the poverty reduction approach for the new decade, focusing on the various dimensions of poverty, and on strategies of how to create a better world free of poverty. Furthermore, the international dimensions are indicated including global actions to fight poverty, analysing global trade, capital flows, and how to reform development assistance in order to improve the livelihood of poor people.

The *World Development Report* (WB 2004), entitled *Making Services Work for Poor People,* provides a practical framework for making basic services such as water, sanitation, health, education, and electricity available for poor people. The framework provides governments and donors with useful tools to reach the common objective of poverty reduction.

The most recent *World Development Report* (WB 2016a), entitled *Digital Dividends*, shows that the current digital revolution offers new opportunities for the promotion of development through three mechanisms of the internet: inclusion, efficiency and innovation. The poor in particular can benefit from digital technologies in their access to markets and services. On a wider scale, digital technologies are instrumental in the accumulation and storage of different knowledge systems for development, not least indigenous knowledge systems.

In the meantime, the United Nations became intensely involved in the global strategies of poverty reduction, which was expressed in the successive *Human Development Reports* annually published since 1990 by the *United Nations Development Programme* (UNDP). In highlighting its constructive contribution to the international development debate, UNEP became a pioneer in highlighting the human dimension of development in the international debate on human development and related issues, including poverty reduction, over the past few decades. Starting with the first *Human Development Report* (UNDP 1990), entitled Concept and Measurement of Human Development, the attention was drawn on people, and how development extends their choices, which goes further than the economic growth of the Gross National Product (GNP), income and wealth. Embarking on this perspective, human development was measured by a more comprehensive index as mentioned above—known as the human development index—reflecting life expectancy, literacy and command over the resources to enjoy a decent standard of living.

The Human Development Report (UNDP 1991), entitled ‘Global Dimensions of Human Development’, analysed the global markets in relation to their ability to meet, or fail to meet, the needs of the world’s poorest people. The following *Human Development Report* (UNDP 1993), entitled *People’s Participation*, underscored the importance of peoples' participation in the events and processes which shape their lives. If participation is operationalised in an appropriate national and global framework, it could become a significant source of vitality and innovation for the creation of new and more just societies.

Special attention for global poverty eradication transpired through the *Human Development Report* (UNDP 1997), entitled *Human Development to Eradicate Poverty*, which embarks on the notion that the world does have the resources and the know-how to create a poverty-free world in less than a generation. In the *Human Development Report* (UNDP 1997), entitled *Human Development to Eradicate Poverty,* the strategy to eradicate poverty is made rather explicit to involve a number of activities, including: (a) removing barriers which deny choices and opportunities for living a tolerable life; (b) safeguarding people from the new global pressures which create or threaten further increases in poverty; (c) building assets for the poor; (d) empowering men and women to ensure their participation in decisions which affect their lives; (e) investing in human development such as health and education; and (f) affirming that the eradication of absolute poverty in the first decades of the 21st century is not only feasible and affordable, but also morally imperative.

As regards the position of indigenous people, the Report underscores that in many parts of the world disparities in income and human poverty affect the indigenous people disproportionately, as they are in general poorer than most other groups of the society. Important evidence is provided of the generally deplorable position of the indigenous peoples, where in developing countries the poorest regions are those in which most indigenous peoples are living. As the *Human Development Report* (UNDP 1997) also documents: "*In Australia, for example, aboriginals receive about half as much income as non-aboriginal. In Mexico, for example in municipios where less than 10% of the population is indigenous, only 18% of the population is below the poverty line. But where 70% of the population is indigenous, the poverty rate rises to 80%.*"

The *Human Development Report* (UNDP 1999), entitled *Globalisation with a Human Face*, draws attention to the era of globalisation which could benefit the lives of people everywhere, but which also poses a challenge to ensure that the benefits are shared equitably and that the related interdependence works for the people, and not for profits. Among the arguments put forward by the Report, that globalisation requires leadership, are the fact that poor people and poor countries are running the risk of being pushed to the margin by globalisation which controls the world's knowledge, and that narrowing the gap between rich and poor should become more explicit global goals.

The subsequent Human Development Report (UNDP 2000), entitled *Human Rights and Human Development*, signalled the importance of respecting human rights in human development, which heralded the launching of the *Millennium Development Goals* (MDGs) by the United Nations in 2000. The declaration of the MDGs encompassed eight international goals which had to be realised by 2015.

In this first major concerted action at the global level, all 189 United Nations Member States committed themselves to the fulfilment, supported by about 22 international organisations of the eight Millennium Development Goals by 2015, of which the first goal was to eradicate extreme poverty and hunger. This was sub-divided into 3 targets: (a) halve, between 1990 and 2015, the proportion of people living on less than $1.25 a day; (b) achieve decent employment for women, men, and young people; and (c) halve, between 1990 and 2015, the proportion of people who suffer from hunger.

The following *Human Development Report* (UNDP 2003), entitled *Millennium Development Goals: A Compact Among Nations to End Human Poverty,* further underscored the United Nations' highest priority of poverty eradication at the beginning of the new century. The unique declaration of solidarity and determination to reduce poverty in the world in 15 years encompassed the international efforts by all member states: "*to eradicate poverty, promote human dignity and equality and achieve peace, democracy and environmental sustainability.*"

The *Human Development Report* (UNDP 2011), entitled *Sustainability and Equity: A Better Future for All*, emphasises that the urgent global challenges of sustainability and equity must be addressed together in both national and international policies in order to sustain the human development progress for most of the world's poor majority, not only for future generations, but also for those living today.

The United Nations Report of the Millennium Development Goals (UN 2015a) documents the results in terms of the largely successful completion of the MDGs, as Ban Ki-Moon, the Secretary-General of the United Nations contended that: "*The MDGs helped to lift more than one billion people out of extreme poverty, to make inroads against hunger, to enable more girls to attend school than ever before and to protect our planet.*" However, as the overall progress has also bypassed minority groups such as women, the lowest on the economic ladder, and the disadvantaged because of age, disability or ethnicity, the inequalities continue to persist while progress has been uneven.

Indeed, as mentioned above, the proportion of people living in extreme poverty —on less than $1.25 a day—in the developing countries had dropped to 14% in 2015, while globally, the number of people living in extreme poverty has declined to 836 million in 2015, which, however, remains an unacceptably high number. Another difficulty which emerged in the strategy is that about half of the 155 member states lack adequate data to monitor poverty and, by consequence, the poorest people in these countries tend to remain invisible. During the 10-year period of time between 2002 and 2011, as many as 57 countries (37%) had none or only one poverty rate estimate available (cf. UN 2015a).

The reduction of the number of people living in extreme poverty on the globe in 2015 is shown in Fig. 4.1, documenting, that during the time of the implementation of the MDGs, i.e. from 2000 up to 2015, the absolute number of people living in extreme poverty globally fell from 1.75 billion 1999 to 836 million in 2015. The United Nations *Millennium Development Goals Report* (2015) not only shows that the number of people worldwide living on less than $1.25 a day has been reduced by half from its 1990 level, but also that the world's extremely poor people are distributed very unevenly across regions and countries.

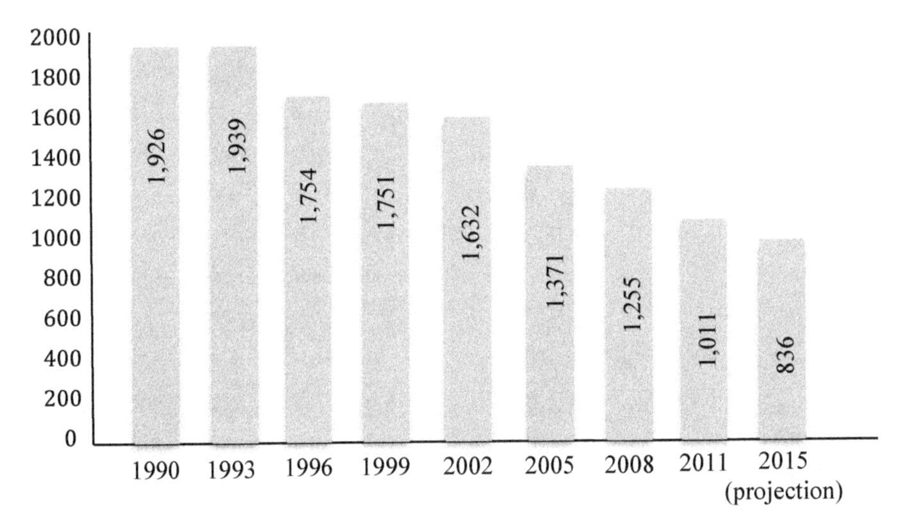

Fig. 4.1 The number of people living in extreme poverty has declined by more than half since 1990. Number of people living on less than $1.25 a day worldwide, 1990–2015 (millions). *Source* The Millennium Development Goals (UN 2015c)

The overwhelming majority of people living on less than $1.25 a day are only living in two regions, Southern Asia and Sub-Saharan Africa, while they account for about 80% of the global total of extremely poor people. In addition, nearly 60% of the world's 1 billion extremely poor people lived in just five countries in 2011: India, Nigeria, China, Bangladesh and the Democratic Republic of the Congo (ranked from high to low; cf. UN 2015a). The lessons learned from the MDGs experience are also reflected in the new United Nations *Post-2015 Agenda for Sustainable Development and Sustainable Development Goals* (SDGs): "*We need to tackle root causes and do more to integrate the economic, social and environmental dimensions of sustainable development.*" (cf. UN 2015b).

The recent *Human Development Report* (UNDP 2015), entitled *Work for Human Development*, draws attention to a broader view of work which goes beyond jobs, also taking into account unpaid care work, voluntary and creative work (Table 4.2).

The Report underscores the important fact that work enables people to earn a livelihood and become economically secure, forming one of the basic requirements for equitable economic growth, poverty reduction and gender equality. Following the completion of the United Nations' eight *Millennium Development Goals* (MDGs) in 2015, in which the alleviation of poverty had already been ranked as the number one goal, the United Nations recently listed poverty eradication as the foremost of the 17 objectives in its new *Post-2015 Agenda for Sustainable Development Goals* (SDGs). The highest priority of poverty reduction has also been recognised by most world leaders at the World Summit in September 2015. Adopted in the *Post-2015 Agenda for Sustainable Development*, poverty

Table 4.2 Countries with the most people in multidimensional poverty

Population in multidimensional poverty			
Country	Year	Millions	Percentage
Ethiopia	2011	78.9	88.2
Nigeria	2013	88.4	50.9
Bangladesh	2011	75.6	49.5
Pakistan	2012/2013	83.0	45.6
China	2012	71.9	5.2

Source Human Development Report Office (2015) calculations using data from Demographic & Health Surveys, Multiple Indicator Cluster Surveys & National Household Surveys

eradication in all its forms and dimensions, including extreme poverty, has become today's ultimate challenge for the next one-and-a-half decades, and an indispensable requirement to attain sustainable development around the globe (cf. UN 2015a, b).

4.4 The 'Missing Link' in the UN/WB Poverty Reduction Policies

For many observers, the problem of the disgraceful condition of poverty still affecting such large proportions of the world's population today seems to have no solution and they see previous efforts for its reduction as only worsening the position of the poor in the end. Some have interpreted the transfer of poverty from one generation to the other even as a manifestation of a persistent *'culture of poverty'*, where the poor are blamed for their problems of human suffering and wasted lives (cf. Lewis 1959). Although pessimism has been growing that at a certain point, the global problem of poverty can hardly be solved, the reasons to remain optimistic and contribute to the realisation of the first of the *Millennium Development Goals* of poverty reduction by the Year 2015 are manifold as they are not only intended for the poor, but also to the non-poor people of the world (UN 2000).

Among these reasons are the fact that poverty basically refers to wasted lives: lives of people who could have realised their potential and contributed to the improvement of the society and its people. Also, poverty often leads to desperation, begging and crime. In many cases, poor people are usually more susceptible for problems of health and disease, including epidemics, bad employment hazards, HIV/AIDS and malnutrition-related high rates of morbidity and mortality. Another reason for the reduction in the poverty of large segments of the population refers to their potential of untapped resources, as Prahalad (2004) shows, that the poor are not a burden, but rather create an opportunity for simple production models, low costs and vast marketplaces. A more critical reason for the developed nations is that the growing number of the poor eventually could cause the breakdown of the 'failed states' which cannot anymore cope with the demands of growing numbers of poor people who resort to political violence, conflict and civil war. Posing a threat to the national security of American and European states, such crises have recently even led to military interventions by the USA, UN or NATO forces such as in Honduras, Serbia, Somalia and Libya. In this context, also the growing numbers of—legal or illegal—immigrants as poor fortune hunters from poor Latin-American and African countries into respectively North America and Europe taking unskilled jobs and living in slums form an additional reason to reduce the poverty in their homelands.

Over the past decades, specific ad hoc actions have been undertaken by international organisations to provide support to the poor and low-income families, often focused on specific target groups which were in need of emergency aid as the result of natural disasters, civil wars or regional conflicts. In this context, Gunatilaka & Kiriwandeniya (1999) refer to the *Triple R Framework*: *relief, rehabilitation* and *reconciliation* as part of the introduction of social safety nets.

The intimate interconnections between development and poverty, widely observed in the practical setting of development programmes, required the construction of a broader framework of sustainable development and poverty reduction for international development cooperation. Back in 1987, the *Brundtland Report* (WECD 1987) paved the way for such a holistic development paradigm which surpassed the limited economic approach towards the exploitative use of natural resources, then dominant among international organisations including the World Bank. The *Brundtland Report* (WECD 1987: 40) clearly indicates that in order to achieve sustainable development, there is a need for: *"a type of development that integrates production with resource conservation and enhancement, and that links both the provision for all of an adequate livelihood base and equitable access to resources."* The implication of this new way of development thinking is, as Mestrum (2003: 50) contends, that: "*Sustainable development, then, is the process for meeting all people's needs, for today and tomorrow. Poverty eradication will be its outcome, due to the equitable distribution of the available resources.*"

The combined approach towards poverty reduction for sustainable development has further been adopted in international strategies which have operationalised the direct relationship between poverty alleviation and sustainable development, most manifest in biodiversity conservation, such as underscored by the *Convention on Biological Diversity on Traditional Knowledge, Innovations and Practices* (UN CBD 1992) and the Millennium Ecosystem Assessment (2005).

Previously, the United Nations Conference on Environment and Development—Agenda 21 (WCED 1987) had already been implementing such a holistic view of development: "*whose primary goals include the alleviation of poverty; secure livelihoods; good health; quality of life; improvement of the status and income of women and their access to schooling and professional training, as well as fulfilment of their personal aspirations; and empowerment of individuals and communities.*"

Since then, several United Nations strategies started to focus on human development, in which poverty eradication is conceptualised as an important step towards sustainable development in which human rights are directly involved. Structural, multilateral strategies have predominantly been designed and implemented by international organisations such as the United Nations, the World Bank, and a number of NGO's. Among the major strategies which these organisations implement to try to reduce poverty on a worldwide scale are the joint efforts to achieve socio-economic growth, to increase international development cooperation, to induce a redistribution of wealth, to promote birth control, or to provide microcredit to the poor and low-income families.

As mentioned before, in September 2000, at the beginning of the Third Millennium, the Member States of the United Nations unanimously adopted the *Millennium Development Goals* (MDGs) (UN 2000). These goals emerged from the agreements and resolutions of various development conferences organised by the United Nations in the course of the 1990s, committing the international community to a common vision of development as a strategy in which human development and poverty reduction receive the highest priority. Poverty reduction became an important component of a global package for human development. In general, the objective of the MDGs was to serve as a guidepost and focus the efforts of the world community on achieving significant, measurable improvements in poor people's lives.

Among the 8 MDGs, the eradication of extreme hunger and poverty by halving the proportion of people living on less than $1 a day and halving malnutrition by 2015 became the prime aim of the United Nations, embedded in subsequent goals of the realisation of improvements in education, gender equality, maternal and child health, treatment of major diseases such as HIV/AIDS and malaria, environmental stability, and fostering a global partnership for development, especially with the poorest countries.

By the end of the period of 2015, as set by the MDGs, a first evaluation of the realisation of these goals has been provided by the *Millennium Development Goals Report* (UN 2015c). The Report provides a final assessment of global and regional progress towards the MDGs since their endorsement in 2000. Although it seeks to show that significant progress has been made across all goals, and that the global efforts to achieve the MDGs have saved the lives of millions and improved conditions for many more around the world, the report has to acknowledge an uneven progress and shortfalls in many areas, which are in urgent need to be addressed without delay.

According to the estimate used by the United Nations, the number of people in extreme poverty, defined as having less than $1.25 a day to live on, has fallen from 1.9 billion in 1990 to 836 million in 2015, i.e. somewhat more than the halving called for in the first MDG on poverty reduction. In proportional terms, that corresponds to a drop from nearly half the population of developing countries living in extreme poverty to only 14% remaining below the $1.25 a day line.

In the process, the Administrator of UNDP, Helen Clark (2014: iv) conceded that: "*...overall global trends are positive and that progress is continuing. Yet, lives are being lost, and livelihoods and development undermined, by natural or human-induced disasters and crises*". In her opinion, eradicating poverty will be a central objective of the new UNDP Agenda for 2030. In this context, however, some scepticism has been expressed about the objectivity of the measurement of 'progress' against the different goals, given the difficulties encountered in gathering and comparing the statistical data over the past 15 years from among the participating countries. In the same way, the publication of the United Nations Millennium Development Goals Report (2015a) has also given rise to several expressions of disappointment and criticism, focused on the limitations of the realisation of its goals, as the approach has generally been dismissed as a mere continuation of 'top-down' as opposed to 'bottom-up', largely directed by the donor countries, and as such subsequently implemented by most national planning agencies. Although some trends indicate that progress has indeed been made in most countries, especially with regard to the goals of eradicating poverty and improving access to education, these trends have been uneven across countries and regions, as well as among social groups.

In addition, the inclusion of the objectives for political and cultural rights of the target population, such as those contained in the *Millennium Declaration,* have so far largely been ignored. Moreover, the rights of the people, as adopted by the United Nations General Assembly back in 1948, and reaffirmed by UN Member States many times since, have also been missing in the MDGs' approach. As Article 1 of the Universal Declaration of Human Rights (1948) states: "*All human beings are born free and equal in dignity and rights. They are endowed with reason and conscience and should act towards one another in a spirit of brotherhood.*", principally, the local peoples' indigenous knowledge systems as well as their indigenous institutions should have received much more attention in the global development framework of the MDGs. Although a reference is made to the 'importance' of the Universal Declaration of Human Rights and 'respect for all human rights', some observers such as the Yale Campuspress (2015) are concerned that the newly adopted Sustainable Development Goals (2015) promote a false sense of success, giving room to governments to go slow on the realisation of the human rights of their population groups.

Closely related are the critical observations that the MDGs have focused attention on average progress, and in doing so have left the persistent inequalities virtually unnoticed. In some cases, the positive assessment of national intervention programmes has blurred the attention for the needs of disadvantaged groups of local communities, including the structurally poor and low-income families.

When in September 2015, the new global *Sustainable Development Goals* (SDGs) of the *Post-2015 Sustainable Development Agenda* of the United Nations (2015a: 1) were adopted by a large number of Heads of States, all member states committed themselves to working vigorously for the full implementation of this Agenda by 2030. The related publication of the UN Report *Transforming our World: The 2030 Agenda for Sustainable Development* (2015b): "*a new plan of action for people, planet and prosperity*", recognises that: "*eradicating poverty in all its forms and dimensions, including extreme poverty, is the greatest global challenge and an indispensable requirement for sustainable development.*" The huge ambitions of the *2030 Agenda for Sustainable Development* encompass 17 Sustainable Development Goals (SDGs) and 169 targets which seek to build on the previous *Millennium Development Goals*, and "*complete what they did not achieve*" The new SDGs are set to integrate and balance the three dimensions of sustainable development: the economic, social and environmental, over the next 15 years. Following the completion of the MDGs in 2015, other international organisations including the World Bank and the Oxford Poverty & Human Development Initiative (OPHI) provided a follow-up to the new targets of the MDGs by making an effort to reconsider their approaches towards the continuing struggle against the global problem of relative poverty, still rampant among about 4 billion people around the world. In its *Report of the Commission on Global Poverty* (World Bank 2016d), the World Bank joins the new *Sustainable Development Goals* of the *Post-2015 Agenda* (United Nations 2015) in the formulation of their primary development goal, being the eradication of poverty by 2030, announcing its two overarching goals: "*the end of chronic extreme poverty by 2030; and the promotion of shared prosperity, defined in terms of economic growth of the poorest segments of society*." In its ambitious objective to measure global poverty over time, the World Bank also seeks to explore alternative approaches to reach more realistic estimates, also using non-monetary indicators (cf. World Bank 2016d).

In this context, it is interesting that the position of one of the main target groups in global poverty reduction, i.e. the indigenous people, receive at least some—but still marginal—attention, albeit only when it comes to sub-national poverty measurement, encapsulated in the concept of *'within-country disaggregation'*. A recent report of the *United Nations Permanent Forum on Indigenous Issues* (UNPFII) (2016) estimates that there are still some 370 million indigenous peoples living in 70 countries across the world. In view of the historical process of exploitation and marginalisation of these indigenous groups, it is not surprising that evidence underscores that they are suffering most from much higher poverty rates. The recent study of the World Bank (2016e) in Latin America, where an estimated 42 million indigenous people are living, similarly documents that they face poverty rates which are: "*on average twice as high as for the rest of Latin Americans*." Measured in terms of the percentage of people living on less than the International Poverty Line ($1.25 PPP in the late 2000s), 9% were below, compared with only 3% for non-indigenous people, based on a weighted average for Bolivia, Brazil, Ecuador, Guatemala, Mexico, and Peru (cf. World Bank 2016a; Calvo-Gonzãlez 2016).

Meanwhile, the *Oxford Poverty & Human Development Initiative* (OPHI) has extended its input to the controversial measurement of global poverty by research which seeks to contribute to an integrated poverty reduction framework, based on Amartya Sen's capability approach. As indicated in the *Introduction*, this framework incorporates multiple interconnected dimensions of poverty and wellbeing, which seeks to provide policy-making with adequate data and foster the international debate on poverty reduction and development. In 2010, UNDP had decided to introduce the *Multidimensional Poverty Index* (MPI) which uses micro-economic data to assess the percentage of households which are confronted with overlapping deprivations in three dimensions—education, health and living conditions. In the same year, Alkire and Santos (2010) had analysed poverty across 78% of the world's people in 104 developing countries using the MPI, releasing the results in advance of the *Human Development Report* (UNEP 2010). In a recent approach,

Alkire et al. (2016) developed a new methodology to measure multidimensional poverty in conjunction with chronic poverty, wellbeing and inequality, useful not only for targeting and monitoring social policies, but also for measuring poverty. In addition, brief survey modules have been developed for the five 'missing dimensions' of poverty data which appear to be important to deprived people, but have so far been overlooked in large scale surveys: quality of work, empowerment, physical safety, without shame and psychological wellbeing.

Despite the above-mentioned renewed and extended approaches to the eradication of global poverty before 2030, promoted by the United Nations (2015b), and the World Bank (2016a), based on 'lessons learned' and supported by the confident statements of the United Nations and its Member States (UN 2015b): "*We are determined to ensure that all human beings can enjoy prosperous and fulfilling lives and that economic, social and technological progress occurs in harmony with nature*", all the related strategies show a serious, and for some a rather incomprehensible 'Missing Link' in the well-intended efforts to eradicate poverty in the course of the next one-and-a-half decades*: the integration of the target group par excellence, i.e. the indigenous peoples and their systems of knowledge, beliefs, practices and institutions.*

Indeed, a closer review of the above-mentioned Reports focused on the 17 SDGs and the related 169 targets, from a neo-ethnoscience emic perspective on the position of indigenous people in the developing countries with regard to achieving poverty reduction for all by 2030, reveals hardly any substantial reference beyond some general lip-service, while these groups together are currently estimated to make up some 370 million indigenous peoples living in 70 countries across the world, as estimated by the United Nations *Permanent Forum on Indigenous Issues* (UNPFII 2016). Their substantial number, their vast body of indigenous knowledge and their significant place in human history merits the recognition and integration of their indigenous systems and institutions into a more participatory, 'bottom-up' approach required to attain the reduction of poverty on a global level.

On the contrary, hardly any special reference is made to the considerable group of the indigenous peoples and their knowledge systems around the world within the context of global poverty reduction, neither in the *Millennium Development Goals Report* (UN 2015a), nor in the *subsequent Post-2015 Agenda for Sustainable Development and Sustainable Development Goals* (UN 2015b). Although the previous World Bank Report (1998/1999) entitled *Knowledge for Development* did recognise the significant role of indigenous knowledge for development, and the recent Report of the World Bank (2017), entitled *Monitoring Global Poverty,* also makes a reference to the position of indigenous peoples with regard to the issue of rural/urban disaggregation, no further indication of specific strategies is mentioned for either the integration of local and global knowledge systems or the realisation of the 10 recommendations proposed for monitoring extreme poverty in the coming years up to 2030.

The conclusion is that despite their position as an important and considerable target group of stakeholders and participants in the concerted efforts to achieve poverty reduction, the indigenous peoples and their knowledge, beliefs and practices continue to be largely left outside the poverty reduction approaches of the major international organisations and agencies, rendering them a *'Missing Link'* in the design and implementation of strategies to achieve sustainable community development for all by 2030. In this respect, the paradigm shift in global development cooperation would still need a further elaboration of its principles towards achieving a community-based knowledge integration strategy with special attention for poverty reduction among the indigenous peoples and their knowledge systems. The following paragraphs seek to further substantiate the efforts to bridge this gap by the IKS-based integration strategy of local and global systems in terms of *Integrated Microfinance Management* (IMM) and *Integrated Community-Managed Development* (ICMD) through the indigenous institutions in order to provide an IKS-based contribution to the realisation of global poverty reduction as today's highest challenge for all people worldwide in the next decade.

4.5 Neo-ethnoscience: Indigenous Knowledge Systems and Development

After the period of the beginning of the second half of the 20th century, marked by a growing interest in the cultures and worldviews of indigenous peoples and their knowledge systems which culminated in the emergence of ethnoscience as the study of indigenous peoples from an emic, i.e. cultural relativist's point of view, the 1980s witnessed a functionalist approach to study the role of indigenous knowledge systems within the context of socio-economic development in the developing countries. This radical reorientation emerged from the practical field of international development aid and cooperation, where in a growing number of cases the imported systems of knowledge and technology from the West often showed not to catch on or to link up well with indigenous cultures and their knowledge systems. The failure of the Transfer-of-Technology (TOT) development paradigm became particularly visible in the 'mismatch' between endogenous and indigenous systems of knowledge and technology, where indigenous knowledge had been operational over many generations in several sectors of the community.

Following such reassessment and revaluation of indigenous knowledge in the field in various dynamic sectors of the communities, in which indigenous knowledge and practices—albeit often ignored in the past—eventually proved to provide valuable contributions or sometimes even better alternatives for imported knowledge and technology, a growing number of ethnoscientists started to study, analyse and promote indigenous knowledge systems in the development process, often with remarkable success in various sectors. The new, dynamic field of the study of transcultural development from the participant's point of view soon started to question the effectiveness of the out-dated Transfer-of-Technology paradigm (cf. Chambers et al. 1989; Titilola 1990; Reijntjes et al. 1992; Warren et al. 1995).

The academic interest in indigenous knowledge systems, particularly among the so-called 'primitive' peoples in the tropics, however, had initially been dominated by cultural anthropologists who tended to highlight and document the exotic, non-Western aspects of indigenous cultures, far removed from practical interventions of development and change, often in an effort to 'protect' indigenous cultures from Western modernisation. Coming to terms with the principle of cultural relativism—the outsider's understanding of an individual's beliefs and behaviour in terms of that individual's own culture—the field of cognitive anthropology had evolved into ethnoscience, which in turn transformed into the more development-oriented field of neo-ethnoscience of the 1990s (cf. Warren et al. 1995).

While initially, anthropologists facing the dilemma of their own bias towards the right of self-determination of indigenous peoples—and indigenous knowledge systems—while at the same time protecting them as it were against influences from outside, where local people should continue to focus on their own culture for their life and livelihood, the reality of recent global processes of communication, acculturation and globalisation have rendered such views rather outdated. As mentioned in the Introduction, this irreversible process formed the basis for the neo-endogenous approach to development. Although some anthropologists have recently been actively engaged in efforts to develop 'applied anthropology', their work has mainly been focused on securing a niche for themselves in the current international debate on culture and development (cf. Sillitoe 1998; Sillitoe et al. 2002; Purcell and Onjoro 2002; Halani 2004). In this context, Ellen (2002) rightly refers to: "*Anthropology's unresolved relationship with development*".

The misguided critique expressed by anthropologists Purcell and Onjoro (2002), that ethnoscientists involved in development programmes, such as Warren et al. (1980), Warren (1991) and Warren et al. (1995) would be biased towards the promotion of a technological form of development focused on material progress, not only denies the indigenous peoples' own right of self-determination to opt for the integration of particular knowledge and technology—material or immaterial—pertaining to their improvement in health, agriculture and natural resources management at their convenience, but also reflects the standpoint of theorists who tend to overlook the practical aspects of expressed needs in international development cooperation in the field. The privileged role which Purcell and Onjoro (2002) claim in their model for the ethnographer as 'agent and facilitator' further reveals their presumed position. In this context, Crossman and Devisch (2002) contend that the weight of the imposed Western rational scientific tradition and the Rostovian development model have prevented the development of endogenous, context-specific systems of knowledge. Indeed, in their view: "*anthropology failed to legitimise indigenous knowledge and avoided dealing with the whole issue of plural or alternative knowledge when it had intimate access to local communities the world over.*"

While other anthropologists such as Clammer (2005) and Kassam (2009) recognise that even anthropologists cannot fully comprehend and appreciate the holistic system of indigenous knowledge, mainly as they are trained in the rationalistic Cartesian philosophy of Western science, the new ethnoscientists—either from the communities or from outside—have further developed a holistic and emic perspective on indigenous peoples and their knowledge systems with a view to constructing a less normative, but more realistic picture of the complex process from a multidisciplinary point of view. Such a new ethnoscience approach would also support the indigenous peoples' right to engage in the international development process on their own terms, referred to as '*ethno-development*' (cf. Stavenhagen 1986).

As transpires through the above-mentioned studies by Warren et al. (1980), Warren (1991), Warren et al. (1995) and Toledo (2001), sustainable community development does not merely involve the provision of consumer goods and services to local communities, but rather aims at enabling indigenous people to use and control their own resources and determine their own life on the basis of their choice of material as well as immaterial elements. Notwithstanding their limitations in dealing with processes of development and change, most anthropologists agree that despite the historical, theoretical and methodological obstacles, development cannot be effective unless indigenous knowledge is part of the development process (cf. Sillitoe et al. 2003). Indeed, historically, indigenous knowledge as 'local knowledge' has been, and largely continues to be regarded as inferior to global knowledge being the subject matter of modern science, still reflecting the remnants of the former Western colonial view on indigenous peoples of the 18th and 19th centuries.

In view of the spiritual factors as significant aspects of the indigenous knowledge systems as part of the indigenous cosmologies—or worldviews—and philosophies of nature and the environment, which have largely been left out of the area of interest by modern science, primarily since appropriate methodologies have still been incapable to comprehend such often 'invisible' factors, the integration models should accommodate a holistic approach in order to include all relevant categories of factors. In contrast to the concept 'cosmology' used in the science of physics and astronomy as the study of the origins and evolution of the universe, here, cosmologies specifically refer to sets of indigenous knowledge, beliefs, interpretations and practices of cultures related to explanations about the role and the meaning of humans, life, and the world within the universe or cosmos in the past, present and future.

The cosmologies of indigenous societies are generally characterised by respect for nature and for human wellbeing, and there is often an appeal to keep a balanced coexistence between all three worlds in the universe, because people, ecosystems, the biosphere and cosmos are defined as a network, composed of common components of matter, energy and spirit. Most relevant for development is the fact that cosmologies and indigenous knowledge systems are used as key references for local decision-making concerning matters such as the use of natural resources, the achievement of sustainable management of forests, the extent of human demographic levels and bio-social synergies, and also to establish peace among and within neighbouring communities (cf. Reichel-Dolmatoff 1996; Millar 1999; Kearney 2008).

In a recent study of the concept of cosmovision in Africa, Millar et al. (2008) shows an elaborated representation of the constellation of forms of indigenous

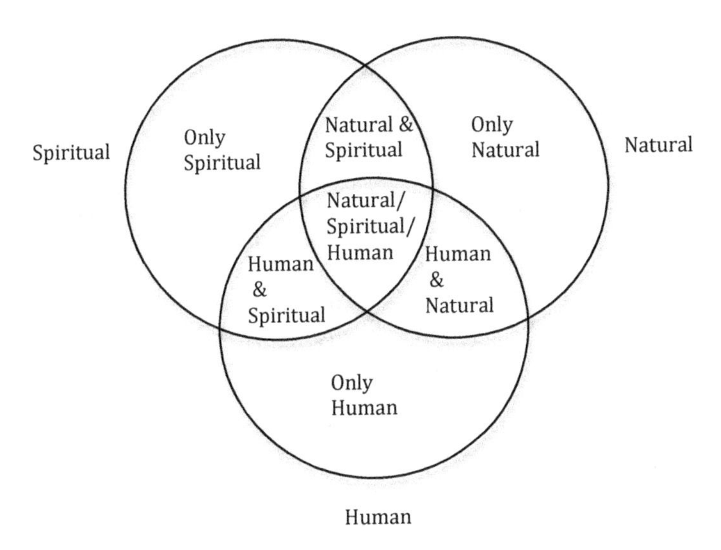

Fig. 4.2 Constellations of knowing from a cosmovision perspective. *Source* Millar et al. (2008)

bodies of knowledge, which centres mainly on the human, spiritual and natural world, as depicted in Fig. 4.2.

In Fig. 4.2, the three (3) circles are depicting the African worldviews which centre mainly on the human, spiritual and natural world. The interaction of the three worlds implies the following constellations of knowledge:

- knowledge resulting from social interactions only;
- combination between the social and natural;
- combination between the social and spiritual;
- knowledge resulting from natural interactions only;
- combination of the natural and spiritual;
- knowledge resulting from spiritual only; and
- combination of social, spiritual, and natural.

According to Millar et al. (2008): "*these constellations highlight the heterogeneity and complexities of African Sciences and therefore engendering different bodies of knowledge and sciences that underscore the development of Africa and this contrasts the western science.*" His conclusion that research of the heterogeneity and complexities of African Sciences should not only focus on the horizontal level of material and social phenomena, but especially on the 'vertical' level of the higher order discourses of the spiritual aspects, rendering a holistic approach necessary, and linking up well with the concept of the LEAD Programme of the comprehensive IKS-based model of integration of local and global knowledge systems, as elaborated in the next Paragraph.

Similarly, Naamwintome and Millar (2015) describe how the historical process of indigenous worldviews or cosmologies in African society have resulted in the development of useful knowledge, knowing and their epistemologies. Nature, peoples, and the spiritual world are prominent features within the African traditional worldviews, which not only guides the sustainable use of natural resources, but also prescribes how community decisions are taken, local problems and conflicts are solved and in what way the rural people organise themselves.

In their study, Naamwintome and Millar (2015) argue, that: "*if Africa is to make a significant presence in the 'knowledge arena' (if not dominate it), the strength of it is in the indigenous knowledge and the cultures of Africa.*" The authors substantiate their pragmatist position that it is necessary not only to harness indigenous knowledge in research, teaching, development, and policy-making, but also that indigenous knowledge is a political instrument requiring the development of a critical mass calling for a new paradigm for higher educational visions for tomorrow's Africa.

Interestingly, there is a striking resemblance in the conceptualisation and representation of these indigenous cosmovisions which have been studied and documented in Sub-Saharan Africa and South-East Asia. These cosmologies are generally depicted in a configuration of three partly overlapping worlds—the human, spiritual and natural world—in which humans are taking a central position. The relations between humans and each of these worlds are basically sacred and harmonic—and have to remain or be restored in order to maintain the cosmic balance, often by the performance of rituals (cf. Millar 1999; Agung 2005; Saefullah 2019).

The importance of the interaction and relationships between humans and their diverse environment in terms of the social, natural and spiritual worlds, which have to be brought continuously into a harmonic balance, has been studied in South-East Asia by Agung (2005) among the indigenous Balinese people and their communities, where the Balinese cosmology of *Tri Hita Karana* is playing a key role in the peoples' conservation behaviour with regard to the island's rich bio-cultural diversity. Similarly, the study by Saefullah (2019) documents the cosmology of the Sundanese people in West Java, known as *Tri Tangtu*, which is guiding the local peoples' ways of life. In Chap. 10, a comparison is made between the representation of the Balinese and Sundanese cosmologies, which reveals certain universal characteristics, most relevant for the process of sustainable community development.

At the beginning of the 21st century, a new impetus was given to the development, promotion and protection of indigenous knowledge systems within an international political framework of transformation and democratisation, particularly with regard to the indigenous communities in developing countries. By that time, the potential of indigenous knowledge for sustainable development had already gained particular attention in health, agriculture and natural resources management, and bio-cultural diversity conservation, where indigenous knowledge, beliefs and practices showed to be complementary to Western knowledge systems, or sometimes even provided more suitable alternative solutions.

As mentioned above, at the international level, organisations including the *World Bank*, the *United Nations Development Programme* (UNEP), the *United Nations Educational, Scientific and Cultural Organisation* (UNESCO) and the *World Health Organisation* (WHO) started to adopt several declarations on the unique position of indigenous peoples and the appropriate protection and use of their knowledge systems in development around the globe. These leading organisations have now entered the new era by promoting a process of sustainable development, of which the practice has shown new opportunities to integrate the great potential of indigenous knowledge systems, still functioning at the community level. Odora Hoppers (2002a) refers to the present 'African Renaissance', which: "*in particular sets forth an agenda that combines identity reconstruction and innovation, human rights, sustainable development and democratisation in South Africa and throughout the African continent.*"

4.6 The Integration of IKS for Sustainable Community Development

The theoretical impediments to approach indigenous and modern knowledge and science on an equal basis—also referred to as 'parity'—have been dominating the discourse for a long time, leading to a general consensus that both philosophical and hierarchical power relations have been blocking a balanced, equal position of both in a universal framework of the philosophy of sciences (cf. Agrawal 1995; Posey 2002). Another theoretical problem is that in contrast with modern knowledge systems on the basis of which different monodisciplines have been structured in separate components, indigenous knowledge systems are built up of multidisciplines, generally embedded in a web of interlinked elements of knowledge, beliefs and practices, which constitutes the local holistic cosmological framework.

Similarly, the methodological difficulties to study, analyse and fully understand indigenous knowledge systems from an outsider's point of view—such as predominantly present among Western-trained scientists—have attracted much attention in the social sciences. As mentioned above, recently, the Leiden ethnosystems approach to the in-depth study, analysis and understanding of indigenous knowledge systems has been elaborated to encompass three methodological principles of respectively the 'Historical Dimension' (HD), the 'Participant's View' (PV) and the 'Field of Ethnological Study' (FES).

Embarking on the basic premise that every indigenous culture has a unique orientation to knowledge, beliefs and practices, represented in the way of life and survival of the community and its members, its language, and its conception of the relationship to its natural, social and spiritual environment, it seems a matter of course that a synthesised form of local and global science would be needed to construct a cross-cultural discipline capable to respond to complicated questions and problems, not least concerning poverty reduction for sustainable development at the community level, which are here at stake. While most Universalists contend that modern Western science would be superior to indigenous perspectives on the natural world, largely because of its predictive and explanatory capabilities, multiculturalists maintain that science is not universal, but rather locally and culturally determined. In his contribution to the establishment of an effective science education in South Africa, Le Grange (2007) proposes a fresh look at the kind of science which is taught to South African school learners, in which the debate moves beyond the binary of modern science/indigenous knowledge to: "*Ways in which Western science and indigenous knowledge might be integrated are explored.*"

Various attempts have been undertaken to compare, validate and integrate the various knowledge systems in an approach of endogenous, 'bottom-up' development in several sectors by scientists and experts working in the field of sustainable development, such as Reijntjes et al. (1992), Warren et al. (1995), Millar (1999), Haverkort et al. (2003), Slikkerveer (2012), and Naamwintome and Millar (2015). These sectors include not only health, education, agriculture, forestry, natural resources management, biodiversity conservation, disaster management and climate change, but also the sectors of financial, medical, and educational services, most relevant to poverty reduction and development. Authors, including Cash et al (2003), contend that the capacity of mobilising and using science and technology (S&T) is increasingly recognised as an essential component of strategies for promoting sustainable development, where the integration of indigenous knowledge and modern breeding methods, termed 'participatory plant breeding', also seek to overcome the boundaries which tend to hinder the integration of long-term knowledge accrued by farmers over many generations with the insights and methods developed by modern plant breeders.

In line with these academic efforts, a growing number of national and international policy processes for protection of traditional knowledge of indigenous peoples and local communities have been designed and implemented in various fora: at the international level, these include the UN Convention on Biodiversity (1992), the UNESCO *Convention on the Protection of Intangible Cultural Heritage* (2003), the FAO *International Treaty on Plant Genetic Resources* (2009), and the WIPO Traditional Knowledge and Intellectual Property Protection (2016).

Neo-ethnoscientists including Warren et al. (1995) have amply illustrated that embarking on the integration of indigenous knowledge systems in various settings promotes local participation as a major prerequisite for attaining sustainable development, which, in turn, is intimately interrelated with community-based poverty reduction. Some interesting models of knowledge integration from the grassroots, relevant for sustainable development at the community level, have been developed by several researchers, including Johannes (1993), Millar (1999), Odora Hoppers (2002b) and Oguamanam (2004). Since recently several attempts have been undertaken to further conceptualise, locate and integrate *Traditional Ecological Knowledge* (TEK) as one of the salient areas of integration with modern ecological science in development programmes of sustainable resource management; it is illustrative to assess the experience as an example of IKS integration. According to Berkes (1993: 3) Traditional Ecological Knowledge (TEK) is: "*…a cumulative body of knowledge and beliefs, handed down through generations by cultural transmission, about the relationship of living beings (including humans) with one another and with their environment process between local and global knowledge systems.*"

Conceptualised in this way, much research has been dedicated to TEK, primarily with a view to 'extracting' TEK, validating it against Western scientific ecology and 'integrating' it with dominant Western science and management systems. However, as Casimirri (2003) contends, indigenous knowledge and systems of management have largely been marginalised.

The *Assembly of First Nations* (1995) described indigenous knowledge as consisting generally of four interlinked components including:

1. creation stories and cosmologies explaining the origins of the earth and its people;
2. codes of ritual and behaviour that govern peoples' relationships with the earth;
3. practices and seasonal patterns of resource utilisation and management, that have evolved as expressions of these relationships; and
4. a body of factual knowledge accumulated in connection with these practices.

Because of its practical potential for natural resources management and bio-cultural diversity conservation, recent TEK-research has particularly focused on the last two components of practices and factual knowledge. Figure 4.3 represents the various elements of indigenous knowledge, showing that TEK is part of a web of indigenous knowledge, in the center of which the spiritual elements of the local culture are located. The representation of the embeddedness of indigenous knowledge in a larger 'web' also identifies the 'invisible' factors of wisdom, beliefs, norms and values, as well as the indigenous institutions which guide human behaviour as key elements in the overarching, holistic worldview. While these 'invisible' elements are often overlooked by outsiders, the actual 'facts' form the components of TEK which can be understood by outsiders. In the integration process, these 'facts' are validated and subsequently removed as 'data' from this web to be applied in Western resource management programmes.

If the analogy is taken one step further, as Casimirri (2003: 3) argues, it becomes possible to see: "*that removing these data points would weaken the structural integrity of the web. Likewise, the data points are interconnected to the web and cannot be fully understood when they are removed from their context.*" In other words, indigenous knowledge systems have to be considered as interconnected with the worldview concerned and as one whole phenomenon.

By consequence, knowledge integration models designed for providing a contribution to sustainable development should be holistic in their approach and also include the invisible factors represented in connection with cosmologies, such as

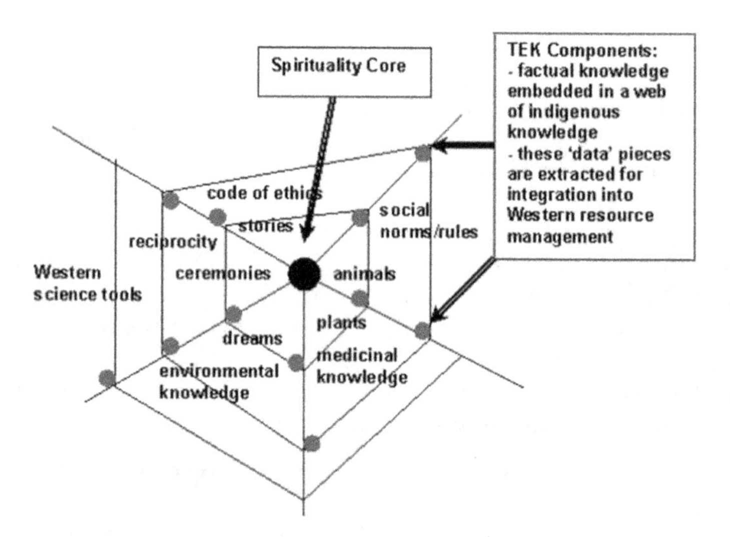

Fig. 4.3 Conceptualisation of Traditional Ecological Knowledge (TEK) within an Indigenous Knowledge web. *Source* Casimirri (2003)

spirituality, sanctity and morality, highly relevant for local peoples' perceptions, activities and behaviour.

In the course of the previous century, the concept of indigenous culture, and in particular its specific components of knowledge, beliefs and practices, has experienced a drastic U-turn in its role in the socio-economic development of the developing countries. After a rather neo-colonial conception of culture as an 'obstacle to development', gradually the recognition evolved of the great potential of culture to guide and enhance the development process in several sectors of the society. Such a radical change of the direction of development cooperation is, as Kendie and Guri (2000: 332) rightly notice: "*largely due to the resilience of culture and its institutions, despite the imposition of Western worldview*". Other scientists have brought into the debate the role of the indigenous cosmologies and indigenous philosophies of nature and the environment, which have enabled and continue to enable the communities in developing countries to retain their culture and related institutions towards the influx from outside of influences of modernisation, westernisation and globalisation (cf. Millar 1999; Agung 2005; Saefullah 2019).

In this context, the argument is made that, because of their sustainable nature, where resources are largely utilised only for strict use and maintenance of present-day communities in order to leave a share for future generations, the indigenous philosophies of subsistence could eventually provide humankind with a non-exploitative, sustainable tool to survive on Planet Earth for a prolonged period of time (cf. Slikkerveer 1999a). Such indigenous philosophy is, as Weber (1905) has shown, in contrast with the capitalist philosophy of commercial surplus-building, characteristic for most Western nations, exploiting and exhausting the natural resources in the shortest possible period of time without taking into account the position of resources and the needs of future generations.[3]

4.7 A New Role for Indigenous Institutions in Human Development

As mentioned in the *Introduction*, institutions are generally referred to as regularised practices or patterns of behaviour structured by rules and norms of the society which are widely used, and are either formal or informal. Formal institutions are rules, laws and constitutions legalised by members of the society, while informal institutions include social norms of behaviour, and conventions which prohibit or permit individuals to undertake certain activities within their social settings (cf. Metha et al. 1999; Hembram 2007; Slikkerveer 2012). Most of these authors view this timely perspective on indigenous institutions and organisations as complementary to the discourse on indigenous knowledge systems and development as the 'Cultural Dimension of Development', introduced by Warren et al. (1995). The *indigenous institutions* are often grounded in strong principles of ethno-cultural affiliation, community cooperation and organised community work over many generations, and often represent indigenous associations based on the local philosophy of mutual aid, cooperation and reciprocity. Examples have been mentioned to include indigenous village associations, farmers' associations, traditional medical associations, traditional legal councils, councils of village elders, mutual aid and reciprocity associations, informal cooperative associations, collective action groups, neighbourhood groups, community water management groups and women's groups.

In his classical study on *Local Institutional Development: An Analytical Sourcebook with Cases*, Uphoff (1986) had already made a distinction of various levels of development-related decision-making, ranging from the international level to the individual level where the middle—or local—level included three local-level institutions: the locality level, the community level and the group level. The author distinguished between an *institution* viewed as a complex of norms and behaviours which persists over time by serving some socially valued purpose, and an

[3]In his book *The Protestant Ethic and the Spirit of Capitalism* (1905), Weber wrote, that capitalism in Northern Europe evolved when the Protestant (particularly Calvinist) ethic influenced large numbers of people to engage in work in the secular world, developing their own enterprises and engaging in trade and the accumulation of wealth for investment. In other words, the Protestant work ethic was an important force behind the unplanned and uncoordinated emergence of modern capitalism. Apart from Calvinists, Weber also discussed Lutherans (especially Pietists), but also noted differences between traditional Lutherans and Calvinists, Methodists, Baptists, Quakers, and Moravians, specifically referring to the Herrnhut-based community under Count von Zinzendorf's spiritual lead.

organisation as a structure of recognised and accepted roles. A crucial element is that indigenous institutions generally refer to local-level institutions with a socio-cultural, endogenous base, rather distinct from exogenous institutions operational through external forces.

In order to avoid possible confusion between the terms *institution* and *organisation*, which tends to be used interchangeably, Huntington's (1965: 378) classical clarification remains valid today: "*institutions are stable, valued, recurring patterns of behaviour. Organisations and procedures vary in their degree of institutionalisation. Institutionalisation is the process by which organisations and procedures acquire value and stability.*"

Thus, organisations which have acquired status and legitimacy for having satisfied peoples' needs and expectations over time have become 'institutionalised'. Blunt and Warren (1996) further clarify the distinction: marriage is an institution which is not an organisation, while a particular family is an organisation (with roles) but not an institution (with longevity and legitimacy). Following this useful differentiation, this Volume will mainly be concerned with long-standing indigenous institutions with an organisational basis in the community being endogenous as opposed to exogenous and operating at the community level.

In his analysis, Uphoff (1986: 5) also observed six categories of 'local institutions' with different advantages and disadvantages for supporting rural development, but he left out of his analysis not only 'local political institutions', since in his opinion: "*...external agencies are expected to avoid getting involved in domestic policies*", but also the indigenous institutions, to which he refered to as '*traditional*' or '*informal*' institutions. Although he recognised that they have evolved and been supported by the local people in dealing with economic, social, cultural, religious and political problems, he contended, that: "*...such institutions almost always do exist, though they may be hard to find or to work with.*" Uphoff (1986: 6) also calls these 'local' institutions: "*pre-existing institutions being often parallel to the above-mentioned categories.*" Although he mentioned many kinds of these 'pre-existing' indigenous institutions, such as *age grade systems*, *women's secret societies*, *craftsmen's guilds* etc., he illustrated his focus on the *formal* 'local' institutions by asserting that certain administrative roles, such as tax collector or land registrar, may have existed for hundreds of years and have later been incorporated into the formal contemporary local administration.

This Volume, however, seeks to pay special attention to these '*informal*' institutions within the context of sustainable development, because of the growing evidence of their crucial role in the development-related community-level decision-making processes, which have become institutionalised over many generations. By consequence, the use of the term *indigenous institution* in this Volume refers specifically to those local-level institutions—informal and sometimes invisible to the outsider—which are rooted in the history of the community and based on strong local philosophical principles of cooperation, mutual aid, and collective action, where the interests, resources and capacities of many community members are structurally joined together in order to achieve common goods and services for the entire community in a non-commercial way.

This perspective links up with the substantial work carried out on *indigenous knowledge systems* in various sectors of the society, which are forming the base for local-level decision-making processes, and as such essential for attaining sustainable community development. Although some authors tend to use terms such as 'traditional', 'informal' or 'customary' with regard to such 'pre-existing' institutions at the community level, this Volume prefers to adhere to the concept of *indigenous institution* as part of the new development paradigm of *indigenous knowledge systems and development*, and as such seeks to avoid previous misconceptions about indigenous knowledge as being 'static', 'geographically distant', 'something from the past', 'old-fashioned', or 'not adaptable to development and change'.

On the contrary, according to a growing number of studies, these indigenous institutions are increasingly regarded by most experts and development agencies as having specific and unique qualities which not only provide the participants with a value of self-respect, but also render their policies most contributive to achieving the objectives of sustainable community development in many sectors of the community (cf. Richards 1985; Chambers et al. 1989; Swift 1991; Warren et al. 1995; Blunt and Warren 1996; Slikkerveer 1999b; Watson 2003; Slikkerveer 2012). As mentioned above, most of these authors view this timely perspective on indigenous institutions and organisations as complementary to the discourse on indigenous knowledge systems and development as substantiating the 'cultural dimension of development', introduced by Warren et al. (1995).

As mentioned before, the informal indigenous institutions and related indigenous knowledge systems are inextricably interrelated, and as such grounded in the local cosmology since many generations. The holistic cosmologies also encompass indigenous institutions of self-help, mutual aid, cooperation and reciprocity. As these institutions have existed for many centuries in different forms, especially among rural communities, they have been responsible for guiding the individuals' behaviour within their communities and their socio-culturally sanctioned access and use of their natural resources.

The philosophy of people-based development from below assumes that participation is not only an end in itself, but also a fundamental precondition and a tool for any successful development process (Oakley 1991). Development is principally considered to be about culture and institutions are the important components which enforce cultural rules, norms and values (van Arendonk and van Arendonk-Marquez 1988). Culture is that whole complex of distinctive spiritual, material, intellectual, and emotional features which characterise a society or social group. It includes not only arts and letters, but also modes of life, the fundamental rights of the human being, value systems, traditions and beliefs (Awedoba 2007). It is evident that culture has steadily, but gradually, made inroads into the governance process and it is serving as the entering point for achieving sustainable development. In a later study, Uphoff (1992) places indigenous institutions within the context of sustainable development, involving many factors. Uphoff (1992: 3) contends that: "*One contributing factor that deserves more attention is local institutions and their concomitant, local participation.*". Other case studies in which indigenous institutions have eventually shown to possess great potential for socio-economic development, notwithstanding certain difficulties emerging between indigenous institutions and outside institutions in the form of development agencies, pave the way for re-activating their role as powerful resources for poverty reduction pertaining to sustainable development.

In this way, Slikkerveer (2012) is further extending a more dynamic conceptualisation of indigenous institutions into a rather functional aspect of the newly-developing field of *neo-ethnoscience*, particularly *ethno-economics and development,* in which the key roles of these indigenous institutions in the local-level decision-making processes are operationalised for the realisation of sustainable community development. In this view, the particular position of poor and low-income members of the community as subjects of community solidarity and cooperation, expressed in an institutionalised form of local-level support—based on respect, equal opportunities and shared benefit from common land, natural products, and cultural goods and services—is providing an important participatory stepping stone for development-related poverty reduction activities.

While Slikkerveer (1990) documented his confrontation with dominant power structures in the Horn of Africa, in which the institutionalised Western medicine had been heavily politicised at the cost of pre-existing institutions of traditional and transitional medicine, recently Chirangi (2013) documents similarly constrained relationships which used to form an obstacle for the well-planned integration of institutions representing the traditional, transitional and modern medical systems in Tanzania. Watson (2003) documents the difficulties encountered by the German Technical Cooperation Agency (GTZ) in natural resources management among the Borana in Southern Ethiopia living on the border with Kenya, largely because of the political embeddedness of the *gada* system, an institutionalised age-grade system which has been guiding and organising the Borana society over many generations. These case studies also show that such often politically-driven obstacles can be overcome through the intermediary and input from the local-level decision making processes, facilitated by indigenous institutions and the worldviews in which they have been grounded for many generations: solidarity, mutual aid, cooperation and respect for the connected worlds of fellow humans, nature and the spirits.

The conclusions of a recent Seminar on the *Role of Local Communities and Institutions in Integrated Rural Development* held in Iran underscore that the role of local institutions, such as local government units, formal and informal local organisations including cooperatives, culture groups, and NGOs, is becoming more important for the realisation of the integration of various rural development efforts (cf. Wijayaratna 2004).

As mentioned above, the concept of *Integrated Rural Development*, introduced in the 1970s, refers to a rural development approach of the integration of a number of different, sometimes overlapping 'target' approaches, which not only suffered from a vague description of what components were actually 'integrated' in the IRD-strategy, but also the problem that most programmes were unable to solve urgent rural problems of achieving and distributing reliable food resources. A major underlying cause of the decline of the approach has been the attitude of many development experts that their 'scientific' knowledge is superior to the indigenous knowledge of the local people who have not received formal education. The negative attitudes have been enforced by the 'top-down' bureaucratic system built up of external institutions and development agencies, which were organised along hierarchical lines hardly engendering local participation, and generally in contrast with the pre-existing indigenous institutions which have been guiding the local peoples' behaviour and way of life on an equal basis for many generations. (cf. Slikkerveer 1995; Kendie and Guri 2005; Naamwintome and Millar 2015).

However, in view of the recent, more positive experience with the careful integration of indigenous institutions into IKS-based development programmes on the basis of lessons learned in the above-mentioned cases, the new concept of *Integrated Sustainable Community Development* has shown to meet the challenge of poverty reduction for sustainable community development in Africa and elsewhere around the globe.

4.8 IKSIM: A New Model for Integrated Community-Managed Development (ICMD)

The international interest in local poverty reduction and sustainable development has largely been focusing on cross-sectorial socio-economic development programmes in health care, food and nutrition, environmental conservation, agricultural production, as well as in labour and employment, governance, equality and democratic representation. As mentioned before, recent changes in rural areas have led to the development of new models, such as the 'neo-endogenous development model', introduced by Ray (2001), based on the utilisation of endogenous knowledge and capacity-building, and democratisation at the community level (Ray 2001; Cabus and Vanhaverbeke 2003; Ward et al. 2005; Tolón-Becerra and Lastra-Bravo 2009).

The conceptual framework which forms the basis for the development of the model of integration of indigenous knowledge with global knowledge, known as the *Indigenous Knowledge Systems Integration Model* (IKSIM), specifically developed by the LEAD Programme for poverty reduction within the context of sustainable community development, embarks on its functionality in several approaches for the poor and low-income families as target groups, operationalised at the community level. Such a utilitarian approach seeks to functionalise indigenous institutions to overcome the conventional dualism between local and global knowledge by operating at a higher level of abstraction, where the cognitive, spiritual and moral dimensions are similarly integrated as important albeit yet 'invisible' factors, which have shown to play a strong, evidence-based role in the holistic body of indigenous knowledge, beliefs and practices.

As a result, the basic building blocks in the framework of the IKSIM model include the following nine (9) principles:

- *'target group' perspective* on the urgent challenge of poverty reduction within the context of sustainable development, specifically on the poor, low-income and marginalised groups among the indigenous peoples in the developing countries;
- *parity-oriented approach* towards indigenous and modern knowledge systems recognised as components of a body of synthesised science on an equal basis—also referred to as 'parity'—being relieved of the burden of philosophical and hierarchical power relations;
- *multidisciplinary perspective* on the integration of knowledge, beliefs and practices for achieving poverty reduction as a multidimensional configuration of humanitarian, economic, social and cultural factors operational in the context of sustainable community development;
- *multi-sector and cross-sectional approach*, focused on the operationalisation and integration of potential indigenous systems of knowledge and technology;
- *holistic approach* towards the interaction between the human, natural and spiritual worlds, conceptualised in the local cosmology as a comprehensive network of knowledge, predominately manifest in such sectors as health, education, agriculture and nutrition, natural resources management and bio-cultural diversity conservation;
- *institution-based strategy* on the functionalisation and input of indigenous institutions into integrative sustainable development programmes and projects in order to secure local participation and governance;
- *humanitarian orientation* towards all stakeholders involved, including local people, leaders and specialists, educators, ethnoscientists, development experts, government agents, etc., who are expected to make joint efforts to reduce poverty for sustainable development at the community level, in which communication, discussion and cooperation are focused on equal opportunities, benefit sharing, mutual respect, dignity and the observance of human rights in order to incite truly participatory development programmes;
- *life-long learning attitude* among all participants involved of mutual exchange of knowledge, experience and opinions contributing to the ongoing dialogue on relevant matters of integration and functionalisation of local and global knowledge systems; and
- *'bottom-up' approach* in policies and programmes towards integrated sustainable community development as a contribution to the Sustainable Development Goals of the United Nations (2015c).

After prioritising the poverty-related key components at the community level in terms of cross-sector problems, factors, aspects, opinions and procedures in close cooperation with the community representatives, the process of integration involves the participatory balancing, harmonisation and fine-tuning of relevant data collected through the execution of the participatory comparative study of the implementation of the 'Leiden Ethnosystems Approach' as the methodological basis for the documentation, analysis, understanding, validation and selection of relevant local and global knowledge of the target community. The activities are routinely followed by the joint assessment, selection and formulation of appropriate recommendations for integrated policy planning and implementation, primarily focused on poverty reduction and sustainable community development.

In Fig. 4.4, the *IKSIM Model* is depicted as a build-up of two oval circles depicting the interacting local and global knowledge systems; this process is based

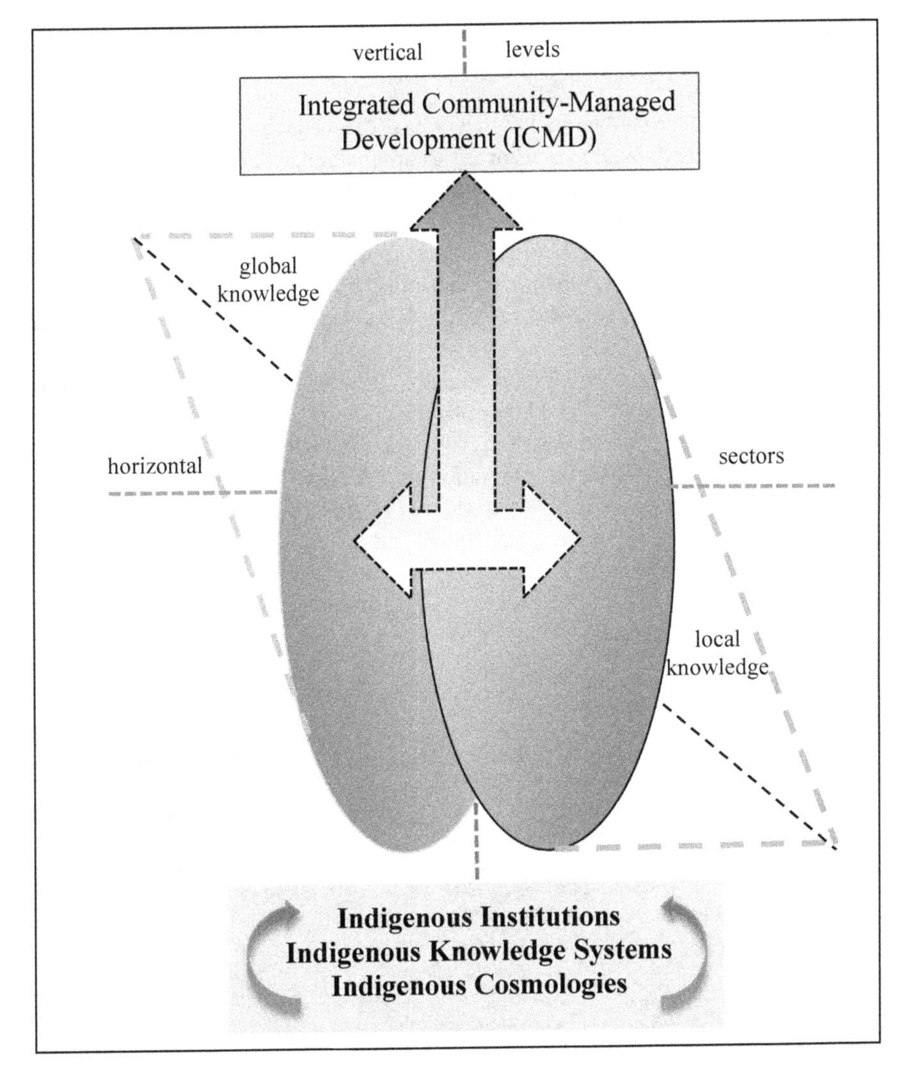

Fig. 4.4 Representation of the *IKS-Based Integration Model* (IKSIM) of Integrated Community-Managed Development (ICMD). *Source* Slikkerveer (2014)

on the complex of the indigenous cosmology, encompassing the three interrelated worlds as the foundation of both indigenous knowledge systems and indigenous institutions, which, in turn, centers on the decision-making processes of the community members, developing the vertical 'bottom-up' approach. The model shows the key position of the indigenous institutions in the strategising process of the local participants in the implementation of the Integrated Community-Managed Development approach (ICMD). As a central concept in the lack of local peoples' participation in development programmes and projects has been the absence of interest or even ignorance among development experts and policy planners towards the target population's perspectives on their culture and their position in life; the overarching, holistic worldview has been situated at the base of the model as the stepping stone for the integration of local and global knowledge systems and institutions.

Since some recent studies in Indonesia have been documenting that for many generations, local groups are accustomed to approaching their existing indigenous institutions of cooperation, mutual aid and neighborhood support at the community level, such as *Gotong Royong*, *Silih-Metulung*, *Gintingan*, *Desa Adat*, *Banjar* and *Arisan*, the IKSIM model seeks in particular to express the key role which these indigenous institutions are playing in the daily life of the local population (cf. Agung 2005; Slikkerveer and Agung 2010; Djen Amar 2010; Ambaretnani 2012; Erwina 2019; Saefullah 2019).

In line with the above-mentioned considerations of the dynamic neo-ethnoscience approach and the neo-endogenous orientation of mid-level development planning in conjunction with the strategic involvement of indigenous institutions, the IKS-based integration model (IKSIM) seeks to contribute to the timely introduction of a new strategy of *Integrated Community-Managed Development* (ICMD) for poverty reduction and sustainable community development in Indonesia and beyond.

References

Agrawal, A. (1995). Indigenous and scientific knowledge: some critical comments. *Indigenous Knowledge and Development Monitor, 33*(3), 3–6.

Agung, A. A. G. (2005). *Bali Endangered Paradise? Tri Hita Karana and the Conservation of the Island's BioculturalDiversity*. Ph.D. Dissertation. *Leiden Ethnosystems and Development Programme (LEAD) Studies*, No. 1. Leiden:Leiden University. xxv + ill., pp. 463.

Agung, A. A. G. (2007). *Microfinancing: Its development to date and prospects*. Contribution to the Cleveringa Lecture. Presented by L. J. Slikkerveer, Jakarta, Indonesia: The Trisakti School of Management and the Sekar Manggis Foundation.

Alkire, S., & Santos, M. E. (2010). *Acute multidimensional poverty: A new index for developing countries*. Working Paper 38. Oxford Poverty & Human Development Initiative (OPHI), Oxford: Oxford Department of International Development.

Alkire, S., & Foster, J. (2009). *Counting and multidimensional poverty measurement.* Oxford Poverty & Human Development Initiative (OPHI), Oxford: Oxford Department of International Development.

Alkire, S., Jindra, C., Robles, G., & Vaz, A. (2016). *Multidimensional poverty index—Briefing 42.* Oxford Poverty & Human Development Initiative (OPHI). Oxford: Oxford Department of International Development.

Anand, S., & Sen, A. (1997). *Concepts of human development and poverty: A multidimensional perspective.* Human Development Report Office Occasional Paper. New York, United Nations Development Programme.

Austin, M. J., Lemon, K., & Leer, E. (2005). Promising practices for meeting the multiple needs of low-income families in poverty neighborhoods. *Journal of Health & Social Policy, 21*(1), 95–117. http://www.haworthpress.com/web/JHSP.

Ambaretnani, P. (2012). *Paraji and Bidan in Rancaekek: Integrated Medicine for Advanced Partnerships among Traditional Birth Attendants and Community Midwives in the Sunda Area of West Java, Indonesia.* Ph.D. Dissertation. Leiden Ethnosystems and Development Programme. LEAD Studies No. 7. Leiden University. xx + ill., 265 pp.

Awedoba, A. K. (2007). *Culture and development in Africa with special reference to Ghana.* Historical Society of Ghana.

Berkes, F. (1993). Traditional ecological knowledge in perspective. In J. T. Inglis (Ed.), *Traditional ecological knowledge: Concept and cases.* Ottawa, ON: International Development Research Centre.

Blunt, P., & Warren, D. M. (1996). *Indigenous organisations and development: A training manual for non-government organisations.*, IT Studies on Indigenous Knowledge and Development London: Intermediate Technology Publications Ltd.

Cabus, P., & Vanhaverbeke, W. (2003). *Towards a neo-endogenous rural development model for the Flemish countryside.* Paper presented at the Regional Studies Association International Conference, April 12–15 2003, Pisa, Italy.

Calvo-González, O. (2016). *Economic slowdown puts the brakes on middle class growth in Latin America.* Washington, D.C.: World Bank. http://blogs.worldbank.org/opendata/economic-slowdown-puts-brakes-middle-class-growth-latin-america.

Cash, D. W., Clark, W. C., Alcock, F., Dickson, N. M., Eckley, N., Guston, D. H., et al. (2003). Knowledge systems for sustainable development. *Proceedings of the National Academy of Sciences of the United States, 100*(14), 8086–8091. http://www.pnas.org/content/100/14/8086.full.pdf.

Casimirri, G. (2003). *Problems with integrating traditional ecological knowledge into contemporary resource management.* Paper submitted to the XII World Forestry Congress, Quebec, Canada.

Chambers, R., Pacey, A., & Thrupp, L. A. (1989). *Farmer first: Farmer innovation and agricultural research.* London: Intermediate Technology Publications Ltd.

Clammer, J. (2005). Culture, development and social theory: On cultural studies and the place of culture in development. *The Asia Pacific Journal of Anthrolopogy, 6*(2), 100–119.

Cohen, J. M. (1987). *Integrated rural development: The Ethiopian experience and the debate.*, The Scandinavian Institute of African Studies Motala, Sweden: Motala Grafiska.

Crossman, P., & Devisch, R. (2002). Endogenous knowledge in anthropological perspective. In C. A. Odera Hoppers (Ed.), *Indigenous knowledge and the integration of knowledge systems.* Claremont, S.A: New Africa Books.

Cull, R. J. (1997). *Financial sector adjustment lending: A mid-course analysis.* Policy Research Working Paper No 1804. Washington, D.C.: World Bank.

Chirangi, M. M. (2013). *'Afya Jumuishi': Towards Interprofessional Collaboration between Traditional & Modern Medical Practitioners in the Mara Region of Tanzania.* Ph.D. Dissertation. Leiden Ethnosystems and Development Programme. LEAD Studies No. 8. Leiden University. xxvi + ill., 235 pp.

Deubel, T. (2008, January). Persistent hunger: Perspectives on vulnerability, famine, and food security in Sub-Saharan Africa. *Annual Review of Anthropology, 35*(1).

Djen Amar, S. C. (2010). *Gunem Catur in the Sunda Region of West Java: Indigenous Communication on the MAC Plant Knowledge and Practice within the Arisan in Lembang*. Ph.D. Dissertation. Leiden Ethnosystems and Development Programme. LEAD Studies No. 6. Leiden University. xx + ill., 218 pp.

Ellen, R. (2002). Reinvention and progress in applying local knowledge to development. In P. Sillitoe, A. Bicker, & J. Pottier (Eds.), *Participating in development: Approaches to indigenous knowledge*. ASA Monographs No 39. London: Routledge.

Erwina, W. (2019). *Iber Kasehatan in Sukamiskin: Utilisation of the Plural Health Information & Communication System in the Sunda Region of West Java, Indonesia*. Ph.D. Dissertation. *Leiden Ethnosystems and Development Programme (LEAD) Studies*, No. 10. Leiden: Leiden University.

Gunatilaka, R., & Kiriwandeniya, P. A. (1999). *Protection for the vulnerable*. Workshop Presentation for a Policy Framework for Poverty Reduction in Sri Lanka, October 1999.

Halani, L. (2004). Book Review: Paul Sillitoe, Alan Bicker & Johan Pottier (eds.) 2002. *Participating in Development: Approaches to Indigenous Knowledge*. (ASA Monographs No 39. London: Routledge.) *Anthropology Matters Journal*, 6(1).

Haughton, J., & Khandker, S. R. (2009). *Handbook on poverty and inequality*. Washington, DC.: World Bank.

Haverkort, B., van't Hooft, K., & Hiemstra, W. (eds.) (2003) *Ancient roots, new shoots: Endogenous developmentin practice*. Leusden, Netherlands: ETC/Compas. London: Zed Books

Hembram, D. (2007). *Use of nontimber forest products by households in the Daniel Boone National Forest region*. Ph.D. Dissertation. West Lafayette, IN: Purdue University.

Hermes, N., & Lensink, R. (2011, June). *Microfinance: Its impact, outreach, and sustainability. World Development, 39*(6), 875–881.

Huntington, S. P. (1965). Political development and political decay. *World Politics*, 17(3) (April 2065)

Jalilian, H., & Kirkpatrick, C. (2001). *Financial development and poverty reduction in developing countries: Finance and development research programme*. Working Paper Series No 30. Bradford Centre for Development.

Johannes, R. E. (1993). Integrating traditional ecological knowledge and management with environmental impact assessment. In J. T. Inglis (Ed.), *Traditional ecological knowledge: Concept and cases*. Ottawa, ON: International Development Research Centre.

Kassam, K.-A. (2009). *Biocultural diversity and indigenous ways of knowing: Human ecology in the Arctic*. Calgary: University of Calgary Press/Arctic Institute of North America.

Kearney, E. (2008). *Cosmology and anthropology: Towards a definition: An interdisciplinary dialogue*. The Cosmology and Cosmologies blog. http://timeo-habla.blogspot.nl/.

Kendie, S. B., & Guri, Y. B. (2005). Indigenous institutions as partners for agriculture and natural resource management. In D. Millar, S. K. Kendie, A. A. Apusigah, & B. Haverkort (Eds.), *African knowledges and sciences: Understanding and supporting the ways of knowing in sub-Saharan Africa* (pp. 106–128). BDU: Barneveld.

Le Grange, L. (2007). Integrating western and indigenous knowledge systems: The basis for effective science education in South Africa? *International Review of Education, 53*, 577–591.

Lewis, O. (1959). *Five families: Mexican case studies in the culture of poverty*. New York: Basic Books.

Mestrum, F. (2003). Poverty reduction and sustainable development. *Environment, Development and Sustainability, 5*, 41.

Metha, L., Leach, M., Newell, P., Scoones, I., Sivaramakrishnan, K. & Way, S. A. (1999). *Exploring understandings of institutions and uncertainty: New directions in natural resource management*. IDS Discussion Paper, No. 372.

Millar, D. (1999). Traditional African world views from a cosmovision perspective. In B. Haverkort & W. Hiemstra (Eds.), *Food for thought: Ancient visions and new experiments of rural people*. London: Zed Books.

Millar, D., Apusigah, A. A., & Boonzaaijer, C. (Eds.). (2008). *Endogenous development in Africa: Towards a systematisation of experiences*. Leusden: COMPAS/UDS.

Millennium Ecosystem Assessment. (2005). *Ecosystems and human well-being: Wetlands and water synthesis*. Washington, D.C.: World Resources Institute (WRI).

Morris, M. W., Leung, Kwok, Ames, D., & Lickel, B. (1999). Views from inside and outside: Integrating emic and etic insights about culture and justice judgment. *Academy of Management Review, 24,* 1781–1796.

Murali, V., & Oyebode, F. (2004). Poverty, social inequality and mental health. *Advances in Psychiatric Treatment, 10,* 216–224.

Myrdal, G. (1968). *Asian drama: An inquiry into the poverty of nations*. New York: Twentieth Century Fund.

Naamwintome, B. A., & Millar, D. (2015). Indigenous knowledge and the African way forward: Challenges and opportunities. *Open Access Library Journal, 2,* e1295. https://doi.org/10.4236/oalib.1101295.

Oakley, P. (1991). *Projects with People: The Practice of Participation in Rural Development*. Geneva: International Labour Office (ILO).

Odora Hoppers, C. A. (2002a). *Indigenous knowledge and the integration of knowledge systems*. Claremont, S.A: New Africa Books.

Odora Hoppers, C. (2002b). Indigenous knowledge and the integration of knowledge systems: Towards a philosophyof articulation. In Odora Hoppers, C. (Ed.), *Indigenous knowledge and the integration of knowledge systems:Towards a philosophy of articulation* (Vol. 1, pp. 2–22). New Africa Books.

Oguamanam, C. (2004). Localizing intellectual property in the globalization epoch: The integration of indigenous knowledge. *Indiana Journal of Global Legal Studies, 11*(2), Article 4.

Pogge, Th. (2012). *Poverty, Human Rights and the Global Order: Framing the Post-2015 Agenda*. Global Justice Program, Yale: Yale University.

Posey, D. A. (2002). Upsetting the sacred balance: Can the study of indigenous knowledge reflect cosmic connectedness? In P. Sillitoe, A. Bicker, & J. Pottier (Eds.), *Participating in development: Approaches to indigenous knowledge*. ASA Monographs No 39. London: Routledge.

Prahalad, C. K. (2004). *Fortune at the bottom of the pyramid: Eradicating poverty through profits*. Upper Saddle River, NJ: Prentice Hall.

Purcell, T., & Onjoro, E. A. (2002). Indigenous knowledge, power and parity: Models of knowledge integration. In P. Sillitoe, A. Bicker, & J. Pottier (Eds.), *Participating in development: Approaches to indigenous knowledge*. ASA Monographs No 39. London: Routledge.

Ray, C. (2001). *Culture economies: A perspective on local rural development in Europe*. Newcastle upon Tyne, UK: Centre for Rural Economy.

Reichel-Dolmatoff, G. (1996). *The forest within: The worldview of the Tukano Amazonian Indians*. London: Themis Press.

Reijntjes, C., Haverkort, B., & Waters-Bayer, A. (1992). *Farming for the future: An introduction to low-external-input and sustainable agriculture*. London: Macmillan.

Richards, P. (1985). *Indigenous agricultural revolution: Ecology and food production in West Africa*. London: Hutchinson.

Robinson, M. (2001). *Microfinance revolution: Sustainable finance for the poor*. Washington, D.C.: World Bank.

Saefullah, K. (2019). *Gintingan in the Sunda Region of West Java: The role of traditional institutions in sustainable community development in Indonesia*. Ph.D. dissertation. Leiden Ethnosystems and Development Programme (LEAD) Studies, No. 11. Leiden: Leiden University.

Schaffer, P. (2001). *New thinking on poverty: Implications for poverty reduction strategies*. Paper Prepared for the Expert Group Meeting on Globalisation and Rural Poverty. United Nations Department for Economic and Social Affairs (UNDESA). United Nations, 8–9 November 2001.

Sen, A. K. (1990). Development as capability expansion. In K. Griffin & J. Knight (Eds.), *Human development and the international development strategy for the 1990s* (pp. 41–58). London: Macmillan.

Shah, A. (2011). *Making Federalism Work for the people of Pakistan: Reflections on the role of the 18th constitutional amendment*. World Bank Policy Paper Series on Pakistan. PK 03/12. http://econ.worldbank.org.

Sillitoe, P. (1998, April). The development of indigenous knowledge: A new applied anthropolog. *Current Anthropology, 39*(2), 223–252.

Sillitoe, P., Bicker, A., & Pottier, J. (eds.) (2002). *Participating in development: Approaches to indigenous knowledge*. ASA Monographs No 39. London: Routledge.

Sillitoe, P., Bicker, A., & Pottier, J. (2003). *Participating in development: Approaches to indigenous knowledge*.London: Psychology Press.

Slikkerveer, L. J. (1990). *Plural medical systems in the Horn of Africa: The legacy of 'Sheikh' Hypocrates*. London: Kegan Paul International.

Slikkerveer, L. J. (1995). INDAKS: A bibliography and database on indigenous agricultural knowledge systems and sustainable development in the tropics. In D. M. Warren, L. J. Slikkerveer, & D. W.Brokensha (Eds.), *The cultural dimension of development: Indigenous knowledge systems* (pp. 512–516). London:Intermediate Technology Publications Ltd.

Slikkerveer, L. J. (1999a). Traditional ecological knowledge (TEK): Practical implications from the African experience. In K. Adimihardja, & M. Clement (Eds.), *Indigenous knowledge systems research & development studies*. No. 3. Leiden: LEAD Programme.

Slikkerveer, L. J. (1999b). Ethnoscience, 'TEK' and its application to conservation. In D. A. Posey (Ed.), *Cultural and spiritual values of biodiversity: A complementary contribution to the global biodiversity assessment* (pp. 169–177). London/Nairobi: Intermediate Technology Publications Ltd./United Nations Environment Programme (UNEP).

Slikkerveer, L. J. (Ed.). (2012). *Handbook for lecturers and tutors of the new master course on integrated microfinance management for poverty reduction and sustainable development in Indonesia (IMM)*. Leiden/Bandung: LEAD-UL/FEB-UNPAD.

Slikkerveer, L. J. (2014). Lectures notes. In *International workshop on integrated microfinance management for poverty reduction and sustainable development in Indonesia (IMM)*. Bandung: UL/FEB-UNPAD.

Slikkerveer, L. J., & Agung, A. A. G. (2010). *Integrated microfinance management and development in Indonesia*.Leiden: LEAD Programme Publication No. 2/2010.

Stavenhagen, R. (1986). Ethnodevelopment: A neglected dimension in development thinking. In R. Apthorpe & A. Kráhl (Eds.), *Development studies: Critique and renewal* (pp. 71–94). E. J. Brill: Leiden.

Swift, A. (1991). Local customary institutions as the basis for natural resource management among the Boran pastoralists in Northern Kenya. In M. Leach & R. Mearns (Eds.), *Environmental change: Development challenges*. Sussex: IDS Bulletin.

Titilola, S. O. (1990). *The economics of incorporating indigenous knowledge systems into agricultural development: A model and analytical framework*. Ames, IOWA: Iowa State University.

Toledo, V. M. (2001). Biocultural diversity and local power in Mexico: Challenging globalization. In L. Maffi (Ed.), *Biocultural diversity: Linking language knowledge and the environment*. Smithsonian Institution Press.

Tolón-Becerra, A., & Lastra-Bravo, X. (2009). *Planning and neo-endogenous model for a sustainable development in Spanish rural areas*. Paper presented at the INRA SFER CIRAD Conference. Montpellier.

Turner, K., & Lehning, A. (2006). Psychological theories of poverty. In M. J. Austin (Ed.), *Understanding poverty from multiple social science perspectives: A learning resource for staff development in social service agencies* (pp. 1–21). Berkeley, CA: School of Social Welfare, University of California.
United Nations (UN). (1992). *Convention on Biological Diversity (CBD)*. Montreal: SCBD. https://www.cbd.int/doc/legal/cbd-en.pdf.
United Nations (UN). (2000). *Millennium Development Goals (MDGs)*. New York: United Nations.
United Nations (UN). (2008). *Human development statistics*. Statistics Division (UNSD), New York: United Nations Department of Economic and Social Affairs (DESA).
United Nations (UN). (2015a). *The millennium development goals report*. New York: United Nations.
United Nations (UN). (2015b). *Transforming our world: The 2030 agenda for sustainable development*. New York: United Nations.
United Nations (UN). (2015c). *Post-2015 agenda for sustainable development and sustainable development goals(MDGs)*. New York: United Nations
United Nations (UDBHR). (2005). *Universal declaration on bioethics and human rights*. New York: United Nations.
United Nations (UDHR). (1948). *Universal declaration of human rights*. United Nations Charter, Chapter IX, New York: Department of Economic and Social Affairs (UNDESA).
United Nations Development Programme (UNDP). (1990). *Human development report: Concept and measurement of human development*. New York: Oxford University Press.
United Nations Development Programme (UNDP). (1991). *Human development report*. NewYork: United Nations.
United Nations Development Programme (UNDP). (1993). *Human development report: People's participation*. New York: Oxford University Press.
United Nations Development Programme (UNDP). (1997). *Human development report: Human development to eradicate poverty*. New York: Oxford University Press.
United Nations Development Programme (UNDP). (1999). *Globalization, with a human face*. New York: Oxford University Press.
United Nations Development Programme (UNDP). (2000). *Human development report: Human rights and human development*. New York: Oxford University Press.
United Nations Development Programme (UNDP). (2003). *Human development report: The millennium development goals: A compact among nations to end human poverty*. New York: Oxford University Press.
United Nations Development Programme (UNDP). (2010). *Human development report: The real wealth of nations, pathways to human development*. New York: Oxford University Press.
United Nations Development Programme (UNDP). (2011). *Human development report: Sustainability and equity: A better future for all*. New York: Oxford University Press.
United Nations Development Programme (UNDP). (2014). *Human development report: Sustaining human progress-reducing vulnerabilities and building resilience*. New York: Oxford University Press.
United Nations Development Programme (UNDP). (2015). *Human development report: Work for human development*. New York: Oxford University Press.
United Nations Educational, Scientific and Cultural Organization (UNESCO). (2003). *Convention on the protection of intangible cultural heritage*. Paris: UNESCO.
United Nations Educational, Scientific and Cultural Organization (UNESCO). (2005). *Universal declaration on bioethics and human rights*. Paris: UNESCO. http://portal.unesco.org/en/ev.php=201html.
United Nations Environment Programme (UNEP). (2010). *Human development report 2016: Human developmentfor everyone*. New York: United Nations.
Uphoff, N. T. (1986). *Local institutional development: An analytical sourcebook with cases*. West Hartford: Kumarian Press.

Uphoff, N. T. (1992). *Local institutions and participation for sustainable development*. Gatekeeper Series No. 31. Ithaca, USA: International Institute for Environment and Development.

van Arendonk, J., & van Arendonk-Marquez, S. (1988). *Direct to the poor: Grassroots development in Latin America*. Boulder, CO: Lynne Rienner Publishers.

Ward, N., Atterton, J., Kim, T,. Lowe, P., Phillipson, J., & Thompson, N. (2005). *Universities, the knowledge economy and neo-endogenous rural development*. Discussion Paper Series No. 1. Centre for Rural Economy.

Warren, D. M. (1991). *Using indigenous knowledge in agricultural development*. Discussion Paper 127. Washington, D.C.: World Bank.

Warren, D. M., Slikkerveer, L. J., & Brokensha, D. (Eds.). (1995). *The cultural dimension of development: Indigenous knowledge systems. IT Studies on Indigenous Knowledge and Development*. London: Intermediate Technology Publications Ltd.

Warren, D. M., Werner, O., & Brokensha, D. (1980). *Indigenous knowledge systems and development*. New York:Latham Publishers.

Watson, E. E. (2003). Examining the potential of indigenous institutions for development: A perspective from Borana. *Development and Change, 34*(2), 287–310.

Wiggins, S., & Higgins, K. (2007). *Pro-poor growth and development Linking economic growth and poverty reduction*. ODI Briefing Paper No. 33. London: Overseas Development Institute.

Wijayaratna, C. M. (2004). *Role of local communities and institutions in integrated rural development*. Report of the Seminar of the Asian Productivity Organization (APO). Islamic Republic of Iran, 15–20 June 2002.

Weber, M. (1905). Die Protestantische Ethik und der Geist des Kapitalismus (The Protestant ethic and the spirit of capitalism). *Zeitschrift für Sozialwissenschaften und Sozialpolitik, 20*(1), 1–54 and *21*(1), 1–110.

Williamson, J., & Maher, M. (1998). A survey of financial liberalisation. *Essays in International Finance,* No. 211, November 1998.

Woodley, E., Crowley, E., de Pryck, J. D., & Carmen, A. (2006). *Cultural indicators of Indigenous Peoples' food and agro-ecological systems*. SARD Initiatives. Paper on the 2nd Global Consultation on the Right to Food and Food Security for Indigenous Peoples. Nicaragua, 7–9 September 2006.

World Bank (WB). (1978). *World development report*. Washington, D.C.: World Bank.

World Bank (WB). (1980). *World development report*. Washington, D.C.: World Bank.

World Bank (WB). (1985). *World development report*. Washington, D.C.: World Bank.

World Bank (WB). (1989). *World development report*. Washington, D.C.: World Bank.

World Bank (WB). (1990). *World development report: Poverty*. New York: Oxford University Press.

World Bank (WB). (1997). *World development report: The state in a changing world*. New York: Oxford University Press.

World Bank (WB). (1998/1999). *World development report: Knowledge for development*. New York: Oxford University Press.

World Bank (WB). (1999/2000). *World development report: Entering the 21st century*. New York: Oxford University Press.

World Bank (WB). (2000/2001). *World development report attacking poverty*. New York: Oxford University Press.

World Bank (WB). (2004). *World development report: Making services work for poor people*. New York: Oxford University Press.

World Bank (WB). (2005a). *World development report: A better investment climate for everyone*. New York: Oxford University Press.

World Bank (WB). (2005b). *Introduction to poverty analysis*. Washington, D.C.: World Bank Institute.

World Bank (WB). (2012). *World development report: Gender equality and development*. New York: Oxford University Press.

World Bank (WB). (2014). *Global monitoring report*. Washington, D.C.: World Bank.

World Bank (WB). (2016a). *Poverty and shared prosperity: Taking on inequality*. Washington, D.C: World Bank.
World Bank (WB). (2016b). *Inequality and shared prosperity*. Washington, D.C.: The World Bank.
World Bank (WB). (2016c). *World bank annual report: Digital dividends*. Washington, D.C.: World Bank.
World Bank (WB). (2016d). *Report of the commission on global poverty*. Washington, D.C.: World Bank.
World Bank (WB). (2016e). *Indonesia economic quarterly*. Pressures Easing. October, 2016. Washington, D.C.: WorldBank
World Bank (WB). (2017). *Monitoring global poverty. Report of the commission on global poverty*. Washington, D.C.: World Bank.
World Commission on Environment and Development (WCED). (1987). *Our common future—The Brundtland report*. New York: United Nations.
World Intellectual Property Organization (WIPO). (2016). *Traditional knowledge and intellectual property background briefs*. No. 1-10. Geneva: WIPO. http://www.wipo.int/publications/en/details.jsp?id=3858&plang=EN.
Yale Campuspress. (2015). *A critique of the sustainable development goals' potential to realise the human rights of all: Why being better than the MDGs is not good enough*. Yale Campuspress, New Haven: Yale University. https://campuspress.yale.edu/thomaspogge/files/2015/10/SDG-HR_Rev-Jan-25-uugh97.pdf.

L. Jan Slikkerveer is Professor of Applied Ethnoscience and Director of the Leiden Ethnosystems and Development Programme, Faculty of Science, Leiden University, Leiden, The Netherlands. He received his Ph.D. on his fieldwork in the Horn of Africa from Leiden University in 1983 and an Honorary Degree from the Faculty of Medicine of Universitas Padjadjaran in Bandung, Indonesia in 2005. He further extended advanced training and research in the newly-developing field of applied ethnoscience in several sub-disciplines, including ethno-economics, ethno-medicine, ethno-biology/botany, ethno-pharmacy, ethno-communication, ethno-agriculture and ethno-mathematics in combination with a focus on three target regions of South-East Asia, East-Africa and the Mediterranean Region. He is the supervisor of 25 Ph.D. students at Leiden University and has published more than 100 books and 300 articles on the subject. While he has also received a number of substantial subsidies from the European Union in Brussels for international development programmes in South-East Asia, East-Africa and the Mediterranean Region, he also conceptualised both the IMM & ICMD approaches in Indonesia, and is Senior Lecturer in the International Master Course on Integrated Microfinance Management (IMM) at the Faculty of Economics and Business of Universitas Padjadjaran in Bandung, Indonesia.

Part II
Global Versus Local Economic Development Processes

Chapter 5
Macro- and Microeconomics and Social Marketing

Tati S. Joesron and Alkinoos Nikolaidis

Policymakers throughout the world have actively tried to improve financial markets in poor regions, but often with disappointing results.
Beatriz Armendariz de Aghion and Jonathan Morduch (2010)

5.1 The Macro- and Microeconomic Context of Development

Environmental analysis of community development plays an important role in identifying various institutions which are related to the community and its development. The micro- and macroeconomic analysis of the environmental factors of the community supports the understanding of micro-pictures and macro-portraits of the community. Figure 5.1 summarises the interrelated factors, institutions, and policies in the micro- and macroeconomic levels, which are related to the community.

Basically, an analysis of the macro-economic environment deals with an aggregate level of all activities done by all people and institutions at regional or national levels. It deals with government policies in both monetary and fiscal policies to stimulate real sectors of the economy. Development policies from the top are also determined in the analysis. However, to deal with a community level where various institutions existed and operated differently, an analysis of the microeconomics environment is more important to identify specific characteristics of available stakeholders including institutions in the community, and how a particular

Alkinoos Nikolaidis is deceased.

T. S. Joesron (✉)
FEB, Universitas Padjadjaran, Bandung, Indonesia
e-mail: tati.joesron@fe.unpad.ac.id

A. Nikolaidis
CIHEAM-MAICh, Chania, Greece

L. J. Slikkerveer et al. (eds.), *Integrated Community-Managed Development*, Cooperative Management, https://doi.org/10.1007/978-3-030-05423-6_5

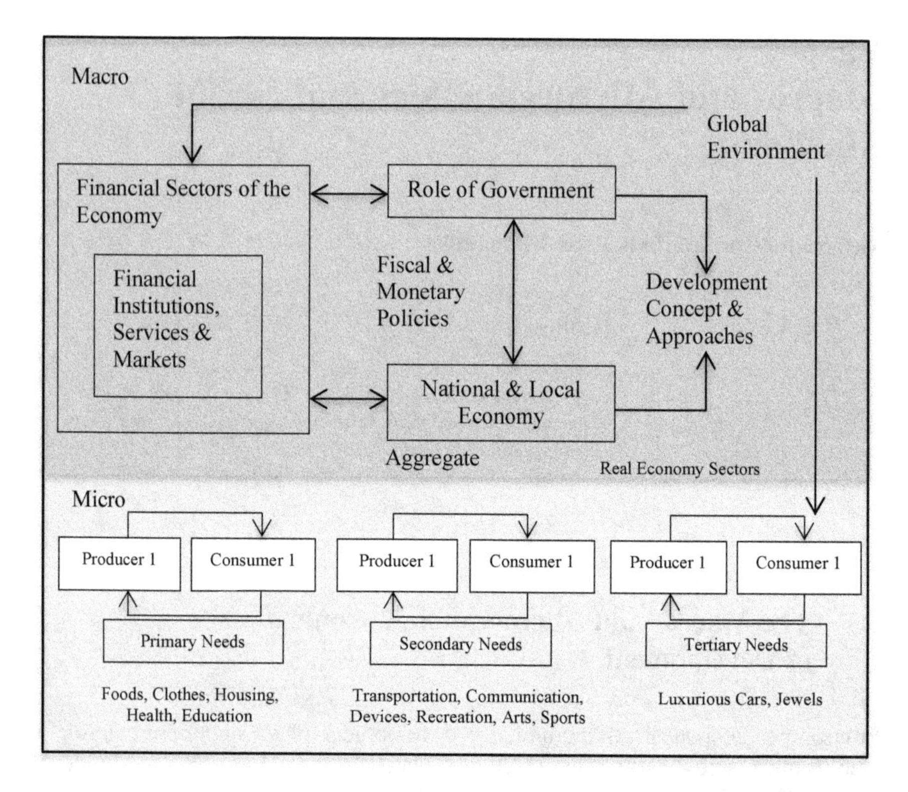

Fig. 5.1 Macro-micro economics framework. *Source* Joesron and Nikolaidis (2012)

development policy should be approached. In community-managed development, this microeconomics analysis will support the government in the identification of local institutions and development agents, and how to empower them to participate in the development process. The analysis will also support the government of how public-private institutions should be integrated in development plans, policies and implementations.

5.2 Macroeconomic Approaches to the Struggle Against Poverty

One of the major challenges for macroeconomic policy makers today is to alleviate poverty. However, since macroeconomic policy planning involves basic aggregate magnitudes, such as total production, employment, consumption and investment, which are determined by the national economy of a particular country—which may be influenced by economic and political decisions in the rest of the world—the realisation of poverty reduction as a goal involves a complicated process.

In general, the challenges in the macro-economy of a nation can be divided into two time frames, i.e. the short term and the long term. Both time frames require different approaches in facing their challenges. Short-term macroeconomic subjects involve the fluctuation of output and employment. The fluctuation of output will have an impact on the employment rates, and thus, will result in fluctuating poverty rates. Policymakers need to address these fluctuations, usually through monetary and fiscal policies. The main question is, however, whether either the monetary or the fiscal policies will be able to realise the reduction of poverty.

Unfortunately, the answer is no. For example, a theory in macroeconomics known as the 'Phillips Curve' explains the trade-off between inflation and unemployment (cf. Mankiw 2015). The increase in the aggregate demand will lead to lower unemployment as people will find more job openings to cope with the increase in demand of output and higher inflation rates. Thus, if the government tries to lower the inflation, it will do so only at the cost of higher unemployment. The government may reduce its spending and reduce aggregate demand, which will pertain to lower inflation and a lower employment rate. Although this process represents the general trend, the specific conditions of a particular country may diverge from the theory. By consequence, poverty reduction may sometimes be sacrificed in order to achieve another objective, rendering the challenge for the government to manage the delicate balance of these trade-offs.

In the long term, the main concerns are about long-term national growth rates and their costs. Going back to the discussion on the balance between inflation and unemployment, in the long term, the trade-off may result in a permanently higher inflation rate and a lower, or stagnant unemployment rate. Hence, the cost of maintaining a low rate of unemployment results in a higher price. One of the challenges in the long term is that almost all factors become variables. An accurate prediction is not possible of what will happen in the next ten or twenty years. By consequence, the development of a policy today which will last for a long-term period of time will rather serve as a guide, albeit prone to changes, but it will not be possible to accurately determine the course of the economy itself in the long term.

Poverty itself is both a short- and a long-term problem. A part of it requires immediate action (short-term), while another part needs a more sustainable and permanent action (long-term). In the short term, poverty reduction becomes the domain of development economics, while in the long term, it is analysed in theories of economic growth. Although the two time frames cannot really be separated in real life situations, such as whether the provision of nutrition to the poor today will have an impact on poverty in the long term, careful consideration should be made when a policy is designed for each time frame. The consequences of a short-term policy can take a very different direction, or create unforeseen effects in the long run such as for instance the occurrence of political turmoil, or when a natural disaster strikes.

Fiscal and monetary policies

Macroeconomic policies in general are divided into several policies, two of them being fiscal and monetary policies. Fiscal policy is an option for poverty elimination at the supply side, and/or the demand side. The supply side defines the sources of funding. Funding is largely generated by implementing progressive tax on income and assets. It is evident that every fiscal budgetary policy to assist the poor will require funding. In general, such funding will be generated by tax revenues. Developing countries are known to be weak in the ability to tax their rich citizens and corporations adequately, and to administer the tax laws, not least because of 'leakages' due to corruption. Hence, many of the poverty elimination programs have to be financed by foreign credit, which as a national debt may become a burden in the long term.

The progressiveness of a tax code can be measured by two factors: one is the rate or percentage of the tax, and the other, less commonly known, is the tax base. The tax base in developing countries tends to progress too slowly, compared to the more developed states. For instance, the 30% rate will be effective at the annual income per capita of 1000 times the national average in some developing countries, while in some Western states this rate becomes effective at 100 times. So, the high tax rate is effective only when income is extremely high in the developing countries, while this is not the case in developed countries.

The approach from the demand side is related to the use of funding. Subsidies for education, health and low-cost housing are examples of government expenditure, known as 'pro poor'. Assisting the poor with education and health services means a focus on poverty reduction at the level of the parents. Poor children with good health and nutrition, and education will grow up to become productive citizens. They are expected to be able to liberate themselves from poverty with adequate employment. Low-cost housing is necessary, as housing expenses take up a large share of the income of the poor, besides the cost of food and transportation. In order to further reduce the expenses of poor households, several municipalities and governments are planning to supply free school buses and school lunches to the children. Another form of such 'pro poor' policies includes cross-subsidies, whereby pricing goods or services for the rich is higher than for the poor. These so-called 'disguised taxes' are created to redistribute income, e.g., the reduction in the general subsidy on fuel in Indonesia, of which 70% was enjoyed by the rich, enabled a reallocation of the saved funds to direct subsidies for the poor. An example is the *Bantuan Langsung Tunai* (BLT)/Direct Cash Assistance and the provision of cheap Liquefied Petroleum Gas (LPG). The government can also allocate a certain budget for *Research and Development* (R&D) on *Appropriate Technology* (AT). The task is then to develop a form of technology which is cheap, simple and non-capital intensive, and as such useful for remote poor rural regions in offering local employment opportunities. Such efforts could also stem the tide of over-urbanisation, which has recently overloaded the urban infrastructure of many cities in developing countries by overcrowding, traffic congestion, slums and urban unemployment[1].

---

[1] *Appropriate Technology* (AT) emerged in the course of the 1970s as an alternative development approach, triggered by the publication of the economist Schumacher's (1973) *Small Is Beautiful: A Study of Economics as if People Mattered.* The new orientation became originally known as 'intermediate technology', focused on an alternative application of a small-scale, decentralised, labour-intensive, energy-efficient, environmentally sound, and locally autonomous process, which embarked on the notion of the introduction of new technologies to developing countries as people-centered. Recently, the concept of appropriate technology is also promoted as a new model of facilitating the introduction of innovations for sustainable development.

One of the most potent fiscal policy tools to eradicate poverty is the allocation of national funds on the improvement and extension of the infrastructure. The rationale is not only that such funding immediately offers employment, which, in turn, is a direct action which will reduce poverty, but an adequate infrastructure will also benefit other sectors of the society. Businesses will grow, more firms and plants will be created and employment will expand. Such use of the budget is a truly long-term solution, offering sustainable expansion of employment with economic growth. The recent miracle of economic growth of the People's Republic of China had started with a large investment in the infrastructure in the late 1990s.

Another way to create employment as a means to reduce poverty is by market expansion. The support of the government for marketing research, efficient logistics and swift documentation contributed to sustaining the recent Chinese export offensive, showing that even micro- and small firms can take advantage of the export market. The availability of foreign demand will also help micro- and small firms to grow bigger and create more employment which pertains to lower poverty rates.

Monetary policies can also serve as a means of poverty alleviation. At the present time, a popular policy to try to eradicate poverty is through microfinance. While initially cheap microcredit for small businesses has been rather effective in expanding employment, the shift to more expensive forms of microfinance since the early 1990s has not been successful in lifting the poor people from the trap of poverty.

Devaluation is a monetary intervention which reduces the value of the currency in a country in relation to foreign exchange. The effect is that imported goods become more expensive, encouraging an increase in the consumption of domestic goods. In addition, cheaper domestic goods relative to foreign products will drive the national export of goods. Such interventions from the government will eventually lead to increased employment. After the national infrastructure had been improved and extended, the Government of the People's Republic of China devalued their currency, the *Yuan*, with a view to starting their export offensive. Devaluation, however, will usually be followed by short-term inflation, if it is not managed in a proper way.

High interest rates are usually viewed as unhealthy for the business sector. So, when the business is not healthy, then employment will be jeopardised. High interest rates may also deter the capital-intensive production process. Counter-intuitively, high interest rates may induce more employment. However, the marginal benefit and marginal cost of such a policy has to be very carefully considered. For instance, a relevant consideration is if labour-intensive domestic goods are competing with imported goods. If these goods are more expensive than imported goods, then the domestic firms cannot survive in the face of imports without trade protection, which, however, will be against globalisation. The above-mentioned examples show that there are several macroeconomic approaches to alleviate poverty, but the key to the struggle against it is to select the maximum combination of policies which would result in a more effective and sustainable reduction of poverty.

5.3 Microeconomic Development for a Better Livelihood

The development of microeconomic research

Microeconomics has developed since the first introduction of classical economics in the 1700s by Smith (1776). Microeconomics has since then evolved from a single science to a number of interdisciplinary fields of study, including health economics, environmental economics and the economics of education. The trend simply stems from the fact that the problems faced by economists today involve mostly cross-cutting subjects which need an integrated approach to be solved. At the same time, the basic principles of economics are related to the topics also discussed in other disciplines. It is currently well recognised that there is no single science which can provide solutions to today's local and global problems, and that the collaboration between sciences has become an important key component in today's policy-making process.

In light of this current trend in the development of microeconomics, the discussions within microeconomics have increased in the last decades on the very essence in which economics are involved: the concept of utility or satisfaction. Thus, researchers who aim at studying or optimising utility are currently rather successful. One of the branches of economics which is entirely focused on this topic is behavioural economics. This field of economics applies the insights of psychology in the analysis of their subjects of research in economics (cf. Oxford Dictionaries 2015). Other branches of economics which are also analysing subjects which are indirectly related to utility include health and education economics. Both of these branches study the issues in the field of health and education by using the principles of economics. Consequently, collaboration with experts from health and education is essential.

It is clear from the increasing trends of utility-related research in microeconomics that the livelihood of the society consists of more than income alone, but also on access to better health and education, and on the optimisation of local knowledge and wisdom. Utility-related research will provide new insights into the understanding of microeconomics, especially if the objective of the research is to contribute to poverty reduction through the study of microeconomics, where not only financial factors, but also other factors such as behaviour, health, and education are playing a significant role.

Microeconomic development and better livelihood for the poor

In line with the emerging trend of utility-related research in microeconomics, the approach to microeconomic development is also changing. Just like the development in microeconomic research during the past decades, microeconomic development also recognises factors other than the income and revenue of people as important for achieving a better livelihood. The recent development has significant implications, especially for the design of interventions to alleviate poverty within a particular community. An understanding of the culture and behaviour of the poor at the community level will provide important material for the development of specific intervention programmes.

A recent study by Kamenica (2012) shows that monetary incentives may not always be useful in an optimal strategy to influence human behaviour. Furthermore, Spears (2011) and Mani et al. (2013) found that poverty also has an impact in reducing the cognitive capacity of people. Poverty can also cause stress and a negatively affected status, leading to short-sighted and risk-averse decision-making, potentially caused by favouring habitual behaviours compared to goal-directed behaviours (Haushofer and Fehr 2014). These findings imply that there is a need for specific strategies in approaching the poor in different settings.

Looking at specific regions, recent research in the Bale Highlands of Southern Ethiopia has found that the most important determinants of choices of livelihood strategies are 'age of household head', 'possession of cropland', 'distance from the market', and 'altitude' (cf. Tesfaye et al. 2011). Interestingly, 'income' is not part of the determinants in the community studied in this area. In Peru, 'ethnicity' and 'location' are correlated with factors such as livelihood strategies, and poverty and asset levels (Porro et al. 2015). The study by Pica-Ciamarra et al. (2015) in twelve developing countries has shown that for a community which relies on livestock, the way to attain a better livelihood is to design specific multiple policies, focused on farming systems, species, and different target groups. The *South Asia Council for the Community and Children in Crisis* (SCA-CCC) (2014) has taken its microeconomic development programmes a step further by helping the target community to solve domestic violence and family problems, recognising that these factors are important for the success of their programmes.

With regard to microfinance as a widely recognised form of microeconomic development, the commercialisation of microfinance as represented by micro-banking and its impact on poverty alleviation are questioned by Bateman (2010) for further analysis both in terms of *displacement effects* and *client exit* behaviour. The effect of displacement occurs in the event that, when microfinance may be able to assist a group of poor people, the resulting increased competition may just displace another group of poor people. The latter is crowded out of the market, and becomes as such marginalised. So, after all, the actual number of poor people is not reduced. In economics this phenomenon is known as the 'fallacy of composition'. It means that what is good micro-economically for a household or unit of business may not necessarily be good macro-economically on a larger scale, i.e. on the regional or national level. The criticism is based on the assumption that micro-business depends on the local market with limited capacity to absorb increased supply. In other words, indiscriminate credit supply to enhance the productive capacity of micro-business pertaining to a microeconomic improvement will not reduce poverty on a macroeconomic scale. The solution is provided by an integrated approach, where—besides capital supply—selected micro-business units are trained and supported to expand their market in order to serve demand beyond the local market.

Client exit, or *client failure*, refers to criticism, which highlights the fact that most microfinance research suffers from a 'survivor bias', being the focus of research on the success stories, and showing less consideration for the businesses which have failed. One reason for such bias is related to the difficulty of tracing businesses which have failed, while the successful businesses are still around and less difficult to study through easier data collection. Among the general population the amount of entrepreneurs is seldom higher than 2%. In logical terms, only this low number of the poor can potentially become successful entrepreneurs. The rest of the population are entrepreneurs out of need, also known as 'survival mode'.

Many studies have corroborated the fact that, in general, micro-businesses suffer from a high rate of failure. However, since several microfinance programmes have reported a high rate of repayment, the conclusion must be that the poor would go to great lengths to repay their loans, regardless of the cost involved. Yet, there are very few data available on these failures and their terrible toll on the poor. The solution refers to another aspect of the integrated approach to microfinance, which is education and health for the poor and their children. It means that in the long term, even if microfinance fails to help the parents, the children can improve their destiny with good education and health. In this respect, the aim is to stop poverty at the parental level, and prevent poverty perpetuation in their children.

5.4 The Social Marketing Approach to Consumer Behaviour

Social Marketing is obviously part of marketing, but more than just the promotion of social ideas as it combines psychology, economics, sociology and anthropology, with the aim of understanding how people behave. Once this understanding is achieved, messages, products and services can be developed which will entice people to change their behaviour. Using a consumer-oriented approach, social marketing can affect a sustainable behavioural change (cf. Dann 2010). In general, the discipline of economics holds the view that individuals will maximise utility, firms will maximise profit and states will maximise welfare, which is conceptualised as more than just an aggregation of the individual utilities of all citizens together. According to the consumer behaviour theory, the standard maximising behaviour can be represented formally as *Maximise Total Utility* (U) subject to *Budget* (Y), where the utility function depends on the goods consumed (Q), or U (Q)[2]. As the science of economics progresses, the awareness that people value goodness grows, such that consumers will be willing to pay more for goods which have social benefit. For example, the *Body Shop* states that their soaps and lotions

[2]Several models of consumer behaviour have been developed which focus on the factors which tend to influence behaviour. According to Kotler and Armstrong (2008), they can be classified as: (a) *psychological factors* (motivation, perception, learning, beliefs and attitudes); (b) *personal factors* (age and life-cycle stage, occupation, economic circumstances, lifestyle, personality and self-concept); (c) *social factors* (reference groups, family, roles and status); and (d) *cultural factors* (culture, subculture, social class system).

are environmentally friendly as they are using non-polluting chemicals, and *Illy Coffee* claims that they purchase their beans from non-exploitative farmers' cooperatives. Similarly, *Aqua* bottled water contributes one liter of clean water for each liter which is sold commercially to the dry regions in Eastern Indonesia. Although these products may cost slightly more than the regular products, the consumers believe that their money will also be allocated to socially desirable goals.

Theoretically, it means that the set of 'goods' consumed (Q) includes items which are not merely consumer goods, but also incorporate social benefits as part of their attributes, as shown in the above-mentioned examples. Furthermore, the 'goods' can also directly include social goals, such as 'not littering' or 'giving alms or charity'. When people purchase goods with socially desirable goals, such as charity, there may be many factors influencing their decisions other than altruism. As Olson (2002) notes, people are sometimes motivated by psychological objectives, called by Andreoni (1990) the 'warm glow effect'; hence acquiring socially desirable objectives does not contradict with standard consumer behaviour since there is an increase in the utility derived from 'socially desirable goods'. In this way, a 'warm glow' effect represents 'feeling good after doing good'. In other words, pursuing social goals is not mutually exclusive to a commercial orientation.

The role of social marketing is to encourage consumers to pay extra for these altruistic goods. So, the objective of the social marketing campaigns is not to gain profit, but to trigger behavioural change, where consumers will be convinced to adopt the view that it is better to consume goods which have additional social values as part of their attributes. A social marketing approach to consumer behaviour seeks also to 'educate' consumers to place value on goods with a social benefit, to see them as something more than simple consumption goods. The short-term success of this approach is shown by the number of people who are willing to pay more for these altruistic goods. In the long term, the approach can be classified as successful if people start to demand—or insist—that all goods should have such additional attributes for social benefit: they are not prepared to purchase goods without such attributes. If people have fully adopted the social marketing approach, the idea might stay, and eventually become sustainable.

Social Marketing as a part of Social Entrepreneurship

Social Marketing is part of Social Entrepreneurship, also known as 'Socio-Preneurship', referring to the use of the general principles of entrepreneurship, such as creativity, innovation and risk taking, for social goals, instead of for profit (cf. Yunus et al. 2010; Kotler and Keller 2012). As for social marketing, it also means using marketing theories and principles to 'sell' ideas in order to promote social change. Instead of the usual legalistic approaches of government decrees, the new ideas could be better accepted by the population if they are popularised with marketing techniques, in the same way as any firm tries to sell its goods or services through advertising.

In general, there are three levels or degrees of commitment to social entrepreneurship, and at each level, social marketing will be a useful tool:

- *Corporate Social Responsibility* (CSR): when a firm does business as usual, but allocates some of its profit towards advancing some social objectives. Most of the time, these 'social objectives' coincide with the long-term goals of the firm, rendering CSR being blamed as another marketing gimmick or propaganda tool: 'Just to make the company look good'. Although it is evident that no firm wants to finance activities against its business interests, some CSR funds are really beneficial to the society, for example by offering scholarships. In the long term, a well implemented CSR policy could only create goodwill for the firm, which eventually could enhance its business.
- *Social Business*: in this case, the firm's objective is not only focused on profit making, but also on a contribution in the social realm, such as for instance the provision of maximum employment. Although efficiency may dictate the use of less labour, the firm would deliberately hire more labour. Since such policy is against economics, the firm must at least aim for a financial break-even position in order to ensure sustainability. Otherwise the firm would need to rely on limitless external donations to sustain its capital or suffer bankruptcy shortly.
- *Social Entrepreneurship*: the use of the techniques by entrepreneurs to develop, fund and implement solutions to social goals, where in its full-fledged version the objective is not the achievement of monetary gains, but sustained adoption of a social initiative. Its success is not measured in money, but, for instance, in how many households adopt its advice, or whether a new regulation is enacted.

Muhammad Yunus, founder of the Grameen Bank in Bangladesh, introduced the term 'Social Business' to describe firms which have other social objectives besides simply making a profit. In his case, the Grameen Bank offers microcredit—and the necessary training to improve the capacity of the poor to manage their business—with the objective of poverty reduction. In this way, profit is accumulated solely to expand loan capacity in order to reach more poor people. Its second objective is the empowerment of women, to the extent that more than 80% of the clients are poor women (cf. Yunus et al. 2010).

Social Marketing and Financial Sector Development

As mentioned above, an integrated approach is required towards microfinance in order to contribute to the solution of the problems of the effect of displacement and client exit, identified as the main factors by the critics of conventional microfinance. In order to be able to promote this new integrated approach to microfinance to the world at large, its proponents—among them development economists and community activists—also have to include social marketing strategies and techniques in their strategies. A relevant question in this context is: "*Why do we need to 'sell' the idea of Integrated Microfinance Management?*", as has recently been introduced effectively in the training at post-graduate level of these new cadres at the *Faculty of Economics and Business of Universities Padjadjaran* in Bandung, Indonesia.

A major reason is that most bankers, as economic stakeholders in the micro-banking sector, are still thinking that the extension of credits through the provision of loans to the poor—whom they perceive as 'unbankable'—will be too risky, especially those people classified as the 'poorest of the poor'. In addition, the integrated approach also advocates loan extension for non-productive, i.e. non-business purposes, including education and health care. Notwithstanding, these integrated policies are stemming from a fundamental difference in orientation: poverty reduction through human community development instead of profit-making through purely commercial financial-economic development.[3]

The proponents of the new approaches of *Integrated Microfinance Management* (IMM) and *Integrated Community-Managed Development* (ICMD) have more work to do in order to convince the following parties and stakeholders of the efficacy of their novel approach towards poverty reduction and sustainable development:

- *Bankers*: they are the majority of micro-banking players and they have the financial means of capital, necessary for any form of microfinance endeavour.
- *Government Officials*: they make the rules and regulations. With the right rules, favourable to the integrated microfinance management approach, the new strategy can be more effective as a tool to eradicate poverty.
- *Social Activists*: they are the experts in the field of community development. They are able to mobilise grassroots support for the new initiative.
- *Local Leaders*: specifically, they are the informal indigenous leaders, because they represent the real behavioural model for local people and their way of life.
- *Local People*: they are the ultimate target of the new approach of integrated microfinance management whose vision and participation is indispensable for attaining the realisation of poverty reduction at the community level.

In order to gain the support of all the parties mentioned above, the social marketing approach should also be included in the new integrated strategies. Instead of relying on legal decrees or regulations, such new ideas may be better accepted after a well-planned approach towards socialisation as a major part of the social marketing strategy.

A successful social marketing campaign will contribute significantly in convincing the local leaders, soon followed by the local population, who eventually may support and adopt the newly-developing integrated initiatives for poverty reduction and sustainable development, including both innovative approaches of *Integrated Microfinance Management* (IMM) and *Integrated Community-Managed Development* (ICMD), further highlighted throughout this Volume.

---

[3]Although many bankers hold the opinion that poor people are being helped by providing them with a bank account in order to get better access to financial services, an increasing number of reports indicate a growing number of people without bank accounts, while those who have accounts, however, continue to use alternative financial services including loans. According to Servon (2013), in the United States the number of people who are 'unbanked' increased nationwide from 10 million in 2002 to 17 million people in 2013, while 43 million became underbanked. She found that banks are often costlier for the poor than check cashers and other alternative services.

References

Andreoni, J. (1990). Impure altruism and donations to public good: A theory of warm-glow giving. *The Economic Journal, 100*(401), 464–477.

Armendariz de Aghion, B., & Morduch, J. (2010). *The economics of microfinance*. Cambridge, MA: The MIT Press.

Bateman, M. (2010). *Why doesn't microfinance work? The destructive rise of local neoliberalism*. London; New York: Zed Books.

Dann, S. (2010). Redefining social marketing with contemporary commercial marketing definitions. *Journal of Business Research, 63*(2).

Haushofer, J., & Fehr, E. (2014). On the psychology of poverty. *Science, 344*(6186), 862–867 http://www.sciencemag.org/cgi/doi/ https://doi.org/10.1126/science.1232491.

Joesron, T. S., & Nikolaidis, A. (2012). Macro- & microeconomics and financial sector development. In Slikkerveer, L. J. (Ed.), *Handbook for lecturers and tutors of the new master course on integrated microfinance management for poverty reduction and sustainable development in Indonesia* (IMM). Leiden/Bandung: LEAD/UL/UNPAD/MAICH/GEMA PKM.

Kamenica, E. (2012). Behavioral economics and psychology of incentives. *Annual Review of Economics 4*(1), 427–452. http://www.annualreviews.org/doi/abs/10.1146/annurev-conomics-080511–110909.

Kotler, P., & Armstrong, G. M. (2008). *Principles of marketing*. Upper Saddle River, NJ: Prentice Hall.

Kotler, P., & Keller, K. L. (2012). *Marketing management*. Upper Saddle River, NJ: Prentice Hall.

Mani, A., Mullainathan, S., Shafir, E., & Zhao, J. (2013). Poverty impedes cognitive function. *Science, 341*(6149), 976–980. http://www.sciencemag.org/cgi/doi/10.1126/science.1238041.

Mankiw, N. G. (2015). *Principles of economics* (7th ed.). Stamford CONN: Cengage Learning.

Olson, M. (2002). *The logic of collective action: Public goods and the theory of groups* (20th ed.) Cambridge, MA: Harvard University Press. http://www.amazon.com/The-Logic-Collective-Action-printing/dp/0674537513 http://homes.ieu.edu.tr/ibagdadi/INT230/MancurOlson-Logicof CollectiveAction.pdf.

Oxford Dictionaries. (2015). *Behavioural economics*. Oxford: Oxford University Press. http://www.oxforddictionaries.com/definition/english/behavioural-economics.

Pica-Ciamarra, U., Tasciotti, L., & Otte, J. (2015). Livestock in the household economy: Cross-country evidence from micro-economic data. *Development Policy Review, 33*(1), 61–81.

Porro, R., Lopez-Feldman, A., & Vela-Alvarado, J. W. (2015). Forest use and agriculture in Ucayali, Peru: Livelihood strategies, poverty and wealth in an Amazon frontier. *Forest Policy and Economics, 51*, 47–56. http://linkinghub.elsevier.com/retrieve/pii/S1389934114002299.

Schumacher, E. F. (1973). *Small is beautiful: A study of economics as if people mattered*. London: Blond & Briggs Ltd.

Servon, L. J. (2013). The high cost for the poor of using a bank. *The New Yorker*. 9 October 2013.

Smith, A. (1776). *An inquiry into the nature and causes of the wealth of nations*. London: W. Strahan and T. Cadell.

Spears, D. (2011). Economic decision-making in poverty depletes behavioral control. *The B.E. Journal of Economic Analysis & Policy, 11*(1). http://www.degruyter.com/view/j/bejeap.2011.11.issue-1/1935-1682.2973/1935-1682.2973.xml.

South Asia Council for the Community and Children in Crisis (SCA-CCC). (2014). *Rural Micro-Economic Development*. http://www.sac-ccc.org/services-view/rural-micro-economic-development.

Tesfaye, Y., Roos, A., Campbell, B. M., & Bohlin, F. (2011). Livelihood strategies and the role of forest income in participatory-managed forests of Dodola area in the Bale highlands, southern Ethiopia. *Forest Policy and Economics, 13*(4), 258–265. http://linkinghub.elsevier.com/retrieve/pii/S1389934111000190.

Yunus, M., Moingeon, B., & Lehmann-Ortega, L. (2010). Building social business models: Lessons from the Grameen experience. *Long Range Planning, 43*(2–3), 308–325. http://linkinghub.elsevier.com/retrieve/pii/S0024630109001290.

Tati S. Joesron is Professor of Economics at the Faculty of Economics and Business, Universitas Padjadjaran, Bandung, Indonesia. She is Senior Advisor of the Study Center for Islamic Economics and Business (PSEBI), and the Center for Entrepreneurship Studies (PSIK), as well as Chair of the Islamic Foundation which manages two Islamic hospitals in Bandung. She is also Senior Lecturer in the Course on *Integrated Microfinance Management* (IMM), Bandung.

Alkinoos Nikolaidis was the Director of the Mediterranean Agronomic Institute of Chania (CIHEAM-MAICh) for more than 25 years until 2012. He passed away in 2016. His legacy includes not only his outstanding leadership of the Institute, but also a number of articles on various applications of Linear Programming and International Political Economy. In 2009 he also joined the Project on *Integrated Microfinance Management in Indonesia* (IMM).

Chapter 6
Microfinancial Sector Assessment and Development

Dimitrios Niklis, George Baourakis and Constantin Zopounidis

> *Microcredit is based on 'trust', not on legal procedures and systems. It is offered for creating self employment for income-generating activities and housing for the poor, as opposed to consumption.*
>
> Muhammad Yunus and Karl Weber (2007)

6.1 Analytical Framework: Concepts, Definitions and Models

A Microfinance Institution (MFI) is an organisation which provides financial services to the poor, particularly in rural areas (Ledgerwood and White 2006). This very general definition includes different providers which vary in their legal structure, mission, and methodology. However, all have as a common theme the provision of financial services to clients who are poorer and more vulnerable than traditional bank clients (clients that can provide collateral to receive a loan).

The key word for the assessment of the regulatory framework of MFIs is 'access'. In order to be able to cope with the multidimensional role of access, the following points should be taken into consideration: (a) the variance of the financial services provided (formal, semi-formal, informal) together with the target groups served; and (b) the degree of interest in financial services, from different parts (households, small and micro-enterprises) with different income. By analysing the above, it can be identified if customers need different products from those already provided, and if the regulatory framework causes obstacles in financing the above categories. Microfinance is built on a compelling logic: hundreds of millions of poor and very poor households seek capital to build small businesses, but their lack of collateral restricts access to loans. But this is only a contributive factor which

D. Niklis (✉) · G. Baourakis · C. Zopounidis
CIHEAM-MAICh, Chania, Greece
e-mail: dniklis@hotmail.com

L. J. Slikkerveer et al. (eds.), *Integrated Community-Managed Development*, Cooperative Management, https://doi.org/10.1007/978-3-030-05423-6_6

seeks to blame the poor: in reality, culture, in terms of ignorance of indigenous knowledge and practices and a top-down approach, is the predominant obstacle.

Innovative '*microbanks*' try to meet the demand with more flexible collateral requirements and thus seek to unleash untapped productive power. The narrative, highlighted by the Nobel Peace Prize committee in awarding the 2006 prize to Muhammad Yunus and the Grameen Bank of Bangladesh, has driven the global expansion of microfinance (cf. Counts 2008). The notion of millions of unbanked households accords with the evidence of most formal banks' shallow outreach to the poor (cf. Armendariz and Morduch 2005). But a lack of use does not imply a lack of access. Some of the 'unbanked' may be excluded, despite having worthy uses for capital; others may simply not be creditworthy or in some cases may be creditworthy but not interested in taking on debt, but most back off as they prefer to continue to use indigenous institutions embedded in their own culture. According to Yunus and Weber (2008), the key features of microcredit include the idea for which the loans are designed: "*to help the poor families to help themselves to overcome poverty*". Moreover, "*it is not based on any collateral or legally enforceable contracts. It is based on 'trust,' not on legal procedures and system*". More specifically, "*it is offered for creating self-employment for income-generating activities and housing for the poor, as opposed to consumption*". In confronting conventional banking practices, the writers define which microcredit "*was initiated as a challenge to conventional banking which rejected the poor by classifying them as 'not creditworthy'*".

6.1.1 Microfinance and Poverty

Since the late 1990s, microfinance has come to be seen by some authors as an integral part of developmental policy and an effective poverty reduction tool (cf. Johnson et al. 2009). Littlefield et al. (2003) argue that microfinance has been shown to have an impact on recipients' income, savings, expenditure, and the accumulation of assets, as well as non-financial outcomes including health, nutrition, food security, education, child labor, housing, job creation, and social cohesion (cf. Chowdhury and Bhuiya 2001; Montgomery 2005; Ghalib et al. 2012; Van Rooyen et al. 2012; Mazumder and Lu 2015). McIntosh et al. (2011) noted that access to credit is associated with moderate increases in variables associated with household welfare. Although microfinance intervention does not directly influence clients' level of education, it has been shown to have a positive impact on the education of clients' children, with children of microfinance clients more likely to go to school and stay there for longer than the children of non-clients (cf. Littlefield et al. 2003; Duvendack et al. 2011).

On the other hand, a growing number of researchers has recently shown that microfinance is not a useful tool for alleviating poverty, and in some cases even aggravates the position of the poor (cf. Adams and Von Pischke 1992; Buckley 1997; Woodcock 2007; Schreiner 1999; Bhatt and Tang 2001; Waller and

Woodworth 2001; Bhatta 2001; Sanders 2002; Karnani 2007; Rao 2012; Bateman 2010, Bateman and Sharife 2014; Slikkerveer 2007; Roodman 2011, 2012).

From the above studies, it is hard to identify if microfinance can individually reduce poverty. Most probably, there is a need for complementary supply-side and demand-side factors. Supply-side factors such as good infrastructure and entrepreneurial skills are necessary in order for micro-enterprises to become more productive. On the other hand, without demand-side factors, few things can be done. A well-established trade and industry framework together with a supportive macroeconomic environment can boost the development of micro-enterprises, creating more job positions.

6.1.2 Microcredit and Risk

Collateral is different between microcredit and regular lending institutions like banks. It is obvious that if physical collateral was a prerequisite for borrowing, the vast majority of MFI clients would not be able to take out a loan due to their poverty level. For the above reason, MFIs focus on group lending, which is based on the principal of joint liability. In essence, the group takes over the underwriting, monitoring, and enforcement of loan contracts from the lending institution (cf. Wenner 1995). In this way, each group member is responsible for the loans of the whole group. If someone cannot repay the loan, the others will have to help, otherwise they lose access to future loans. It is thus in each member's interest to ensure that the other members pay.

6.1.3 Microcredit Programmes

Microloans are, almost by definition, small, and typically relatively short-term (i.e., one year or shorter), and have high repayment rates. A broad vision of the structure of microcredit can be gleaned from the Microfinance Information Exchange (MIX) dataset, which provides comparable data for over 1127 MFIs in 102 countries, totaling $65 billion in outstanding loans and 90 million borrowers in 2009 (MIX 2010).

The theoretical literature has emphasised the aggregate and distributional impacts of financial intermediation in models of occupational choice and financial frictions (cf. Banerjee and Newman 1993; Aghion and Bolton 1997; Lloyd-Ellis and Bernhardt 2000; Erosa and Hidalgo 2008). In these studies, improved financial intermediation leads to more entry into entrepreneurship, higher productivity and investment, and a general equilibrium effect which increases wages. It is shown that the distribution of wealth and often the joint distribution of wealth and productivity are critical.

Related literature has found the impacts of improved financial intermediation in these models on aggregate productivity and income to be sizable (cf. Gine and Townsend 2004; Jeong and Townsend 2007, 2008; Amaral and Quintin 2010). Buera and Shin (2010) and Buera et al. (2011) show the importance of endogenous saving responses and general equilibrium effects through interest rates in quantitative assessment. Microfinance programs are growing around the world, and indeed in some countries, the scale of microfinance is approaching levels where general equilibrium effects should be reckoned with (cf. Buera et al. 2012; Sullivan 2007).

6.1.4 Microfinance and the Macro Environment

Recently, a number of studies (Franks 2000; Gonzalez 2007; Krauss and Walter 2008; Ahlin et al. 2011) have explicitly investigated the relationship between MFIs' performance and changes in the macro environment of the country in which they operate. These studies recognise that the macroeconomic environment is an important determinant for MFI outreach and performance in addition to institution-specific characteristics. More specifically, a country where inflation and unemployment are at low levels, whereas savings and investments are increasing, creates opportunities for the development of the microfinance sector. The above environment enhances the psychology of citizens who therefore want to invest and consume more. This is an area where MFIs play an important role.

However, an important aspect of the macro environment, which is under-investigated in the literature, is the link between the outreach and performance of MFIs and the development of the traditional bank sector (savings or investment institutions). There is a theory that the market failure of microfinance is caused due to the fact that MFIs serve a different purpose from commercial banks and focus especially on the unbanked poor. According to this theory, the inability to access financial resources is presumably caused by information asymmetry and the lack of collateral, and withholds many potential entrepreneurs from starting a business. There are however more factors involved which should be taken into consideration.

Nonetheless, this is an important question given the recent commotion on MFIs' 'mission drift', i.e. the fear that MFIs are shifting away from their original mission as the sector increasingly commercializes (cf. Armendariz and Szafarz 2009; Kono and Takahashi 2010). For example, Cull et al. (2009) show that the more commercially-oriented MFIs focus on a better-off clientele. MFIs seem in this way to act more and more as pure commercial banks. In this process, it has become increasingly unclear which MFIs are actually serving the poor, and which objectives they are pursuing (cf. Fernando 2006). It has even become clear that these commercially-oriented MFIs have recently focused on making more money on the new target group of the poor instead of serving them as previously subsidised microcredit used to do.

Furthermore, commercial banks have also become interested in serving microfinance clients. This has led to competition between banks and MFIs (McIntosh and Wydick 2005). So it is interesting to study the relationship between the two sectors and to assess how this influences the fulfillment of MFIs' mission and performance. *Are MFIs reaching more clients in countries with well or less developed traditional banking systems? Do they stand in direct competition, or do the two sectors complement each other? And how does this influence the number of poor people served?*

6.1.5 Mission Drift of MFIs

MFIs have a dual mission of financial sustainability and serving small enterprises and low income families. In recent times, mission drift in MFIs is being hotly debated. Mission drift is defined as MFIs losing their original mission, i.e. their focus on the poor, and moving up-market (cf. Copestake 2007). An increasing number of studies have indeed demonstrated an important trade-off between financial and social performance. According to Mersland and Strøm (2010), mission drift can be measured by increasing the loan sizes of MFIs. Cull et al. (2009) show that commercialised MFIs offer higher loan sizes, which points towards mission drift. Hermes et al. (2008) demonstrate a trade-off between depth of outreach and efficiency. They define cost efficiency in terms of how close the actual costs of the lending activities of an MFI are to what the costs of a best-practices MFI would have been. They conclude that if MFIs focus on maximising efficiency, mission drift might be stimulated, since MFIs serving the poorer parts of the population are less efficient.

Mersland and Strøm (2010), who study the evolution of average loan sizes offered by MFIs, argue the other way around: MFIs should increase efficiency to offer smaller loan sizes. They claim that cost aspects are crucial in the assessment of mission drift and argue that average loan sizes may be increased due to inefficient management of the organisations and not by a shift in the markets which the MFIs tend to serve.

Gutierrez-Nieto et al. (2007) emphasise that it is important to use different efficiency measures, because the results are dependent on the kind of efficiency measured. MFIs can be seen as a part of the overall financial system which is focusing on the poor unbanked segments of the population (cf. Christen et al. 2004; Richter 2004). Armendariz and Szafarz (2009) claim that, overall, the development of the financial sector is an important factor to take into account when evaluating which MFIs are actually serving the poor. They argue that MFIs offering higher loan sizes, one of the determinants to assess the level of the poorness of the clients served, does not necessarily mean that MFIs are shifting away from their mission. MFIs can simply be cross-subsidising. This is more probable with a larger unbanked population. In this case, the unbanked market is more diverse, including not only the poorest of the poor, but also the better-off middle-class people; then

MFIs may be serving a whole different range of client segments, thus fulfilling the work which commercial banks should be doing.

6.2 The Structure and Components of Financial Systems

Financial sector development is seen as important, because it fosters economic growth (cf. King and Levine, 1993). Levine (2004) and Kono and Takahashi (2010) argue that limited access to financial services is a major bottleneck for people in developing countries wanting to improve their livelihoods. According to Barr (2005), the promotion of MFIs has been viewed as a development policy able to address the market failures of the banking system, and has received increased attention as a tool for poverty reduction (cf. Barr 2005). While most MFIs started as Non-Governmental Organisations (NGOs), they have increasingly transformed into commercial institutions, which influenced the objectives MFIs are pursuing. Currently, MFIs attempt to fulfill a dual objective, i.e. both to reach the unbanked poor as well as to become self-sustainable institutions (cf. Armendariz and Morduch 2010).

In line with the above-mentioned market failure theory of microfinance, one could argue that some MFIs are substitutes for the commercial banking sector. MFIs could solve the shortcomings of the commercial banking sector by using different lending methodologies. Stiglitz and Weiss (1981) first showed that banks often fail to serve a part of the population due to information asymmetries and transformation costs. MFIs have traditionally used group lending to overcome lending constraints (Besley and Coate 1995), but at present use a whole set of different lending methodologies (cf. Armendariz and Morduch 2000). These alternative methodologies gave the opportunity to MFIs to be able to serve clients considered too risky by banks. MFIs thus tend to concentrate on a clientele which is not served by banks. In countries with well-developed financial systems, a high percentage of the population is served by commercial institutions. For example, in rich countries over 80% of households have an account with a financial institution whereas that fraction decreases to 20–40% in developing countries (cf. Demirgüç-Kunt et al. 2008). In countries where financial markets are more developed, the need for microfinance will be less acute and the demand will be smaller. Consequently, the microfinance sector is expected to be less developed in places where the commercial banking system is well established.

Another argument is related to the prediction of a negative relation between microfinance; the development of the formal banking sector relates to competition between the two sectors (cf. Hermes et al. 2009). Specifically, in well-developed banking systems, commercial banks are more efficient and profit from scale advantages and diversification. In that sense, commercial banks can be very active and flexible in adapting credit contracts and serving different groups of people. If MFIs prove that it is possible to serve a specific market, traditional banks could efficiently adapt their credit contracts towards the poorer sections of the population.

Furthermore, commercial banks are able to offer loans at lower costs, which results in lower interest rates. MFIs' interest rates are traditionally higher than the interest rates asked by commercial banks due to the high transaction costs which MFIs bear, and could lead to a crowding-out effect where the MFI clientele substitutes its MFI loans with commercial bank loans at lower interest rates.

6.2.1 Regulatory Framework of Financial Services

The *regulatory framework* of MFIs providing financial services is different from that of formal banking, due to differences in the operational market and client characteristics. The term *financial services* extends beyond the credit products and savings deposit facilities provided to varying degrees by different types of rural finance and MFIs. The term includes payments, money transfer and remittance services, insurance and contractual savings products. It is important to focus on access to payments and savings products by different segments of the population and the supply of those products by different institutions. Payment and savings products are often the most important financial services for low-income households. Improved access to savings products can help households achieve higher returns on their savings and smoother cash flows, as well as reducing vulnerability to external shocks.

The degree and quality of access to financial services available to low-income rural households and their small businesses is influenced by the quality of the legal and regulatory framework. Available data and information show that more efficient financial markets can contribute to accelerated agricultural growth and food security. Scaling up access in rural markets to a wider array of financial services through a varied range of financial intermediaries becomes critical to help low-income rural households' smooth consumption and enhance labor productivity, which is the most important production factor controlled by the poor. Also, agriculture has strong forward and backward multiplier effects for the overall economy (cf. Robinson 2001; Zeller 2003).

There are examples of agricultural development banks, MFIs and credit unions developing strong rural portfolios, while commercial banks do not generally seem to fit this market niche so readily. Some MFIs have tried to transform from non-governmental status to a regulated, supervised financial institution; however, with notable exceptions, this has not proven to be a reliable route for the improved rural outreach of financial services. The rural finance and microfinance sector is small, relative to the commercial financial sector, with limited effect on the overall stability of the financial system. In a large number of developing countries, the total loans outstanding in the rural finance and microfinance sector was about 1% of the broad money supply, with this sector reaching fewer than 1% of the population as clients. A handful of countries stand out from the rest with higher levels of microfinance outreach and penetration, especially in Indonesia (6.5%); Thailand (6.2%); Vietnam and Sri Lanka (4.5%); Bangladesh and Cambodia (3.0%); Malawi

(2.5%); and Bolivia, El Salvador, Honduras, India, and Nicaragua (at 1.0% or slightly more) (cf. Honohan 2004).

During the 1970s and 1980s, the microenterprise movement led to the emergence of NGOs which provided small loans for the poor. In the 1990s, a number of these NGOs transformed into formal financial institutions in order to access client savings and lend these onwards, thus enhancing their outreach. Specialised MFIs have proven that the poor are 'bankable'.

Formal providers are those that are subject not only to general laws but also to specific banking regulation and supervision (development banks, savings and postal banks, commercial banks, and non-bank financial intermediaries). Formal providers may also be any registered legal organisations offering any kind of financial services. *Semi-formal providers* are registered entities subject to general and commercial laws but are not usually under bank regulation and supervision (financial NGOs, credit unions and co-operatives). *Informal providers* are non-registered groups such as rotating savings and credit associations (ROSCAs) and self-help groups. MFIs can be government-owned, like the rural credit co-operatives in China; member-owned, like the credit unions in West Africa; socially-minded shareholders, like many transformed NGOs in Latin America; and profit-maximising shareholders, like the microfinance banks in Eastern Europe.

According to the legal structure of the provider, there is a limitation regarding the types of services which can be offered. For example, non-regulated institutions are not allowed to provide savings or insurance. New players in microfinance such as commercial banks manage to bring financial services closer to poor clients through extended branch networks and automatic teller machines. In many cases there is cooperation between different types of providers in order to extend their clientele. One lasting problem is the existence of private commercial moneylenders posing as social non-profit institutions, especially in urban areas.

According to the Consultative Group to Assist the Poor (CGAP 2004), all providers, over time, have tended to increase their product offerings and improved their methodologies and services, as poor people proved their ability to repay loans and their desire to save. In many institutions, there are multiple loan products providing working capital for small businesses, larger loans for durable goods, loans for children's education and loans to cover emergencies. Safe, secure deposit services have been particularly well received by poor clients, but in some countries, NGO MFIs are not permitted to collect deposits. Remittances and money transfers are used by many poor people as a safe way to send money home. Banking through mobile phones (mobile banking) makes financial services even more convenient and safer, and enables greater outreach to more people living in isolated areas. Innovation in financial services for poor people is still regarded as a powerful instrument for reducing poverty, enabling them to build assets, increase income, and reduce their vulnerability to economic stress. Innovations are demonstrated throughout the world in which the credit facility is built up. Nowadays an MFI can function as an intermediary 'buffer' between the producer and the buyer, with examples taken from the agricultural sector (cf. Miller 2011).

In terms of operations, Ismawan and Budiantoro (2005) divide MFIs into four categories: savings-led microfinance, credit-led microfinance, micro-banking and the linkage model. Savings-led microfinance is a mechanism which is based on the accumulation of money by the people in the community, using it for various kinds of purposes, from consumption to productive economic activities. NGOs, such as credit unions, self-help groups and co-operative organisations, are some of the examples of this type of microfinance. Credit-led microfinance is a mechanism in which the source could be external from contributions such as government funds or private investors. Banks such as the *Badan Kredit Desa* (Village Credit Unit, BKD), *Lembaga Dana dan Kredit Pedesaan* (Fund and Credit Village Institution, LDKP) and *Baitul Maal wat-Tamwiil* (Islamic Co-operative Microfinance Institution, BMT) are among these. Micro-banking is an approach run by a commercial bank which gives services to microenterprises or economically active poor people such as the well-known *Bank Rakyat Indonesia* (cf. Bank of the people, BRI). Another example is Danamon Simpan Pinjam (under the Danamon Bank), and *Bank Perkreditan Rakyat* (Rural Banks or the People's Credit Banks).

Ascarya (2011) who studied MFIs around the world, established 17 categories: (1) Association model; (2) Bank Guarantee model; (3) Community Banking model; (4) Cooperative model; (5) Credit Union model; (6) Grameen Bank model; (7) Group model; (8) Individual model; (9) Intermediary model; (10) Non-Government Organisation (NGO) model; (11) Peer Pressure model; (12) Rotating Savings and Credit Association (ROSCA) model; (13) Small Business model; (14) Village Banking model; (15) Self-Help Group model; (16) Graduation model; and (17) Micro-banking Unit model. He concluded that each type of MFI will require a different approach in the analysis and implementation of its programme, and will follow a different path to become a sustainable MFI.

6.2.2 Islamic Finance

There are many common goals between microfinance and Islamic banking. According to Dhumale and Sapcanin (1999): "*both systems advocate entrepreneurship and risk sharing and believe the poor should take part in such activities*". Moreover, terms such as: social justice, equitable distribution of income and wealth, and fair economic development, are very crucial for Islamic banking (cf. Rosly and Bakar 2003; Wajdi Dusuki 2008; Smolo and Ismail 2011). A common example is the case of loans without collateral, where there is the same vision. Such a relationship could be beneficial for both poor entrepreneurs and investors. The entrepreneurs would have access to credit markets and investors would be able to diversify their investments (Rahman and Rahim 2007). According to Karim et al. (2008), the main Islamic microfinance types of contracts are the following:

- *Murabaha* Sale (cost plus markup sale contract). The most widely offered Sharia-compliant contract is *murabaha*, an asset-based sale transaction used to finance goods needed as working capital;
- *Ijarah* (leasing contract). *Ijarah* is a leasing contract typically used for financing equipment, such as small machinery;
- *Musharaka* and *Mudaraba* (profit and loss sharing). The profit and loss sharing (PLS) schemes are the Islamic financial contracts that are most encouraged by Sharia scholars;
- *Takaful* (mutual insurance). The equivalent of Islamic insurance, *takaful* is a mutual insurance scheme.

Islamic microfinance has the potential to become very popular throughout the Muslim world. The most important is to convince people that it is able to meet their needs on a large scale.

6.3 Research Methods and Techniques for Financial Sector Analysis/Impact Assessment

Financial Sector Analysis is a crucial process in microfinance where updated data should be gathered in order to be assessed in a proper way. In order for this to be feasible, it entails an understanding of basic financial sector attributes like statements; basic accounting; ratio analysis; performance measurement; budgeting systems; cost-benefit analysis; forecasting; risk management etc. Financial statements are important for two reasons: (a) to uncover problems, and (b) to identify appropriate action. The most important financial statements are the balance sheet, the profit and loss (income) statement, and the cash-flow statement.

A balance sheet is a list of the accumulated assets and liabilities incurred by the business. The difference between the two represents the net worth of the business. The profit and loss statement basically provides information about the ability to generate profit by increasing revenue, reducing costs, or both. The cash-flow statement answers the question "*Where was the cash used?*" Understanding these statements is crucial, since they all tell what has happened in the past. But from a management perspective, it is essential to know what is going to happen in the future. Therefore, forecast plans for profit and loss together with cash-flows should be developed.

The analysis of financial statements (known as quantitative analysis) is one of the most important elements in the financial analysis process. For this reason, representative indicators should be well selected and explained, in order to familiarise all the involved parties (investors, clients, managers, debtors, employees etc.). Among the dozens of financial ratios available, 30 indicators are chosen, the most relevant to the investing process, organised into six main categories, as shown in Table 6.1.

Table 6.1 Main categories of ratio indicators

(1) Liquidity measurement ratios – Current ratio – Quick ratio – Cash ratio – Cash conversion cycle	(4) Operating performance ratios – Fixed asset turnover – Sales/revenue per employee – Operating cycle
(2) Profitability indicator ratios – Profit margin analysis – Effective tax rate – Return on assets – Return on equity – Return on capital employed	(5) Cash flow indicator ratios – Operating cash flow/sales ratio – Free cash flow/operating cash ratio – Cash flow coverage ratio – Dividend payout ratio
3) Debt ratios – Overview of debt – Debt ratio – Debt-equity ratio – Capitalisation ratio – Interest coverage ratio – Cash flow to debt ratio	6) Investment valuation ratios – Per share data – Price/book value ratio – Price/cash flow ratio – Price/earnings ratio – Price/earnings to growth ratio – Price/sales ratio – Dividend yield – Enterprise value multiple

Source Authors' own work

The above-mentioned indicators cover all aspects which should be taken into consideration. The aim here is to point out that when someone wants to take an investment decision, all these different aspects should be analysed. A thorough analysis increases the degree of a successful decision.

Another important topic has to do with the selection of an appropriate budgeting system. The system should be easy to work, otherwise it will cause a lot of bureaucracy due to the dozens of forms needed to be filled. Knowledge of the yearly budget is crucial because it simply depicts how much money someone can spend and it assists in the prioritisation of needs. The better organised a budget, the more needs it covers with the same amount of money. Expenses should be first of all covered and the rest of the money should be placed in the most profitable investments.

A cost-benefit analysis should also be done in order to determine how well, or how poorly, a planned action will turn out. Since the cost-benefit analysis relies on the addition of positive factors and the subtraction of negative ones to determine a net result, it is also known as running the numbers.

Forecasting is the process of making statements about events whose actual outcomes (typically) have not yet been observed. A commonplace example might be the estimation of the expected value of interest at some specified future date. Prediction is similar, but in more general terms. Both might refer to formal statistical methods employing time series, cross-sectional or longitudinal data, or alternatively to less formal judgement methods.

Risk management can be considered as the identification, assessment, and prioritisation of risks followed by the co-ordinated and economical application of resources to minimise, monitor, and control the probability and/or impact of unfortunate events, or to maximise the realisation of opportunities. Stress testing can be really helpful here, where many scenarios under different assumptions are examined in order to be able to face any difficulties which may occur.

The next angle of approach is to look at the actual results of the operations labelled as 'Impact Assessment'. Most of the current impact assessments are focusing on outreach and MFI sustainability, which refers to the availability of microfinance. They tend to ignore to what extent microfinance gives a positive impact on sustainable economic and social development. This focus is currently also being questioned. Considering outreach, it can be observed that most microfinance operations tend to exclude the poorest of the poor, while in terms of MFI sustainability, Bateman (2010) concludes that of the 100,000 MFIs currently estimated to be operating in the world, only 3–5% will become financially self-sustainable. This estimate is similar to the findings of Seibel and Agung (2005) in their study of Islamic Microfinance Institutions in Indonesia. Their study concludes that while Islamic banks are able to raise their assets, the performance of microfinance institutions tends to deteriorate. Islamic rural banks tend to be stagnant while *Baitul Maal wat-Tamwiil* (BMT) ('Islamic Co-operative Microfinance Institutions') decrease significantly. Out of 3000 institutions operating in the late 90s, only 20% were sustained in 2005. This criticism brings us to the major problems which are neglected in current research on microfinance impact assessment, including *Randomised Control Trials* (RCT), a prominent method in impact assessment.

The major problems, as underlined by Bateman (2010), are displacement or spill-over effects, and client exit/failure. The displacement effect refers to the loss of jobs and income in non-client microenterprises. While one client is financed by MFIs, which enables him to enter the market, an existing microenterprise which is not dependent on a MFI could be affected by this new competitor. The problem of market saturation can lead to a lower price in the market, and lower margins for small-scale or micro-enterprises; when faced with higher costs, they cannot be sustained in the long term. An example of this displacement effect is the Grameen Bank anti-poverty Grameen Phone programme. This programme stimulated Bangladeshi women to run a phone business, selling mobile phone air time. These women were given $420 loans to run the business. The business was very successful with a dramatic increase in women sellers from 50,000 in 1997 to 280,000 in 2005 since the business was supported by a micro loan from the Grameen Bank. However, the average income of these women dramatically decreased from $750–$1200 in 1997 to around $70 in 2005 because of this phenomenon.

Regarding client exit/failure, this problem refers to their exit from the branch after they cannot maintain a business in a saturated market. The aforementioned could create a more severe situation for the client compared to the condition they were in before they received credit. They will not have an economically viable income anymore, but they will still have the repayment obligation, which has to be settled alternatively, either by relatives, friends or even a second loan, on other collateral. In short, these two major problems in the impact assessment of microfinance, together with the outreach and sustainability limitations as already explained above, need to be evaluated.

The role of microfinance therefore should not only be considered as being limited to financial services to the people; it should be considered as including other aspects of these people's lives. A recursion to the wide range of aspects of human life which are framed in the context of poverty would consider that microfinance's role should be assessed in all these dimensions to ensure that it effectively supports sustainable economic and social development.

The latest research by Duvendack et al. (2011) on the evaluation of various impact assessments generally concludes that there is currently no convincing impact of microfinance on the well-being of poor people, including the position of women, which is found to be methodologically weak and based on inadequate or non-representative data. Their findings again are in line with the criticism of Bateman (2010) on impact assessment. As cited in Bateman, the 2009 World Bank publication *The Moving Out of Poverty* reveals the result of a study across fifteen countries in Africa, East Asia, South Asia, and Latin America involving more than sixty thousand interviews with poor people. The study concluded that "*Microcredit can help the poor subsist from day to day, but in order to lift them out of poverty, larger loans are needed, so that the poor can expand their productive activities and thereby increase their assets*". By relating this conclusion to the work of Rahmat et al. (2006), the question might be extended in a way such as that implied in the following: "What amount of money could lift the poor out of poverty?" According to their study, if the loan was about double or more in size, a situation of decreasing performance would also occur.

Finally, the impact of 'external influences' should be mentioned (i.e. the world economy or financial crisis, or climate). As is commonly believed, the sphere of operations of many MFI institutions is in the twilight zone between formal and informal economies. Especially with regard to rural or suburban microenterprise, it is suggested that the impact of a crisis is less when the distance to the formal economy is larger, but as the business is scaled up and starts to be tied in with the formal economy, the risk of default increases (cf. Gonzalez 2011). Under these circumstances, the MFI industry has to anticipate a situation where its portfolio of clients is susceptible to great fluctuations. Market saturation gets a new definition here, as it does not refer to the number of suppliers but to the limits of growth under the overall circumstances. The indication of sector analysis means that the wider economic context of the sector's operations comes into scope, depending on the trends in that country or region.

6.4 Implementation of Sector Analysis Research: Cases from the Field

Microfinance is mostly known as a tool that lifts poor people (especially women) out of poverty by funding their microenterprises and increasing their income. There are many stories that recite the success of micro-entrepreneurs who started with a small loan and managed to become successful. Unfortunately, this is only the exception. According to many scientific studies, there is no straightforward connection between getting a microloan and becoming wealthier. Although some studies which address this challenge found that microcredit produced certain economic and social benefits, recently a growing number of studies on microfinance found the opposite, indicating that microfinance actually becomes a poverty trap for many low-income families (cf. Bateman 2010; Roodman 2011).

In the last years, researchers have started using *randomised controlled trials* (RCT) to test microfinance's impact. Similar to biomedical studies which test new pharmaceuticals, randomised controlled trials in the field of microfinance isolate the effect of a chosen innovation by assigning a random selection of individuals or villages to the innovation (the treatment group), and another equivalent selection of individuals or villages to maintain the status quo (the control group). They compare the results between the groups, following the control group paradigm. One selects a large enough group of studies so that when it is randomly divided, the two subgroups can be presumed to be statistically identical. The first subgroup gets loans, and the second subgroup does not. If one subgroup experiences better outcomes than the other, the researcher can be reasonably sure that it is due to the loans, because the loans are the only ex-ante difference between the groups. Many RCT studies of microfinance have already been completed (cf. Gine et al. 2006; Karlan and Nadel 2006; Karlan et al. 2009; Karlan and Zinman 2010, 2011; Duvendack et al. 2011; Hermes and Lensink 2011; Karlan and Valdivia 2011; Lund et al. 2011; Augsburg et al. 2012; Pande and Park 2012; Van Rooyen et al. 2012). The researchers that looked at standard microcredit programmes found no evidence of short-term welfare improvements, although they did find some other possible benefits.

Karlan and Zinman (2010) studied the effect of microcredit on small business investment in Manila, Philippines. They claim that profits from business increase, especially for male and higher-income entrepreneurs. Moreover, they find rather striking results showing that businesses substitute away from labour into education, and formal insurance into informal insurance.

Banerjee et al. (2009) evaluated the impact of the opening of MFI branches in the slums of Hyderabad, India. Half the 104 slums were randomly selected for opening a new branch. They found mixed results, but on the whole, the effect of introducing microfinance appears to be very moderate.

McIntosh et al. (2011) develop the so-called Retrospective Analysis of Fundamental Events Contiguous to Treatment. According to the authors, this methodology allows measuring welfare changes—due to a treatment such as, for example, access to microfinance—based upon a single cross-sectional survey in which questions are included on fundamental events in the history of the respondents. These fundamental events are defined as events in a household's history which are discrete, unforgettable, and important to the household's welfare. By using questions which relate to such events, researchers can create a retrospective panel dataset in order to measure the impact of a certain treatment. In particular, analysing the timing of these events within a window around the timing of a treatment allows for statistical tests based on changes in household welfare variables occurring after the treatment. They apply their methodology to a survey among 218 Guatemalan households which have obtained access to microfinance in different years and examine the effects of access to credit on dwelling improvements. The results of their analysis show that access to microfinance increases the probability of dwelling improvements, although the effects are relatively modest. Collins et al. (2009) present the results of year-long financial diaries collected about twice a month from hundreds of rural and urban households in India, Bangladesh, and South Africa. The diaries reveal that financial instruments are critical survival tools for poor households indeed, and that these tools are even more important for the poor than for richer people. Another interesting point is that poor people prefer using informal instruments (informal savings, loans from family and friends) rather than microfinance from formal providers due to the flexibility. On the other hand, there is a drawback that has to do with reliability (e.g. the one to whom money is deposited does not have cash to give it when needed). As they conclude, "*whether or not the microfinance movement was right to stress loans for microenterprises, or has been too slow to embrace savings and other services, its contribution is beyond dispute. It represents a huge step in the process of bringing reliability to the financial lives of poor households*".

A very characteristic example is BancoSol in Bolivia, which started in the mid-1990s with a few million dollars of donor subsidies, and turned into a loan portfolio of over $200 million with services for over 300,000 active clients at the end of 2008. Unfortunately, this is only a part of the story. As researchers such as Schreiner (2004), Brett (2006) and Bateman (2010) document, in contrast to the original promise by Bolivia's microfinance industry, the commercial expansion of microfinance in the late 1990s has in fact driven the country's poor further into poverty. Not only has almost no progress been found in 2006 in reducing income-based poverty in Bolivia for more than two decades, but also the long-term result of commercialised microfinance is found to weaken light industry and agricultural production which had been carefully built up in Bolivia since the 1950s, undermining the country's long fight against poverty.

Innovative schemes, such as the BRAC Ultra-Poor Programme, tried to open up pathways to economic activity and access to financial services for the extreme poor. Recent research shows that while 'start-up' capital is virtually unknown in Bangladesh, it is now extremely difficult for SMEs to access capital for their economic activities. DfID (2008: 2–3) notes in this context: "*Rural areas with the highest potential for lifting low income groups out of poverty are cut off from most financing mechanisms*".

Another example, CGAP, has launched pilots in India, Pakistan and Haiti, which are modelled on the BRAC programme. The programme targets destitute clients through a carefully sequenced combination of livelihood grants and microfinance, with savings playing a critical role, in order for clients to come out of poverty. The role of CGAP as a major World Bank institution promoting the 'new wave' of microfinance as a 'best practice' is highly controversial, particularly in view of CGAP's implicit support for the ultra-high interest-rate policies of its close partner Compartamos IPO Bank of Mexico. It is now generally accepted that despite CGAP's contention that commercialisation lowers interest rates, it is actually, on the contrary, increasing interest rates. Summing up, success in reaching poorer people with microfinance is determined by MFIs' mission, and their ability to translate that mission into effective products and services.

The aim is to see more clients overall, and very poor people in particular, served with appropriate, varied products from a variety of institutions. Unfortunately, the efforts done till now are not sufficient. Most of them start focusing on the poor but eventually transform into typical credit institutions.

6.5 Regulatory Framework and the Government's Role

Microfinance clients are often described according to their poverty level (from very poor to vulnerable non-poor). This description is problematic in the sense that clients have different needs and providers should supply them with different products. In their vast majority, microfinance clients are women (around two thirds). There is one particular category of microfinance clients: if they have an entrepreneurial spirit, it is very likely that they will become successful. On the other hand, there are people who start running a business because they cannot find another job and the future of these firms is uncertain.

Microfinance is a financial impulse for small enterprises to empower them economically, and is expected to result in an increased scale of their business. The discussion on the role of microfinance in poverty alleviation is somehow marginalised since the commercialisation of the sector by banking institutions. Poverty reduction tends to be handled by the government as a scheme of intervention. After the announcement of *Millennium Development Goals* (MDGs) in the early 2000s, the role of microfinance in poverty alleviation has become more prominent in many developing countries (Bhatt and Tang 2001; Kamani 2007). According to Robinson (2002), microfinance can help poor people in three steps:

In the first step, it supports poor people to have access to basic needs such as food, clothing, health and shelter.

In the second step, microfinance not only enables poor people to consume, but also to make them economically active, providing cash which can be used as investment capital.

Table 6.2 Microfinance regulation and deposit collection per country

Regulation status	Countries
Countries with regulated and non-regulated MFIs	Armenia, Bosnia and Herzegovina, Bolivia, Cambodia, Colombia, Haiti, India, Jordan, Kenya, Mexico, Mozambique, Nicaragua, Nigeria, Peru, Philippines, Togo, Uganda
Countries where regulated MFIs collect deposits	Bangladesh, Bolivia, Cameroon, Colombia, Dominican Republic, Ecuador, Ethiopia, Indonesia, Madagascar, Mexico, Mongolia, Nepal, Palestine, Paraguay, Peru, Senegal, Tajikistan
Countries where non-regulated MFIs can collect deposits	Armenia, Cambodia, Honduras, India, Kenya, Mali, Mozambique, Nicaragua, Philippines, Rwanda, Sri Lanka, Togo, Turkey, Uganda
Countries where regulated MFIs do not necessarily collect deposits	Albania, Bosnia and Herzegovina, Benin, Bolivia, Dominican Republic, Egypt, Guatemala, Haiti, India, Jordan, Kazakhstan, Kosovo, Madagascar, Mexico, Mongolia, Morocco, Mozambique, Nicaragua, Pakistan, Palestine, Peru, Philippines, Serbia, Slovakia

Source Hartarska and Nadolnyak (2007)

In the third step, microfinance enables economically active poor people to enlarge their scale of activities. In this step, microfinance is meant to create activities with long-term sustainability. Table 6.2 depicts microfinance regulation and deposit collection among different countries.

Conclusively, access to financial services creates the possibility of improving the economic conditions of the poor. However, what is important to remember is that credit, or debt, is a big responsibility. Incidences of over-indebtedness do occur and clients may end up less well-off, reminding us that microfinance, and credit in particular, must be used judiciously.

6.5.1 What Is the Government's Role in Microfinance?

Governments can play a range of roles—they can promote financial access, protect customers, or even provide financial services directly. Whether this happens by assisting to develop the financial infrastructure (through payments or credit information systems) or by maintaining the country's macroeconomic stability, governments have been instrumental in promoting greater access to financial services to low income people, and improving the quality of those services (cf. Yaron et al. 1997; Marmot et al. 2008; Beck et al. 2009; Deininger and Mpuga 2005).

There is little analysis of the results of national microfinance policies and strategies. The massive numbers of clients—for instance, in Bangladesh, Indonesia, and Bolivia—where there is an absence of such strategies, suggests that microfinance growth does not always keep up with microfinance policies and strategies. There is a positive role for governments to play in adopting appropriate 'light-touch' consumer protection policies and market conduct regulation, such as disclosure requirements, protections against over-indebtedness, and simple accessible recourse mechanisms, coupled with clients' financial education. The challenge here is that the system should secure that clients have full information about the products they choose and their rights and obligations are clear in advance. When microfinance providers offer voluntary deposit services, then there is a role for governments to play in prudential regulation and supervision, in order to protect depositors as well as the stability of the financial system. Even though governments generally have a bad track record with direct provision of microfinance—especially lending—there are still countries where this procedure occurs. There are just a few cases of successful publicly provided micro-lending, which are the exception rather than the rule. Examples of successful direct provision of savings services by government providers are more numerous. Yet funds collected from low income savers are often used for governmental purposes which offer little if any direct benefit to the communities where they were collected.

6.5.2 Assessment of the Regulatory Framework

Assessment of the regulatory framework for the rural finance and microfinance sector covers both the institutional aspects and the benchmarks used to evaluate the sector's performance and soundness. The considerations include: (a) assessing the need for prudential supervision versus non-prudential regulation, and for the technical capacity for supervision, as well as the costs of that supervision; (b) determining which agency should carry out the supervision or regulation, and whether delegated or auxiliary supervision may be warranted or justified; and (c) establishing benchmarks and standards for evaluating outreach and for financial performance and soundness. However, an analytical evaluation of their outreach, operating performance, and financial soundness—as well as the primary problems they face or may pose to the rest of the sector—may be an important aspect of the assessment of adequacy of access.

Some key questions in assessing the regulatory framework of rural finance and MFIs include the following: *Is there a need to regulate (but not prudentially supervise) those other institutions? If so, what is the scope of the regulation?* Very often the distinctions between broad regulatory oversight (sometimes called non-prudential regulation) and detailed prudential supervision are ignored in a number of countries. Inappropriate regulatory approaches have led to the misallocation of scarce supervisory and staff resources in an attempt to impose prudential standards and requirements on rural finance and MFIs which are not engaged in mobilising and intermediating public deposits, a step which poses a systemic risk. Prudential supervision involves the regulatory authorities verifying the compliance of institutions with mandatory standards—such as minimum capital levels and adequacy, liquidity management ratios, and asset quality standards—as measures for financial soundness. A critical topic is the extent to which regulatory authority should be centralised, delegated, or decentralised.

6.5.3 The Financial System

The role of the financial system in every country is quite crucial (Mishkin 2007; Ayyagari et al. 2008). The welfare of a country is based on money circulation. Without money, the economy is like a car without gas. The case of the barter economy is rarely found nowadays. A country without a solid financial system is hard to develop and be competitive. The vast majority of financial sector assets belong to banks. The central bank of each country is responsible for the regulation and supervision of the entire system (cf. Allen and Gale 2000; Chinn and Ito 2002).

Long-term macroeconomic policies, a well-expanded public infrastructure, effective market control and mechanisms for ensuring minimum standards of systemic protection increase the quality of supervision and attract more investments from abroad (cf. Dias 2013).

MFIs, as previously mentioned, seek to provide access to financial services to those who are excluded from the conventional financial sector. On the other hand, the banking system forces MFIs to act in accordance with the typical obligations of commercial financial institutions. Table 6.3 shows how typical banks react towards different characteristics compared with the way that MFIs think and react.

Risk management is a factor which is quite hard to confront. A vast number of individual institutions are not fully prepared to face up to their risk exposures, which is more obvious during severe economic conditions. Broad components of the asset classification and provisioning framework applied to delinquent loans are

Table 6.3 Comparison of banking and microfinance logic

Characteristic	Banking logic	Microfinance logic
Goals	Deriving a rent or profit	Increasing the access of the disenfranchised to financial services while fulfilling fiduciary obligations towards depositors and investors
Target population	Clients are customers and seen as more or less risky sources of income	Clients are customers and seen as micro-entrepreneurs
Management principles	Maximising profit while fulfilling fiduciary obligations not only to investors but also to depositors	Striking a balance between maximising access of the disenfranchised to financial services and fulfilling fiduciary obligations to depositors and investors

Source Battilana and Dorado (2010)

in line with international standards. On the other hand, some shortcomings cause the understatement of non-performing loans and the overstatement of income. The above-mentioned should be reviewed in order to be able to reflect institutional, regional, and global conditions.

6.5.4 Rural Finance and MFIs

Tax services may present obstacles to rural finance and MFIs, preventing them from more effectively providing access to financial services. The legal and non-profit status of non-bank NGO MFIs may be questioned by tax authorities on the grounds that the credit services they are providing to their clientele are priced at commercial rates, rather than at 'charitable' levels. In other instances, licensed specialised banks and non-bank finance institutions may not be permitted by tax accounting laws and regulations to provide for possible loan losses, in spite of prudential regulations issued by the central supervisory authority, which creates an unnecessary but real economic burden to such specialised banks and non-bank finance institutions.

A related problem stems from the requirement by tax authorities that delinquent loans may be written off only when the sale and disposition of collateral securing such a defaulted loan results in recovering a monetary value which is less than the value of the collateral. Credit registries allow borrowers to build up a credit history which can assist lenders in assessing risk, thereby reducing the cost of lending and improving access. Credit registries which give easy and reliable access to a client's credit history can dramatically reduce the time and costs of obtaining such information from individual sources, and can therefore reduce the total costs of financial intermediation. Under this process, borrower credibility is easier to be identified, which locates who has more chances to default. It helps borrowers to build up a credit history and achieve lower lending costs. As is easily observed, this is important for SMEs, because their creditworthiness is more difficult to evaluate and because they have less visibility and transparency relative to large enterprises.

Often, current regulations may provide for the sharing of only negative information (i.e., information on non-performing loans). But this information does not always tell the truth. A company has periods of growth and recession, where some loans may be non-performing. In order to achieve a reliable credit risk evaluation all positive and negative parameters of a company should be taken into account. Reporting positive information significantly increases the predictability of rating and scoring models used by lenders, thereby translating into lower loss rates, higher acceptance rates of credit applicants, or both (cf. Staschen 2003). Sharing positive information will also allow borrowers to build their credit history, which can especially benefit small borrowers, because it will allow them to establish a reliable borrowing reputation and to improve their chances to increase their borrowing as their business grows. Regulations governing information sharing should also allow for adequate consumer and data protection mechanisms. Allowing all finance providers to share both positive and negative information on their borrowers will allow small businesses to participate in the process of reputation building and to generate credit history. All the above will assist the transfer from microfinance to bank finance for successful borrowers. The common share of information among all finance providers could lead to higher competition and less segmentation.

6.5.5 Consensus Guidelines on Regulating and Supervising Microfinance

Where the government does not need to supervise and certify the financial soundness of regulated institutions, it is preferable to deal with commercial and criminal laws rather than apply prudential regulation, with the advantages of lower costs and a higher degree of effectiveness, as indicated below:

- A high level of minimum capital needs could enable the supervisory authority to avert new institutions from entering the system, which would deteriorate the quality of supervision.
- Regulatory reform should preclude financial institutions (banks, finance companies etc.) from being able to offer microfinance services, or make it quite difficult to lend to microfinance institutions.
- Any design of microfinance regulation should take into consideration the realistic estimation of supervision costs and the identification of a sustainable mechanism for this payment. People who encourage governments to supervise new types of institutions should be ready to help finance the initial costs of such supervision.
- 'Self-supervision' of an entity in developing countries will deteriorate the regular flow of the system.
- No license should be given to any microlending institution, unless it is able to cope with its lending profitability and the administrative costs of mobilising the deposits it proposes to capture.
- The large financial cooperatives should be supervised by a specialised financial authority different from the one responsible for all cooperatives (both financial and non-financial).

6.6 Concluding Remarks

The success of microfinance in developing countries is attributed partly to the close relationships among borrowers, which create encouragement and social pressure in order to repay the loan (cf. World Bank 2015). Trust and confidence among group members, as well as views of the lender, serve as an important foundation for successful microcredit lending; the design of information-sharing mechanisms may be guided with this insight in mind (cf. UNDP 2015). Many studies indicate that access to financial services for households with limited income is an important factor in reducing poverty and inequality (cf. Karlan and Morduch 2010; World Bank 2008; Imai and Azam 2012; Mullainathan and Shafir 2009). The recent report of the World Bank shows that by extending beyond the conventional reaches of the markets, MFIs enable the poor to smoothen income shocks (cf. Armendariz and Morduch 2010).

Risk management is a crucial process that is closely related to microfinance. As long as the vast majority of MFI clients are low income clients, there should be some interlocks. Collaboration between MFIs and exchange of client information could be really helpful towards this direction. There are clients who take advantage of the intense competition between MFIs, and get multiple financing without any terms. People forget that taking out a loan is quite easy; the hard part is to repay this loan. On the other hand, regulators must also be interested in the client, and not just giving broad overviews to market operators. Thus, the ability to identify recurrent defaulting creditors in various regions or areas will flag off on their monitoring systems, thereby preventing systemic risks in the industry.

Another problem closely related to risk is the tendency of people to borrow money for consumption rather than income generation, which leads to mounting indebtedness in many countries and even suicides. Unfortunately, there are many cases where loans taken out by one are serviced by another, who in turn has collected a cache of money by persuading several people to become ghost loaners for them (cf. Dogra 2015).

To sum up, microfinance is a really useful tool that has managed to alleviate poverty in many countries. People have been taking out loans in order to invest and improve their lives. This aim seems to be changing in recent years. Many MFI clients spend the money on goods consumption, which makes them more and more mired in debt. This is a problem that all involved parties (industry practitioners, regulators, policy makers and all other concerned stakeholders within the Microfinance Industry) should examine carefully and try to find solutions to it as soon as possible.

References

Adams, D., & Von Pischke, J. D. (1992). Microenterprise credit programs: 'deja vu'. *World Development, 20,* 1463–1470.

Aghion, P., & Bolton, P. (1997). A theory of trickle-down growth and development. *Review of Economic Studies, 64,* 151–172.

Ahlin, C., Lin, J., & Maio, M. (2011). Where does microfinance flourish? Microfinance institutions' performance in macroeconomic context. *Journal of Development Economics, 95,* 105–120.

Allen, F., & Gale, D. (2000). *Comparing financial systems*. MIT Press.

Amaral, P., & Quintin, E. (2010). Financial intermediation and economic development: A quantitative assessment. *International Economic Review, 51,* 785–811.

Armendariz, B., & Morduch, J. (2000). Beyond group lending. *Economics of Transition, 8,* 401–420.
Armendariz, B., & Morduch, J. (2005). *The economics of microfinance*. Cambridge, MA: MIT Press.
Armendariz, B., & Morduch, J. (2010). *The economics of microfinance* (2nd ed.). Boston: MIT Press.
Armendariz, B., & Szafarz, A. (2009). *On mission drift of microfinance institutions*. CEB Working Papers Series No. 09/015. Brussels: Universite Libre de Bruxelles.
Ascarya, P. (2011). The persistence of low profit and loss sharing financing in Islamic banking: The case of Indonesia. *Review of Indonesian Economic and Business Studies, 1*. LIPI Economic Research Center.
Augsburg, B., De Haas, R., Harmgart, H., & Meghir, C. (2012). *Microfinance at the margin: Experimental evidence from Bosnia and Herzegovina*. Available at http://ssrn.com/abstract=2021005 or http://dx.doi.org/10.2139/ssrn.2021005.
Ayyagari, M., Demirgüç-Kunt, A., & Maksimovic, V. (2008). How important are financing constraints? The role of finance in the business environment. *The World Bank Economic Review, 22*(3), 483–516.
Banerjee, A. V., Duflo, E., Glennerster, R., & Kinnan, C. (2009). *The miracle of microfinance? Evidence from a randomized evaluation*. Hyderabad, India: MIT Jameel Poverty Action Lab, Indian Centre for Micro Finance, Spandana.
Banerjee, A. V., & Newman, A. F. (1993). Occupational choice and the process of development. *Journal of Political Economy, 101,* 274–298.
Barr, M. (2005). Microfinance and financial development. *Michigan Journal of International Law, 26,* 271–296.
Bateman, M. (2010). *Why doesn't microfinance work? The destructive rise of local neoliberalism*. London: Zed Books.
Bateman, M., & Sharife, K. (2014). The destructive role of microcredit in post-apartheid South Africa. In M. Bateman & K. Maclean (Eds.), *Seduced and betrayed: Exposing the contemporary microfinance phenomenon*. Santa Fe, NM: SAR Press.
Battilana, J., & Dorado, S. (2010). Building sustainable hybrid organisations: The case of commercial microfinance organisations. *Academy of Management Journal, 53*(6), 1419–1440.
Beck, T., Demirgüç-Kunt, A., & Honohan, P. (2009). Access to financial services: Measurement, impact, and policies. *The World Bank Research Observer, 24*(1), 119–145.
Besley, T., & Coate, S. (1995). Group lending, repayment incentives and social collateral. *Journal of Development Economics, 46,* 1–18.
Bhatt, N., & Tang, S. Y. (2001). Delivering microfinance in developing countries: Controversies and policy perspectives. *Policy Studies Journal, 29*(2), 319–333.
Bhatta, G. (2001). Small is indeed beautiful but…: The context of microcredit strategies in Nepal. *Policy Studies Journal, 29*(2), 283–295.
Brett, J. (2006, March). We sacrifice and eat less: The structural complexities of microfinance participation. *Human Organization, 65*(1), 8–19.
Buckley, G. (1997). Microfinance in Africa: Is it either the problem or the solution? *World Development, 25*(7).
Buera, F. J., Kaboski, J. P., & Shin, Y. (2011). Finance and development: A tale of two sectors. *American Economic Review, 101,* 1964–2002.
Buera, F. J., Kaboski, J. P., & Shin, Y. (2012). *The macroeconomics of microfinance* (No. w17905). National Bureau of Economic Research.
Buera, F. J., & Shin, Y. (2010). *Financial frictions and the persistence of history: A quantitative exploration*. Working Paper 16400. National Bureau of Economic Research.
CGAP. (2004). *Key principles of microfinance*. Retrieved March 6, 2015, from CGAP http://www.cgap.org/publications/key-principles-microfinance.
Chinn, M. D., & Ito, H. (2002). *Capital account liberalization, institutions and financial development: Cross country evidence* (No. w8967). National Bureau of Economic Research.

Chowdhury, M., & Bhuiya, J. (2001). Do poverty alleviation programs reduce inequity in health? Lesson from Bangladesh. In D. Leon & G. Walt (Eds.), *Poverty inequity and health*. Oxford: Oxford University Press.

Christen, R., Rosenberg, R., & Jayadeva, V. (2004). *Financial institutions with a double-bottom line: Implications for the future of microfinance* (pp. 2–3). CGAP Occasional Paper. July, 2004.

Collins, D., Morduch, J., Rutherford, S., & Ruthven, O. (2009). *Portfolios of the poor: How the world's poor live on $2 a day*. Princeton, NJ: Princeton University Press.

Copestake, J. (2007). Mainstreaming microfinance: Social performance management or mission drift? *World Development, 35,* 1721–1728.

Counts, A. (2008). *Small loans, big dreams*. Hoboken, NJ: Wiley.

Cull, R., Demirgüç-Kunt, A., & Morduch, J. (2009). *Microfinance tradeoffs: Regulation, competition, and financing*. World Bank Policy Research Working Paper No. 5086. Available at http://ssrn.com/abstract=1492566.

Deininger, K., & Mpuga, P. (2005). Does greater accountability improve the quality of public service delivery? Evidence from Uganda. *World Development, 33*(1), 171–191.

Demirgüç-Kunt, A., Beck, T., & Honohan, P. (2008). *Finance for all? Policies and pitfalls in expanding access*. World Bank Policy Research Report. Washington, DC: The World Bank.

DfID (Department for International Development). (2008). *The road to prosperity through growth, jobs and skills*. Discussion Paper. Dhaka: DfID, Bangladesh.

Dhumale, R., & Sapcanin, A. (1999). *An application of Islamic banking principles to microfinance*. Technical Note, UNDP.

Dias, D. (2013). *Implementing consumer protection in emerging markets and developing economies*. CGAP/World Bank.

Dogra, C.-S. (2015). *Why microfinance is becoming a bad word all over again*. Available at http://thewire.in/2016/01/15/why-microfinance-is-becoming-a-bad-word-all-over-again-18937/.

Duvendack, M., Palmer-Jones, R., Copestake, J. G., Hooper, L., Loke, Y., & Rao, N. (2011). *What is the evidence of the impact of microfinance on the well-being of poor people?* EPPI Centre, Social Science Research Unit, Institute of Education. London: University of London.

Erosa, A., & Hidalgo, C. (2008). On finance as a theory of TFP: Cross-industry productivity differences, and economic rents. *International Economic Review, 49,* 437–473.

Fernando, J. (2006). Introduction: Microcredit and empowerment of women: Blurring the boundary between development and capitalism. In Fernando, J. (Ed.), *Microfinance: Perils and prospects*. Oxon: Routledge Studies.

Franks, R. (2000). Macroeconomic stabilization and the microentrepreneur. *Journal of Microfinance, 2,* 69–91.

Ghalib, A. K., Malik, I., & Katsushi, S. I. (2012). *Microfinance and its role in household poverty reduction: Findings from Pakistan*. Working Paper No. 173/2012. Brooks World Poverty Institute (BWPI).

Gine, X., Harigaya, T., Karlan, D., & Nguyen, B. (2006). *Evaluating microfinance program innovation with randomized control trials: An example from group versus individual lending*. Asian Development Bank, Erd Technical Note No. 16.

Gine, X., & Townsend, R. M. (2004). Evaluation of financial liberalization: A general equilibrium model with constrained occupation choice. *Journal of Development Economics, 74,* 269–307.

Gonzalez, A. (2007). *Resilience of microfinance institutions to macroeconomic events*. MIX Discussion Paper No. 1. Washington, DC: The MIX.

Gonzalez, A. (2011). *An empirical review of the actual impact offinancial crisis and recessions on MFIs MIX*. GMF Summit.

Gutierrez-Nieto, B., Serrano-Cinca, C., & Molinero, C. M. (2007). Microfinance institutions and efficiency. *Omega, 35,* 131–142.

Hartarska, V., & Nadolnyak, D. (2007). Do regulated microfinance institutions achieve better sustainability and outreach? Cross-country evidence. *Applied Economics, 39*(10), 1207–1222.

Hermes, N., & Lensink, R. (2011). Microfinance: Its impact, outreach, and sustainability. *World Development, 39*(6), 875–881.

Hermes, N., Lensink, R., & Meesters, A. (2008). *Outreach and efficiency of microfinance institutions*. Working Paper. Groningen: Groningen Universiteit. Available at SSRN http://ssrn.com/abstract=1143925.

Hermes, N., Lensink, R., & Meesters, A. (2009). *Financial development and the efficiency of microfinance institutions*. Working Paper, Groningen: Groningen Universiteit. Available at http://ssrn.com/abstract=1396202.

Honohan, P. (2004). *Financial Sector Policy and the Poor: Selected Findings and Issues*. World Bank Working Paper No.43. Washington, DC: World Bank.

Imai, K. S., & Azam, M. S. (2012). Does microfinance reduce poverty in Bangladesh? New evidence from household panel data. *Journal of Development Studies, 48*(5), 633–653.

Ismawan, B., & Budiantoro, S. (2005). Mapping microfinance in Indonesia. *Jurnal Ekonomi Rakyat*. Available at http://www.ekonomirakyat.org/edisi_22/artikel_5.htm.

Jeong, H., & Townsend, R. M. (2007). Sources of TFP growth: Occupational choice and financial deepening. *Economic Theory, 32,* 197–221.

Jeong, H., & Townsend, R. M. (2008). Growth and inequality: Model evaluation based on an estimation-calibration strategy. *Macroeconomic Dynamics, 12,* 231–284.

Johnson, N., Garcia, J., Rubiano, J. E., Quintero, M., Estrada, R. D., & Mwangi, E. (2009). Water and poverty in two Colombian watersheds. *Water Alternatives, 2*(1), 34–52.

Kamani, A. (2007). *Microfinance misses its mark*. Stanford Social Innovation Review.

Karim, N., Tarazi, M., & Reille, X. (2008). *Islamic microfinance: An emerging market niche*. CGAP Focus Note, 49. Retrieved fromhttp://www.cgap.org/sites/default/files/CGAP-Focus-Note-Islamic-Micorfinance-An-Emerging-Market-Niche-December-2014.pdf.

Karlan, D., Goldberg, N., & Copestake, J. (2009). Randomized control trials are the best way to measure impact of microfinance programmes and improve microfinance product designs. *Development and Small Business Finance, 20*(3), 167–176.

Karlan, D., & Morduch, J. (2010). Access to finance. In Rodrik, D., & Rosenzweig, M. (Ed.), *Handbook of development economics* (Vol. 5, pp. 4703–4784). Amsterdam: Elsevier.

Karlan, D., & Nadel, S. (2006). *Evaluating microfinance program innovation with randomized controlled trials: Examples from business training and group versus individual liability*. Workshop: Using Research Findings to Improve Design of Products and Services. Microcredit Summit.

Karlan, D., & Valdivia, M. (2011). Teaching entrepreneurship: Impact of business training on microfinance clients and institutions. *Review of Economics and Statistics, 93*(2), 510–527.

Karlan, D., & Zinman, J. (2010). Expanding credit access: Using randomized supply decisions to estimate the impacts. *The Review of Financial Studies, 23*(1), 433–464.

Karlan, D., & Zinman, J. (2011). Microcredit in theory and practice: Using randomized credit scoring for impact evaluation. *Science, 332*(6035), 1278–1284.

Karnani, A. (2007, Summer). Micro-finance misses its mark. *Stanford Social Innovation Review*, 34–40.

King, R., & Levine, R. (1993). Finance and growth: Schumpeter might be right. *The Quarterly Journal of Economics, 108,* 717–737.

Kono, H., & Takahashi, K. (2010). Microfinance revolution: Its effects, innovations and challenges. *The Developing Economies, 48,* 15–73.

Krauss, N. A., & Walter, I. (2008). *Can microfinance reduce portfolio volatility?* Working Paper. Available at http://ssrn.com/abstract=943786.

Ledgerwood, J., & White, V. (2006). *Transforming microfinance institutions: Providing full financial services to the poor*. World Bank Publications.

Levine, R. (2004). *Finance and growth: Theory and evidence*. NBER Working Papers No. 10766. Massachusetts: National Bureau of Economic Research.

Littlefield, E., Morduch, J., & Hashemi, S. (2003). *Is microfinance an effective strategy to reach the Millennium Development Goals?* FocusNote, CGAP. 24. http://www.cgap.org/docs/FocusNote_24.pdf.

Lloyd-Ellis, H., & Bernhardt, D. (2000). Enterprise, inequality and economic development. *Review of Economic Studies, 67,* 147–168.

Lund, C., De Silva, M., Plagerson, S., Cooper, S., Chisholm, D., Das, J., et al. (2011). Poverty and mental disorders: Breaking the cycle in low-income and middle-income countries. *The Lancet, 378*(9801), 1502–1514.

Marmot, M., Friel, S., Bell, R., Houweling, T. A., Taylor, S., & Commission on Social Determinants of Health. (2008). Closing the gap in a generation: Health equity through action on the social determinants of health. *The Lancet, 372*(9650), 1661–1669.

Mazumder, M., & Lu, W. (2015). What impact does microfinance have on rural livelihood? A comparison of governmental and non-governmental microfinance programs in Bangladesh. *World Development, 68,* 336–354.

McIntosh, C., Villaran, G., & Wydick, B. (2011). Microfinance and home improvement: Using retrospective panel data to measure program effects on fundamental events. *World Development, 39,* 922–937.

McIntosh, C., & Wydick, B. (2005). Competition and microfinance. *Journal of Development Economics, 78,* 271–298.

Mersland, R., & Strøm, Ø. (2010). Microfinance mission drift? *World Development, 38,* 28–36.

Miller, C. (2011). *Microcredit and crop agriculture*. FAO. GMF Summit.

Mishkin, F. S. (2007). Is financial globalization beneficial? *Journal of Money, Credit and Banking, 39*(2–3), 259–294.

MIX. (2010). *Microbanking bulletin.* Number 20. Washington.

Montgomery, H. (2005). *Meeting the double bottom line: The impact of Khushhali Bank's microfinance program in Pakistan*. Tokyo: ADBI.

Mullainathan, S., & Shafir, E. (2009). Savings policy and decision-making in low-income households. *Insufficient Funds: Savings, Assets, Credit, and Banking AMONG Low-Income Households, 121,* 140–142.

Pande, R., & Park, Y. J. (2012). Repayment flexibility can reduce financial stress: A randomized control trial with microfinance clients in India. *PLoS ONE, 7*(9).

Rahman, A. R. A., & Rahim, A. (2007). Islamic microfinance: A missing component in Islamic banking. *Kyoto Bulletin of Islamic Area Studies, 1*(2), 38–53.

Rahmat, T., Megananda, S., & Maulana, A. (2006). *The impact of microfinance on micro and small enterprise's performance and the improvement of their business opportunity*. Working Papers in Economics and Development Studies (WoPEDS), 200601. Department of Economics, Padjadjaran University.

Rao, D. S. (2012). Credit market failures and microfinance. *International Journal of Research in Management & Technology, 2*(3), 294–297.

Richter, P. (2004, September). The integration of the microfinance sector in the financial sector in developing countries: The role that Apex mechanisms play in Uganda. In *WIDER Conference*, Finland.

Robinson, M. (2001). *The microfinance revolution: Sustainable finance for the poor* (pp. 199–215). Washington: World Bank.

Robinson, M. (2002). *Microfinance revolution: Lessons from Indonesia*. Washington, D.C.: The World Bank.

Roodman, D. (2011). *Due diligence: An impertinent inquiry into microfinance*. Washington, D.C.: Center for Global Development.

Roodman, D. (2012). *Microcredit doesn't end poverty, despite all the hype*. The inside track on Washington politics. Washington D.C.: The Washington Post (March 10, 2012).

Rosly, S. A., & Bakar, M. A. A. (2003). Performance of Islamic and mainstream banks in Malaysia. *International Journal of Social Economics, 30*(12), 1249–1265.

Sanders, C. (2002). The impact of microenterprise assistance programs: A comparative study of program participants, nonparticipants, and other low-wage workers. *The Social Service Review, 76,* 321–340.

Schreiner, M. (1999). *A scoring model of the risk of costly arrears at a microfinance lender in Bolivia*. Center for Social Development. Washington University in St. Louis.

Schreiner, M. (2004). *Rural microfinance in Argentina after the tequila crisis*. Lewiston, NY: Edwin Mellen Press.

Seibel, D. H., & Agung, W. D. (2005). *Islamic microfinance in Indonesia*. Economic Development and Employment Division. Financial Systems Development Sector Project. Eschborn: Deutsche Gesellschaft fur Technische Zusammenarbeit (GTZ).

Slikkerveer, L. J. (2007). *The need for integrated microfinance management to reduce poverty and enhance health and education in Indonesia* (p. 34). Cleveringa Lecture Series. Jakarta: School of Management, Universitas Trisakti.

Smolo, E., & Ismail, A. G. (2011). A theory and contractual framework of Islamic micro-financial institutions' operations. *Journal of Financial Services Marketing, 15*(4), 287–295.

Staschen, S. (2003). *Regulatory requirements for microfinance—A comparison of regulatory frameworks in eleven countries worldwide*. Deutsche Gesellschaft für Technische Zusammenarbeit.

Stiglitz, J., & Weiss, A. (1981). Credit rationing in markets with imperfect information. *American Economic Review, 71,* 393–419.

Sullivan, N. P. (2007). *You can hear me now: How microloans and cell phones are connecting the world's poor to the global economy*. Wiley.

UNDP. (2015). *Human development report 2015: Work for human development*. United Nations Development Program.

Van Rooyen, C., Stewart, R., & De Wet, T. (2012). The impact of microfinance in sub-Saharan Africa: a systematic review of the evidence. *World Development, 40*(11), 2249–2262.

Wajdi Dusuki, A. (2008). Banking for the poor: The role of Islamic banking in microfinance initiatives. *Humanomics, 24*(1), 49–66.

Waller, G. M., & Woodworth, W. (2001). Microcredit as a grass-roots policy for international development. *Policy Studies Journal, 29*(2), 267–282.

Wenner, M. (1995). Group credit: A means to improve information transfer and loan repayment performance. *Journal of Development Studies, 32,* 263–281.

World Bank. (2008). *Finance for All? Policies and pitfalls in expanding access*. Policy Research Report. Washington, DC.

Woodcock A. (2007). *Validation of the WeDQoL-Goals-Peru: Goal necessity and satisfaction scales and individualised quality of life scores*. Report to the WeD team. Bath: Wellbeing in Developing Countries Research Group, University of Bath.

World Bank. (2015). *World development report 2015: Mind, society, and behavior*. Washington, D.C.: World Bank.

Yaron, J., Benjamin, M. P., & Piprek, G. L. (1997). *Rural finance: Issues, design, and best practices* (Vol. 14) Washington, DC: World Bank.

Yunus, M., & Weber, K. (2008). *Creating a world without poverty: Social business and the future of capitalism*. New York: Public Affairs.

Zeller, M. (2003). *Models of rural financial institutions*. Lead theme paper presented at the International Conference on Best Practices Paving the Way Forward for Rural Finance. Washington, D.C.

Dimitrios Niklis holds a Master Degree in Business Economics and Management and a Ph.D. in Financial Management. His research interests include financial risk management and analysis, programming of financial investments, bankruptcy prediction, credit ratings, multiple criteria decision making and tourism. He has participated in various Greek, EU and international research projects.

George Baourakis is Director of the Mediterranean Agronomic Institute of Chania (CIHEAM-MAICh) in Crete, Greece, and the Studies and Research Coordinator of the Business Economics and Management Department of CIHEAM-MAICh. He holds a Ph.D. in Food Marketing and is an expert in Behavioural Economics with more than three decades' experience in the management and coordination of educational and research activities, including coordination of the M.Sc. Programme of CIHEAM-MAICh, and organization of international seminars and training courses both at CIHEAM-MAICh and in many other Mediterranean countries. He has co-ordinated a large number of EU projects (FP 4th, 5th, 6th and 7th, INTERREG I, II and III-Archimed, MED, Tempus, Phare, Life, Lifelong Learning, Leonardo Da Vinci, European Social Fund/Operational Sectoral Programme), as well as international and national-regional research projects, and has also supervised their scientific, economic and administrative management. He is a Research Fellow at the Centre of Entrepreneurship, Nyenrode University, The Netherlands Business School, and Senior Lecturer in the International Master Course on Integrated Microfinance Management (IMM) at the Faculty of Economics and Business, Universitas Padjadjaran in Bandung, Indonesia. He also holds the positions of Co-Editor in Chief for the International Journal of Food and Beverage Manufacturing and Business Models (IJFBMBM), IGI Global, and Associate Editor for the Springer Cooperative Management Series.

Constantin Zopounidis is Professor in the field of Financial Engineering and Multicriteria Analysis at the Technical University of Crete (Greece), Distinguished Research Professor at Audencia Business School (France) and Academician at the Royal Academy of Economics and Financial Sciences (Spain). His interests lie in operational research, financial risk management, multicriteria analysis and financial management.

Chapter 7
Integrated Poverty Impact Analysis (IPIA): A New Methodology for IKS–Based Integration Models

L. Jan Slikkerveer

Awareness of the differential distribution of impacts among different groups in society, and particularly the impact burden experienced by vulnerable groups in the community, should always be of prime concern.

Frank Vanclay (2003)

7.1 Recent Development of Policy Impact Assessment Models: EIA and SIA

Since the 1970s, economists working in international development programmes have taken an interest in the assessment of economic growth and the distribution of goods and services among different populations in developing countries. In addition to the political role of the state in the policies of provision and distribution of public resources, the interventions of economic development programmes and projects of international donors and agencies started to require appropriate methodologies for the assessment of their impact on the local population. The first economic impact assessment methods include the *Economic Impact Analysis* (EIA) which examines the impact of a new policy or project on a neighbourhood, community or society, often triggered by public concerns about the impact of the proposed policy.

The need, however, for a systematic method to evaluate economic projects emerged not only from a rather vague process of project planning and management, but also from an adverse process of evaluation, where evaluators tended to use their own subjective judgment as to what they thought were 'good things' and 'bad things' (cf. PCI 1979). EIA has evolved from a simple tool to evaluate the effectiveness of a project, known as a *logical framework* or *logic model.* Table 7.1 shows a general logical model on the basis of a matrix, conventionally implemented

L. J. Slikkerveer (✉)
LEAD, Leiden University, Leiden, The Netherlands
e-mail: l.j.slikkerveer@gmail.com

L. J. Slikkerveer et al. (eds.), *Integrated Community-Managed Development*, Cooperative Management, https://doi.org/10.1007/978-3-030-05423-6_7

Table 7.1 Logic model of a 'community health management project of a non-smoking campaign' (CHMP)

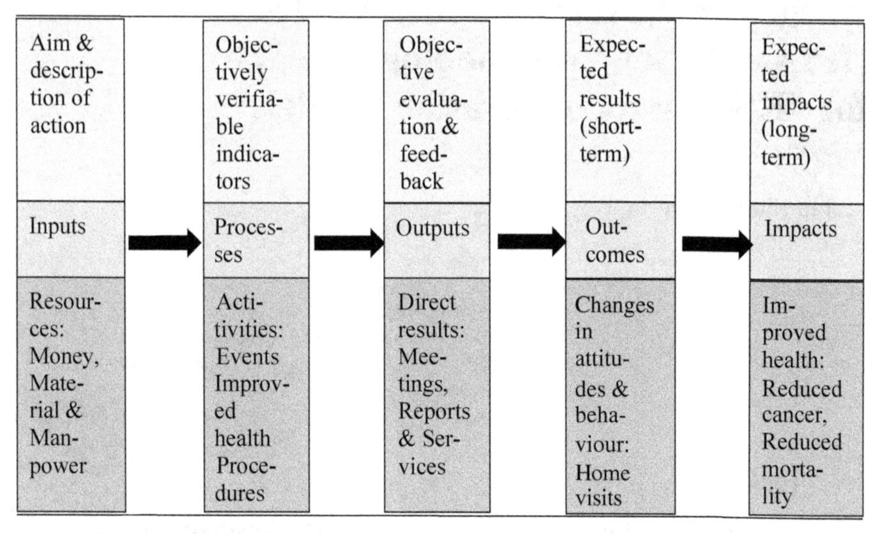

Aim & description of action	Objectively verifiable indicators	Objective evaluation & feedback	Expected results (short-term)	Expected impacts (long-term)
Inputs →	Processes →	Outputs →	Outcomes →	Impacts
Resources: Money, Material & Manpower	Activities: Events Improved health Procedures	Direct results: Meetings, Reports & Services	Changes in attitudes & behaviour: Home visits	Improved health: Reduced cancer, Reduced mortality

Source Slikkerveer (2012)

in project management, to measure the revenues and profits from inputs over a certain period of time.

The related *Logical Framework Approach* (LFA) is a descriptive methodological tool mainly used for designing, monitoring and evaluating development projects, known as a *logic model* or *logical framework*. The *Logical Framework Approach* (LFA) also contributed to the development of public economics by extending the traditional welfare theory to the examination of constraints on policy planning resulting from limited information about the political and financial processes of project management.

Meanwhile, the environmental movement had grown in North America since the 1960s, triggered by the publication of Rachel Carson's book *Silent Spring* in (1962), which marked the beginning of a wider public awareness of the destruction of the natural environment by new technologies, industries and economic power. As human activities were largely seen as responsible for the on-going degradation of the environment, growing concerns led to the demand for reliable measurement of the impact of modern development on the environment, disentangled from particular interest groups. Soon, the methodology of *Environmental Assessment* (EA) was introduced as the assessment of the positive or negative environmental consequences of a distinct policy. The implementation of such an assessment of policies, plans and projects by national governments or international organisations, conducted prior to the decision to put the proposed action into practice, became known as *Environmental Impact Assessment* (EIA). As a policy planning tool, EIA has become an integral component of the decision-making process in public policies, governed by administrative and legal regulations, which usually also includes alternative actions to meet the policy objectives.

As it became clear that interventions of policies, programmes and projects had not only an impact on the natural environment, but also on the population groups living and surviving in the environment, the *Social Impact Assessment* (SIA) was derived from the *Environmental Impact Assessment* (EIA) as a methodology to measure the social effects of development interventions on the population in the target area. In 1992, the 'Interorganisational Committee on Guidelines and Principles for Social Impact Assessment' was established with the aim of outlining a set of guidelines and principles which would assist public- and private-sector agencies and organisations to support their policy planning and implementation efforts, as published in their *Principles and Guidelines for Social Impact Assessment* (IOCGP 2003).

In general, the implementation of most intervention projects tends to follow a series of five main phases, usually starting with initial planning, followed by implementation, operation and maintenance, and concluded by the decommissioning of the project. Since the type and setting of development intervention through a policy, project or programme may differ, it is evident that the type of social impact may differ accordingly. In the case of projects executed in the sequence of the above-mentioned five phases, each phase includes potentially significant social impacts. In order to encourage public participation in the interventions, the method of SIA seeks to involve all affected social groups from the very beginning of the policy planning process and also to indicate possible alternatives.

An important aspect of the *Social Impact Assessment* (SIA) is the identification of those variables which point to measurable change in the human population, the communities, and the social relationships resulting from a proposed action. These SIA variables can be grouped into (a) population change; (b) community and institutional structures; (c) political and social resources; (d) community and family changes; and (e) community resources. The method of the assessment of the social impacts for the different activities in the subsequent stages is generally executed through the construction of a simple matrix, in which the horizontal five phases are indicated, while the above-mentioned groups of variables are listed vertically (cf. IOCGP 2003).

The data needed to complete the SIA matrix are usually collected from the intervention agency in terms of past experience and an assessment of future prospects of the proposed project, documented in records and accounts. In addition, data on the target area, administrative structure and local population are usually available from public sources and statistical reports. Specific research is carried out through qualitative field work including interviews, group meetings and consultation of members of the target population.

The model which the *Social Impact Assessment* (SIA) uses is based on the comparative method to gain adequate *understanding* of the intervention process in an area where a planned intervention has already taken place, and compare the data with the data in an area where the planning process still has to be implemented.

The next critical step, however, is the reliable *prediction* of possible social impacts of the envisaged intervention on the local population, for which only a few conventional techniques have been operationalised. It means that in order to estimate the probable effects of a proposed project in community X, the assessment of the impact of a similar intervention in community Y forms the comparable basis for prediction.

In this way, past experience with similar projects is compared with future possibilities or alternatives. The related technique of trend projections into the future is based on the assumption that the situations are rather similar and that the change will continue at the same rate, enabling a predictive social impacts assessment for the future. Another method to predict the social impact of an intervention refers to the population *multiplier* method, which assumes that every increase in the population implies specific multiple increases of related variables, such as employment, housing, services and infrastructure.

The technique of the use of *scenarios* in SIA refers to the formation of logical imaginations about the future, based on the assumptions about the various variables concerned. Such a technique is closely related to the judgement of experts who are knowledgeable about the study area and are able to assess various scenarios for their implications on the planned intervention.

Most of the methods and techniques discussed above, however, tend to remain at the descriptive and explanatory level of various forms of impact assessment, largely based on the measurement of qualitative criteria.

Despite the recent development of a growing body of environmental and social impact analysis approaches, represented by a range of policy impact tools and methods, there is still a lack of systematic and analytically supported methods which are capable of providing a standardised, objective model to explain and predict the valid impact of distinct policy interventions on the poor and underprivileged segment of the target population. Although most of the assessment methods and techniques—largely confined to various matrix-based models—are labeled under the category of 'impact analysis', it is still unclear to determine to what extent they comply with the concept of analysis as the process of a systematic examination and independent evaluation of data or information, by breaking it into its component parts to uncover their interrelationship and interactions in order to provide an objective basis for problem-solving and decision-making.

From another angle in the current debate on the need for more objective impact analysis methods, the World Bank (2009), in its efforts to help to realise the predominant goal of global poverty reduction in 2030 by introducing particular poverty reduction strategies, is now also paying attention to more objective measurements of the impact of these poverty reduction policies. The related *Poverty Reduction Strategy Papers* (PRSPs) seek to put into practice the *Comprehensive Development Framework* (CDF) of the World Bank (2009), encompassing a set of principles to guide poverty reduction and development as a means to attain its objectives of eliminating poverty, reducing inequity, and improving opportunities for people in low- and middle-income countries. The related *Sourcebook for Poverty Strategies* by Klugman (2002) includes not only some key areas such as understanding the nature and causes of poverty, and identifying obstacles to pro-poor development, but also the assessment of sectoral policies which are involved in poverty reduction. Although Klugman (2002) acknowledges that: "*Assessing the growth and distributional impacts of past policies and programmes is difficult, not least because data and monitoring and evaluation systems are usually weak, and rigorous quantitative assessments are seldom available*", she makes a point for designing these policies within the context of the PRSP: "*The PRSP should estimate the likely effect of its proposed policy measures on the poor and include measures to mitigate any negative impacts.*"

7.2 Towards an Evidence-Based Integrated Poverty Impact Analysis

An interesting, albeit yet underdeveloped methodology of a more standardised research-based explanation and prediction of project impacts is provided by the development of advanced statistical analytical techniques which enable the determination of the evidence-based differences in probabilities between processes which are involved in project interventions and those which are not. Although a standardised and objective methodology is not fully explored, the current analysis is advancing by the use of both qualitative and quantitative measurement of the statistical significance of various groups of relevant variables, operational in the intervention process. Such an analysis, however, starts with the execution of a series of advanced research methods and techniques, followed by appropriate preparation of reliable data in Excel to be transformed into useful SPSS data sets. For some assessors, not only the level, but also the appropriate balance between qualitative and quantitative research methods seems to remain problematic as they question the right 'mix', such as Kanbur et al. (2001).

However, the systematic implementation of the qualitative research of secondary data from the agencies concerned, complemented by primary data from the field with the objective to acquire in-depth knowledge of the complicated processes and configurations, largely as a means to the appropriate execution of subsequent quantitative surveys to measure the spread and coverage of significant phenomena over the target population in a representative mode, has shown to provide a strong contribution to the research-based understanding and prediction of policy impacts on different groups of the community. In this way, the provision of aggregated numerical data for valid and reliable statistical inference to understand not only correlation, but also interaction among significant variables, responds directly to the need for independent and objective statistical analytical models, far removed from the subjective description and personal interest of directly involved stakeholders, particularly in such cases where the position of the poor and underprivileged members of the community, too often subjugated under external interventions, is at stake. Especially the analysis of policy impacts on the local people is most important since the impact of outside interventions from national or international levels may drastically alter their lives and livelihoods, where usually 'scientific assessments' are brought to bear substance to the decision-making process.

While the economic assumptions often show up to be rather arbitrary and implausible, the econometric techniques are sometimes so complex that transparency and usability for a pro-poor assessment have practically gone adrift. As mentioned above, qualitative research is most useful for its explanatory power, while quantitative survey data provide substantial empirical inference with wider coverage and allow for statistical analysis with the comparative advantage of establishing general conclusions about causal impact and covariant change with a high degree of confidence for larger population groups.

The current focus on the position of the poor in international development renders rather timely an appeal to applied-oriented researchers to stretch the responsibility of empirical research in the assessment of impacts of policy and project interventions to building an evidence-based analytical process. Such an evidence-based analytical process is crucial for policy interventions as it also reflects the reality of the changing positions of all participants and stakeholders involved in the proposed intervention, imposing a heavy burden on the objective and trustworthy outcome of the data analysis for the actual decision-making process.

Although Peng (2009: 405) embarks on the premise that: 'the replication of scientific findings using independent investigators, methods, data, equipment, and protocols has long been, and will continue to be, the standard by which scientific claims are evaluated', he also supports the plea for more attention for and development of evidence-based data analysis, where he points directly to related education and training in data analytics techniques and strategies which should be reproducible and replicable. In this context, replicability refers to the likelihood that an independent study on the same research question will provide a result that is consistent with the original study, while reproducibility denotes the capacity to re-analyse given data to obtain a consistent result (cf. Peng 2009).

Writing on reproducible research and biostatistics, Peng (2009) refers to Evidence-Based Data Analysis (EBDA) as an empirically-based approach: "identifying what works in increasing replicability and reproducibility, and then recommending, and expecting, those methods, techniques and tools, to be used in data analysis." In a recent statistics blog called Simply Statistics, Irizarry et al. (2017) assert that the basic idea behind evidence-based data analysis is that for each stage of the operations of the entire research and analysis process—coined by the authors as the pipeline—the best method should be used as the one justified by appropriate statistical research, which provides evidence that favours one method over another: "So a data analysis is really a pipeline of operations where the output of one stage becomes the input of another."

The worldwide adoption of the *Sustainable Development Goals of the 2030 Agenda*, with a high priority on poverty reduction, has further intensified the need for more standardised survey methodologies on poverty and an independent and transparent analysis of project impacts on the poor on a global level (cf. United Nations 2015). Recently, two interesting initiatives concerning advanced research and analysis have emerged in this newly-developing field of poverty impact analysis.

The work of Chambers and Mayoux (2005) reverses the existing research paradigm by launching a model of *Quantification, Participatory Methods and Pro-Poor Impact Assessment.* The authors argue that instead of just being optional labels, participatory approaches, methods and behaviours are indeed essential for the new agendas of pro-poor development. As the impact assessment methodologies are currently at a crossroads, Chambers and Mayoux (2005: 272) signal a critical dilemma in the present debate: "*On the one hand the underlying agendas of pro-poor development and 'improving practice' necessarily require participation by poor women and men in deciding priorities and identifying strategies. On the other hand the sheer numbers of people involved, the potential conflicts of interest and consequently difficulties of decision-making require rigorous quantification and analysis in order minimise domination by vested interests.*"

The attractiveness of the proposed promotion of these participatory methods for the newly developed *Integrated Poverty Impact Analysis* (IPIA) is not only that they have shown to be rather cost-effective for the related surveys, but also that they can generate accurate quantitative data on local priorities, different experiences of poor people, and the potential for innovation in relation to causality and attribution within the context of pro-poor development. In addition, a more direct involvement of the participants in the entire process of research and analysis, in which their voice on their own position is brought forward, adds substantially to the evidence-based orientation of the IPIA approach.

Moreover, the contribution which the IPIA approach seeks to provide for a more systematic poverty impact analysis by including the participatory methods links up well with the overall objective of the integrated community-managed development approach to highlight local participation in its contribution to the *Sustainable Development Goals of the 2030 Agenda* of the United Nations (2015), elaborated in the previous Chapters. Besides, such participatory methods are also implied in the 'Leiden Ethnosystems Approach' developed and refined by the LEAD Programme of Leiden University since the 1980s. As mentioned in Chap. 3, this participatory-oriented research approach is particularly expressed in one of the three basic principles, being the focus of the 'Participants View' (PV), so as to adopt an *emic* perspective of the local people in community-based processes of development and change (cf. Slikkerveer 1991).

Meanwhile, in 2003, the World Bank, in collaboration with the *Poverty Reduction Group* (PRMPR) and *the Social Development Department* (SDV), introduced the concept of *Poverty and Social Impact Analysis* (PSIA) (World Bank 2003). In its efforts to support developing countries in the design and implementation of national development policies, the World Bank (2003) presented a *User's Guide to Poverty and Social Impact Analysis*, based on the application of economic and social tools and methods in policy reforms. More recently, with funding from the Government of Norway, the United Nations Development Programme and the World Bank are conducting this type of analysis in a number of developing countries in order to assess the social impact of the recent economic crisis. Although the World Bank has aggregated important global data on poverty impact and reduction as a basis for policy reforms, and its reports refer to 'analysis', the current execution of its impact assessment in many developing countries still remains confined to the descriptive use and interpretation of largely qualitative data collected from a range of resources including stakeholders, institutions, agencies, transmission channels etc. For instance, the social development tools described in the *PSIA User's Guide* of the World Bank (2003) are rightly focused on key issues: "*non-income dimensions of poverty, such as stakeholder interests, social capital, and vulnerability*", which contribute to the further understanding of the complicated process. The *analysis*, however, remains in fact at the level of *assessment, assumption* and *estimation*, most illustrative in the related estimation of policy impacts. As the World Bank's *PSIA User's Guide* (2003: 7) indicates: "*Forecasting or simulating likely impacts of policy by definition presupposes a view of likely causality and behaviour. Depending on the analyst's information base these can be empirically 'estimated' based on the past, derived on the basis of theory, or assessed on the basis of knowledge of the country context and discussions with key stakeholders and experts.*"

The importance, however, which the World Bank attaches to the role of multi-topic household surveys is transpiring through its *Living Standards Measurement Study* (LSMS), in operation since the 1980s. The World Bank's LSMS (2017) is providing technical assistance to many statistical offices around the world with the objective of: "*generating high-quality data, incorporating innovative technologies and improved survey methodologies, and building technical capacity*", and is as such also providing technical support in the design and implementation of household surveys and in the measurement and monitoring of poverty. The LSMS Programme encompasses various tools to assist researchers and practitioners in the design, implementation and analysis of household surveys, in which some analytical building blocks for the statistical analysis of poverty impacts are also provided. The previous publication by Deaton (1997) of *The Analysis of Household Surveys* elaborates on the use of household survey data from developing countries in several policy issues, and is as such rather conducive to the further development of an evidence-based poverty impact analysis.

In their contribution on *Poverty Measurement and Analysis* in the above-mentioned *Sourcebook for Poverty Strategies* by Klugman (2002), Coudouel et al. (2002) elaborate on the measurement and analysis of poverty, inequality and vulnerability by focusing on two main dimensions of 'poverty', i.e. *income* and *consumption.* Before turning to the analysis of 'poverty', first its measurement has to be computed by: (a) defining the indicator of poverty; (b) selecting the poverty line, being the threshold below which individuals are classified as 'poor'; and (c) choosing the appropriate poverty measures to be operationalised.

The poverty indicators usually include monetary indicators, which include income and consumption, where consumption has shown to be a better outcome indicator than income because of its direct relation to the individual's well-being and standard of living. Moreover, consumption is less fluctuating over time and more behaviour-oriented at the individual level. The non-monetary indicators of poverty include insufficient consumption of nutrition, health, education and social services, where participatory methodologies in the household surveys are crucial for the subjective perception of poverty.

The indication of the poverty line for a particular country or region is often guided at the national level, which is not only based on a minimum standard of living, but also on values and norms regarding the minimum level of meeting basic needs.

The selection of the poverty measure refers to statistical standard measures, such as the incidence of poverty or the headcount index, the depth of poverty or the poverty gasp, and the poverty severity or the squared poverty gap. The three measures are usually calculated on the basis of the proportion of households which are below the poverty line in the case of the headcount index (cf. Coudouel et al. 2002). After the indicators, lines and measurement of poverty have been determined, the next step is the comparison of the various characteristics of the poor and non-poor in the analytical model. In supporting these objectives, the CDF of the World Bank (2009) emphasises the interdependence of all social, structural, human, governance, environmental, economic, and financial elements of development, which links up well with the holistic integration approach of the LEAD Programme.

Before elaborating on the *Integrated Poverty Impact Analysis* (IPIA), first the analytical framework, the preparatory research process and the analytical model are described in the next Paragraphs.

7.3 Analytical Framework for IPIA: Understanding and Prediction

Basically, the new *Integrated Poverty Impact Analysis* (IPIA) encompasses a process of analysing, explaining, predicting, monitoring and managing the positive, neutral and negative impacts of planned interventions of policies, programmes and projects on the condition of the poor and vulnerable members of the target population. IPIA therefore refers to a specific methodology of analysing the impact of a proposed intervention on the poor and low-income families of the community.

The analytical framework of the IPIA model provides a multi-dimensional context of the integrated impact analysis, focused on the comparative understanding and prediction of contemporary and future configurations of the impact of outside interventions, such as policies, projects and programmes on local peoples and communities in developing countries, in particular on the poor and vulnerable members of the indigenous communities. The IPIA analytical framework is based on the following four dimensions:

1. principles (6)
2. levels of stakeholders (3)

3. phases of the project intervention (6)
4. stages of the IPIA process (10)

1. *Principles*

The basic principles of the analytical framework of IPIA include a multidisciplinary strategy, an integrated approach, a combined qualitative, quantitative and participatory research methodology, an evidence-based analytical process, a pro-poor orientation, and a human rights-based perspective, being the six core elements of the underlying *IKS-Based Integration Model* (IKSIM), which also forms the basic tool for the two innovative strategies of IMM and ICMD for sustainable community-managed development (cf. Chap. 3). The evidence-based analytical process is, of course, the core principle in which the analysis reflects reality in an optimum mode by determining not only the appropriate conceptualisation of the objectives and constructing a suitable conceptual model, but also selecting the balanced qualitative and quantitative research methods and techniques, conducting objective measurements and household surveys in order to achieve reliable data sets for the execution of the advanced stepwise analyses.

Before engaging in the execution of the process of integrated poverty impact analysis, the assessor should secure the coverage of the legal and administrative preconditions at the national level of the state in accordance with the prevailing government laws and regulations.

2. *Levels of stakeholders*

The impact of project interventions is usually taking place at three major levels: the macro-, meso- and micro-level, which are generally represented by the international, national, regional and local levels at which agencies, institutions and their staff are designing, implementing and concluding the policy or project interventions.

The definition of the positions of different stakeholders are important since they are active in and between these three levels, and include not only international organisations and donors, national agencies and planners, provincial authorities and administrators, regional administrators and leaders, but also the participants of local communities, households and individuals. As these different categories of stakeholders tend to have different interests in the project interventions, the analytical framework seeks to integrate the assessment of their distinct roles into the overall poverty impact analysis.

3. *Phases of project intervention*

In general, most interventions in the form of policies, programmes or projects proceed through a series of six phases, from identification through implementation to finalisation of the intervention, known as the ‘project cycle’. Usually, project interventions in developing countries include the following six phases:

Phase 1: identification and design
Phase 2: conceptualisation
Phase 3: planning

Phase 4: implementation
Phase 5: operation and maintenance
Phase 6: project conclusion

The subsequent execution of the six project phases takes place at the three above-mentioned levels: the *macro-level* of international and national organisations and agencies (government or non-government); the *meso-level* of interaction between national agencies and local communities; and the *micro-level* of the indigenous communities and institutions, households and individuals.

4. *Stages of the IPIA process*

The Integrated Poverty Impact Analysis (IPIA) refers to an evidence-based analytical process of a series of ten subsequent stages by which assessors are capable of analysing the impact of a project at all these stages, from complementary research to the finalisation of the report, in relation to their different impacts on the poor segment of the target population. In this context, the six-fold aim of IPIA is to provide an evidence-based impact analysis on the basis of: (1) execution of advanced complementary research; (2) provision of empirical data sets in SPSS; (3) construction of the appropriate analytical model; (4) execution of the stepwise analysis; (5) formulation of an objective explanation of the contemporary relations and interactions among relevant factors and a comprehensive prediction of the related future relations and interactions; and (6) presentation of the final report on the impact of the project and the alternative interventions.

Following the above-mentioned sequence of the six phases of the proposed project intervention, the elaboration of the six-fold aim into ten stages of the IPIA process, leading to the report as a basis for an adequate and responsible policy decision-making process, can be described as follows:

Stage 1: a–f Complementary research (6 steps)
Stage 2: Sorting of empirical data
Stage 3: Data input into SPSS
Stage 4: Analytical model
Stage 5: Bivariate analysis
Stage 6: Mutual relation analysis
Stage 7: Multivariate analysis
Stage 8: Multiple regression analysis
Stage 9 a–b: Explanation and prediction (2 steps)
Stage 10: IPIA final impact report

The integration of the above-mentioned 4 dimensions of the IPIA analytical framework of principles (6), levels of stakeholders (3) and phases of project intervention (6), together with the stages of the IPIA process (10), provides an analytical framework for the execution of an evidence-based impact analysis with a focus on the poor and low-income families and individuals at the community level. Figure 7.1 shows a schematic representation of the integrated IPIA analytical

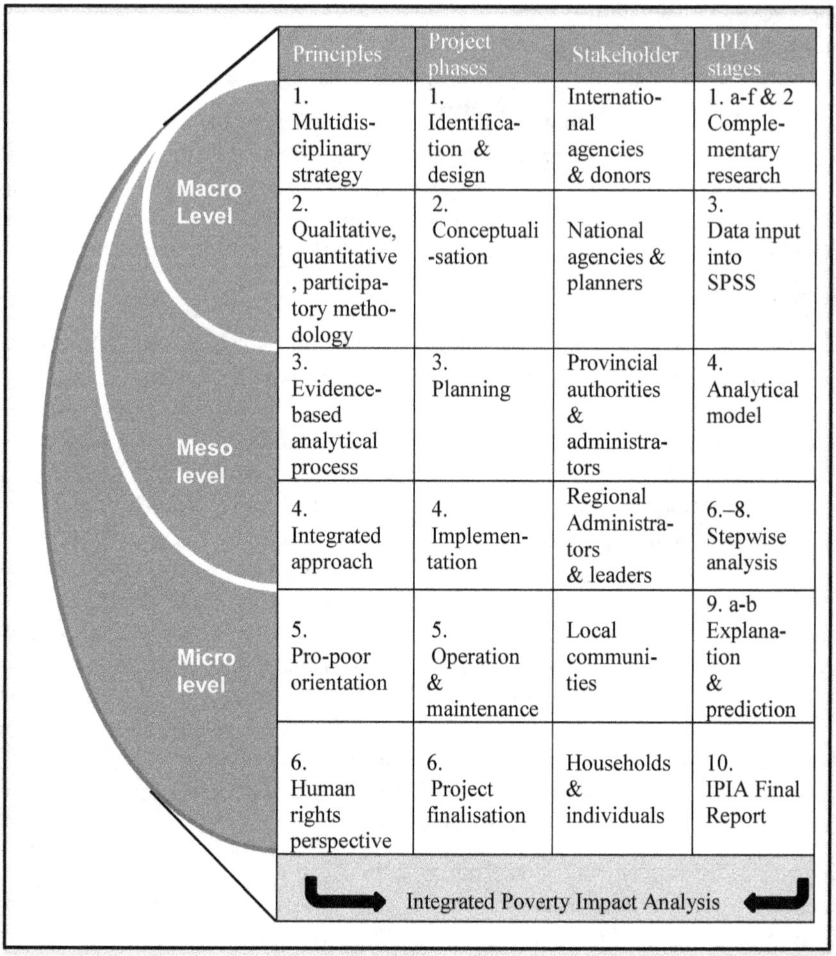

Level	Principles	Project phases	Stakeholder	IPIA stages
Macro Level	1. Multidisciplinary strategy	1. Identification & design	International agencies & donors	1. a-f & 2 Complementary research
	2. Qualitative, quantitative, participatory methodology	2. Conceptualisation	National agencies & planners	3. Data input into SPSS
Meso level	3. Evidence-based analytical process	3. Planning	Provincial authorities & administrators	4. Analytical model
	4. Integrated approach	4. Implementation	Regional Administrators & leaders	6.–8. Stepwise analysis
Micro level	5. Pro-poor orientation	5. Operation & maintenance	Local communities	9. a-b Explanation & prediction
	6. Human rights perspective	6. Project finalisation	Households & individuals	10. IPIA Final Report

Integrated Poverty Impact Analysis

Fig. 7.1 Schematic representation of the IPIA analytical framework, based on the integration of the major principles, project phases, stakeholders, and IPIA stages as the basis of the provision of an evidence-based poverty impact analysis. *Source* Slikkerveer (this chapter)

framework, based on the integration of the major principles, project phases, stakeholders and IPIA stages as the basis for the provision of such an evidence-based poverty impact analysis. The following Paragraphs will further elaborate on the 10 stages of the evidence-based IPIA process, sub-divided into the three core activities of respectively the execution of complementary research, the selection of the analytical model and the implementation of the advanced stepwise analysis, all leading up to the final impact analysis report.

7.4 The Foundation of the IPIA Process: Complementary Research

The build-up of the evidence-based IPIA process encompasses the integrated sequence and interconnection of the subsequent 10 stages, leading up from complementary research for the collection and transformation of empirical data into SPSS datasets, through the implementation of the analytical model and the stepwise analysis for the explanation and prediction of the poverty impact of interventions, to the formulation of the final report. The IPIA process is schematically represented in Fig. 7.2, while a short description of each of the stages is presented below.

Stage 1 a–f: *Complementary research*

Following the general aim of the IPIA process to develop an evidence-based analysis of the impact of a project on the poor individuals and households of the

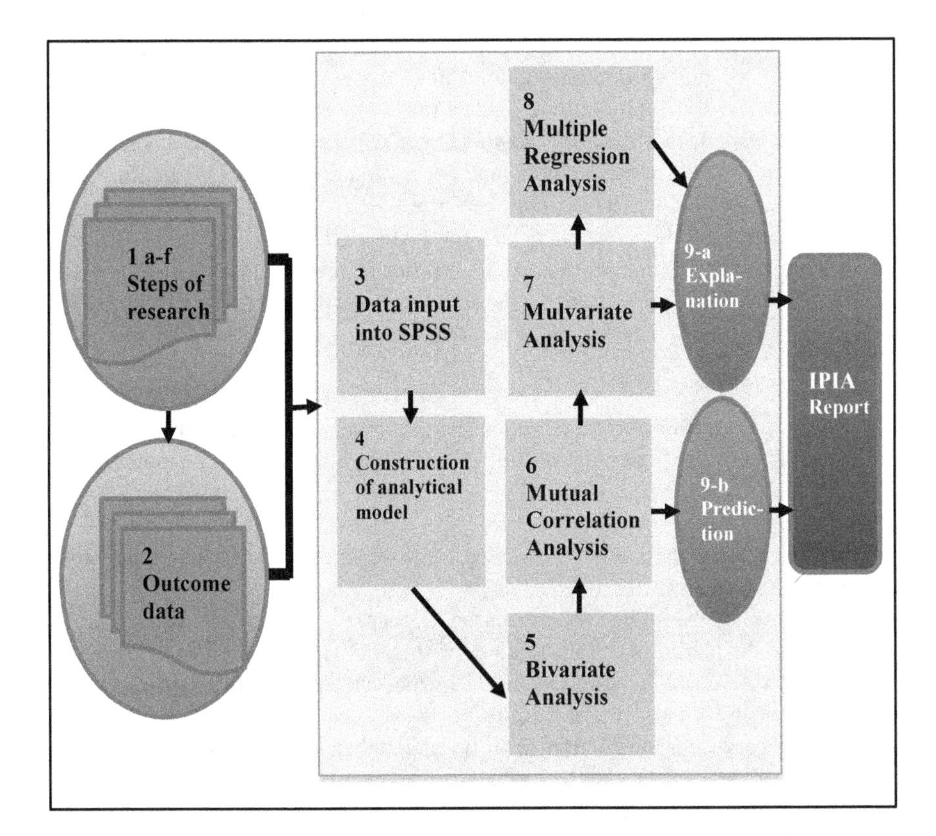

Fig. 7.2 Schematic representation of the 10 stages of the evidence-based IPIA process: Progress from complementary research (1a–f) and output of empirical data (2), through data input into SPSS (3); construction of the analytical model (4) into the stepwise analysis (5–8) to provide explanation (9a) and prediction of impact (9b), to the presentation of the IPIA report (10). *Source* Slikkerveer (this chapter)

target population with a view to reconsidering the proposed project, it starts with the important complementary research stage and its subsequent stage of outcome of empirical data, which can be sub-divided into 6 major tasks, as indicated below.

Following the above-mentioned model of *Quantification, Participatory Methods and Pro-Poor Impact Assessment* of Chambers and Mayoux (2005), the specific IPIA research seeks to include the technique of participatory rural appraisal in order to identify the subjects which are most relevant to the poor within the context of their income and consumption behaviour under impacting conditions.

Although the specific research methodology is building on the current state of development of the appropriate complementary research, i.e. a critical combination of qualitative and quantitative research methods and techniques, particular ways of fieldwork data collection are implemented in order to provide a sound and realistic research base for the policy planning and implementation of the new approaches of the IMM and ICMD models. After clarifying the problem-oriented purpose of the research, in this case in general terms of the description of the final attainment of the development programme concerned, a detailed sequence of research activities has to be prepared which can be conceptualised as a circular process, where the tasks are continuously guided, monitored, controlled and evaluated.

In view of the focus on the intervention impact on specific poverty groups and individuals, respondents from the segment of the poor and socially excluded are central to the research and include lone-parent families and large families with children; ill and disabled individuals, and the elderly; migrants and members of minority groups; and homeless and unemployed individuals.

The first *stage of the complementary research* (1a–e) of the IPIA process includes the execution of 6 major steps, which can schematically be conceptualised to follow a circular process, built up as follows:

Step a: *Identification and planning*

The first step concerns the identification and planning of the research to reach the provision of empirical data, being the foundation of the entire IPIA process, and includes the following activities:

- identification of the problem-oriented purpose of the research, including the general aim and specific objectives;
- formulation of the subject of the research: concepts, definitions, and terms;
- review of the literature to substantiate the theoretical orientation of the research;
- indication of possible ‘missing links’ in the current state of the theories and approaches relevant to the research;
- planning of the various activities to be undertaken in the sequence of phases, stages and steps;
- determination and measurement of the concept and utility of the proposed intervention project;
- review of the relevance of the project to the target population, particularly the poverty groups and individuals in the communities;

- identification of the different positions and interests of various groups of stakeholders and participants involved in the envisaged project;
- determination of the role of the indigenous institutions, local leaders, and participants at the micro-level of communities, households and individuals;
- building a profile of the poor and underprivileged members of the local communities forming the subjects of the IPIA process.

Step b: *Qualitative research*

The second step regards the execution of qualitative research and includes the following activities:

- description of the national context of the study;
- selection and description of the selected geographical research area;
- description of the natural environment: zones, elevation and vegetation;
- providing a geographical research map on the settlement pattern of the target population;
- identification and documentation of secondary research data from the study area;
- preparation of a 'sociography' of the study area, including detailed information on existing general data, the target population, statistical resources and administrative structures[1];
- selection of techniques of qualitative data collection and documentation;
- identification and interviews with key informants, opinion leaders and stakeholders;
- implementation of the specific *Leiden Ethnosystems Approach* encompassing the Historical Dimension (HD), the Field of Ethnographic Study (FES) and the Participant's View (PV) (cf. Slikkerveer 1991).

Step c: *Conceptual model and questionnaires*

The third step concerns the selection of the conceptual model which serves as a guide for the setup of the quantitative household surveys, including the design of the questionnaires, and forms the basis for the later analytical model. It includes the following activities:

- selection of the conceptual model underlying the research;
- identification and categorisation of relevant factors as variables in the model;
- creation of a *codebook* which includes detailed information of the relevant factors operationalised into the variables of the conceptual model;

[1]The description of the natural and cultural research setting often starts with the sociography of the study area. The concept of 'sociography' was introduced by Steinmetz (1913) as a means of describing a particular society in general terms on the basis of existing data, available in statistical accounts and reports, before conducting in-depth research. In addition, the sociography provides an overview of the human settlement patterns, the infrastructure and the natural environment: zones, altitudes and types of vegetation. It is usually supplemented with a detailed geographical research map on the actual settlement patterns of the target population within its natural context, usually available through local libraries, statistical offices or the internet.

- operationalisation of theoretical concepts into measurable variables;
- determination of scales of measurement of selected variables[2];
- identification of levels of expected significant associations among different levels of variables;
- construction of structured questionnaires according to the categories in the conceptual model;
- implementation of the sequence of categories of relevant variables in the paragraphs of the questionnaires;
- choice of methods of sampling, data recording, and data collection.

The operationalisation of concepts into variables, categorised into the 'blocks' of variables in the model, is a deductive process, where general theoretical concepts are 'translated' into measurable factors, which, in turn, have to be 'converted' into variables, which are provided with pre-coded answer categories in the questionnaires. As an example, Table 7.2 shows the operationalisation of the concept of a respondent's 'phase in life' in the category of 'socio-demographic variables' to the variable 'age', and the ten related answer categories.

In this context, it is important to make a differentiation between the various possible classes of measurement of the selected variables, especially since the type of scale is decisive for the later determination of the appropriate significance tests in the various analyses of collected data (cf. Table 7.2).

[2]Measurement scales are used to categorise and/or quantify the values of variables. In general, four scales of measurement are used in statistical analysis: nominal, ordinal, interval, and ratio scales.

Nominal scales: refer to two or more named categories (classes) which are qualitatively different from each other, such as the variable 'marital status': single; married; divorced; widowed.
Ordinal scales: refer to two or more named categories (classes) which are qualitatively different from each other, but have an additional quality put in a rank order or ladder, such as the variable 'socio-economic class': poor; average; well-to-do.
Interval scales: refer to two or more named categories (classes) which are qualitatively different from each other, but have an additional quality where the intervals between the classes are equal, such as the variable 'temperature', measured in degrees Celsius, where the difference between 20° and 25° is the same as between 32° and 37°C.
Ratio scales: refer to two or more named categories (classes) which are qualitatively different from each other, but have an additional quality where zero indicates the absence of the attribute, rendering the ratio between numbers in the scale similar to that between the amounts of the attributes under measurement, such as the variable: 'income': measured in dollars, euros or pounds etc.
Likert scales: refer to a set of usually five collective responses to a question, and formatted in a range of answers, each with a score on the scale of 1–5. The terms of the possible responses used in the scaling are usually collected in a pilot study with the target group in order to reach consensus to safeguard the validity of the measurement. In a Likert scale, the structured answer categories often represent a choice of opinion or attitude of the respondents, in which they indicate their level of agreement or disagreement on a symmetric agree-disagree scale for a series of statements ranging from very positive to very negative.

Table 7.2 Operationalisation of the variable 'age' in a household survey

Concept	Factor	Variable	Indicator	Answer categories	
Phase in age (1)	'age'	Age	Number of years	(1)	0–9
19				(2)	10–19
29				(3)	20–29
39				(4)	30–39
49				(5)	40–49
59				(6)	50–59
59				(7)	60–69
79				(8)	70–79
				(9)	90+

Source Slikkerveer (1990)

Step d: *Pilot study and questionnaire testing*

The fourth step concerns the execution of a pilot study and the testing of the quantitative questionnaires, and includes the following activities:

- introduction of the research to the regional and local authorities, specialists and the study population;
- assessment of the feasibility of the research activities in the field;
- elaboration of the specific research objectives;
- identification of the sector(s) involved in the research;
- implementing the qualitative research: key informants;
- identification of relevant factors and their value categories;
- testing the (draft) quantitative questionnaires: labels, relevance of factors/variables, and answer categories;
- monitoring representation, reliability and validity;
- drawing a simple geographical map of the research area.

Step e: *Quantitative household surveys*

The fifth step includes the execution of the participatory fieldwork and includes the following activities:

- selection and training of teams of interviewers and interpreters;
- identification of participants as co-researchers in the field;
- determination of the selected sample(s) from the parent population;
- identification of the sample households in the survey(s);
- execution of household surveys through respondents,
- monitoring and control of the completed questionnaires of the surveys;
- collecting, ordering and recording of survey data.

Step f: *Sorting of empirical data*

The sixth step includes the *sorting of empirical data* collected during the research and includes the following activities:

- cleaning of the data, which refers to the process of detecting, correcting or removing corrupt or inaccurate records from the collected empirical data;
- sorting of the data, which includes the rearrangement of the rows of the matrix with respect to one or more variables with a view to easy reading of the data;
- ordering the data, which refers to the arrangement of unordered data in visual displays such as tables and graphs;
- subsetting of the data, which refers to the division of the data into separate samples;
- aggregation of the data, based on particular key indicators.

In some surveys, it is useful in this step to construct so-called 'composite variables' on the basis of the measured values of two or more variables. For these composite variables, the measurement of several variables is combined through the statistical technique of *factor analysis* into *one* new variable. The factor analysis examines a set of interdependent relationships among variables in order to identify the underlying dimensions which explain the correlations among a set of related variables. An example is the variable of 'socio-economic status' (SES), which can be recalculated on the basis of a combination of the outcome of a set of single variables such as 'annual income', 'material assets', 'type of house', 'area of arable land' and the 'subjective classification' of both the respondent and the interviewer, etc., which in combination provides a more realistic assessment of the respondent's overall socio-economic status.

The process of the logical sequence of the above-mentioned 6 steps of the complementary research stage of the IPIA process of, respectively, identification and planning; qualitative research; conceptual model and questionnaires; pilot study and questionnaire testing; quantitative household surveys; and sorting of empirical data is represented in Fig. 7.3. In the field of the objective measurement of poverty impacts over time—as is the case in intervention policies, projects and programmes—the research cycle can be extended by the execution of comparative research between a community which has experienced similar intervention and a community which has not experienced such intervention. The outcomes of such parallel studies are conducive not only to the explanation, but also to the prediction of particular impacts on the target population.

After completion of the 6 above-mentioned steps of the first *stage of the complementary research* (1a–f) of the IPIA process, the next stage of the *outcome of empirical data* (2) includes the preparation of the data input into the following phases (3–9) which encompass the actual complex of the evidence-based analysis of intervention impacts on the local people, particularly the poor. The subsequent phases are further elaborated in the next Paragraphs.

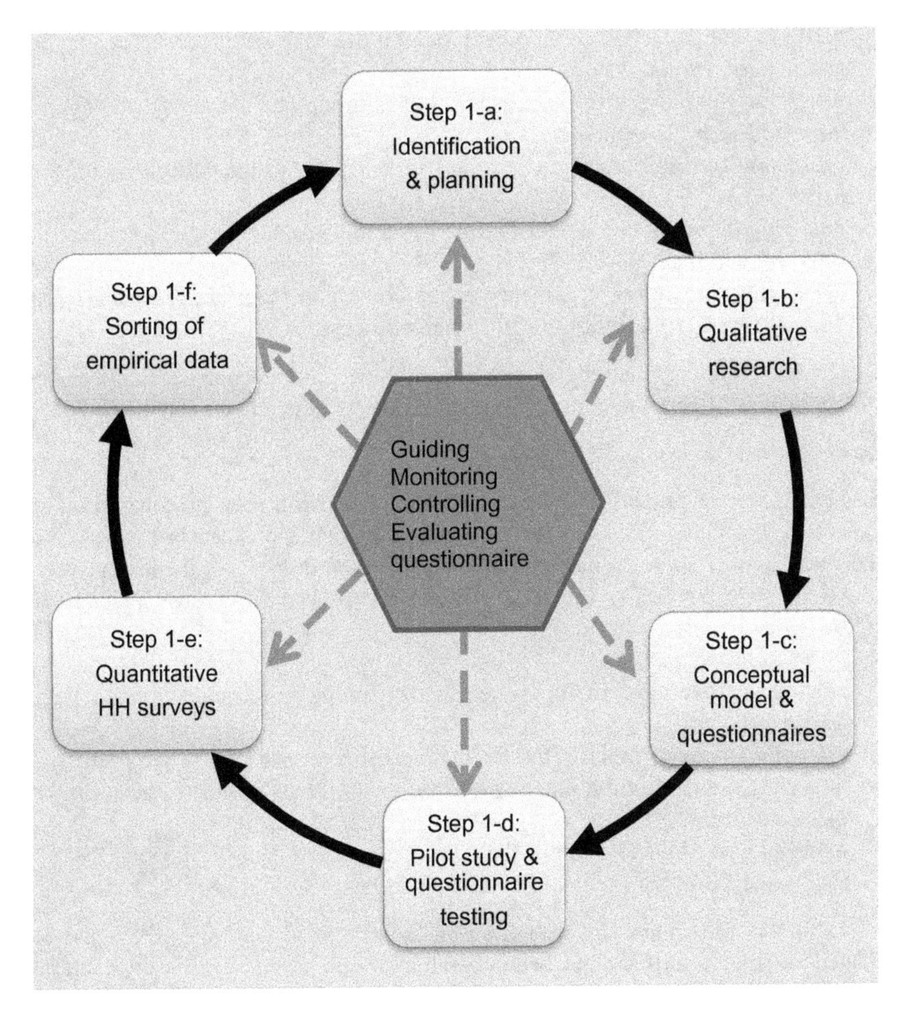

Fig. 7.3 The process of the logical sequence of the 6 steps of the complementary research stage of the IPIA process: (1-a) identification and planning; (1-b) qualitative research; (1-c) conceptual model and questionnaires; (1-d) pilot study and questionnaire testing; (1-e) quantitative household (HH) surveys; and (1-f) sorting of empirical data. *Source* Slikkerveer (author's illustration)

7.5 Outcome of Empirical Data, Input into SPSS and Analytical Model

Stage 2: *Outcome of empirical data*

The second stage of the outcome of empirical data (2) of the IPIA process follows after the completion of the 6 steps of the first stage, and includes the following activities:

- verifying the collection of empirical data on the basis of the accuracy, quality and integrity of the data;
- weighting the balance of the execution of both the qualitative and the quantitative research components;
- examining the unbiased and controlled measurement of the collected empirical data;
- preparing the set of outcome data from the complementary research for the next stage of data input into SPSS;
- identifying the empirical evidence in the collected data in terms of the reproducibility and replicability of the research process.

The last mentioned activity is crucial for the overall evidence-based dimension of the analysis, being the core of the entire IPIA process (cf. Irizarry et al. 2017).

Stage 3: *Data input into SPSS datasets*

The third stage of data input and transformation into SPSS datasets is based on the execution of specific statistical tasks, which are guided and executed by the use of the Excel and SPSS software programmes, using the drop-down menus or syntax (cf. Kent State University Libraries (KSUL) 2016). The third stage includes the following activities:

- creating an overview of the collected data properly formatted in the Excel spreadsheet;
- formatting the data input in the Excel spreadsheet for the SPSS program;
- importing the data in the Excel spreadsheet into the SPSS format according to the software;
- navigating in SPSS to monitor the appearance of the data in SPSS;
- saving the results of the data file after importation into SPSS.

The process of importing data from an Excel Spreadsheet into SPSS and the related criteria are further described in detail by the Kent State University Libraries (KSUL 2016).

Stage 4: *Construction of the analytical model*

Embarking on the holistic IKSIM research model, as described in Chap. 3, the fourth stage of the IPIA process encompasses the crucial construction of the analytical model which underlies the core analysis of the Integrated Poverty Impact Analysis (IPIA).

It is the schematic representation, conceptualised in the previous conceptual model of the configuration of measurable variables in their categories, in which the values of the relationships are eventually conducive to the explanation and prediction of the impact of projects on the target population. Based on a review of the literature and recent case studies in East Africa, the Mediterranean Region and Indonesia, the theoretical and methodological foundations of the IPIA form the basis of the selected analytical model. It encompasses different categories of relevant factors documented in the field and operationalised in the model into variables at different levels, categorised into so-called 'blocks' of variables.

The construction of the analytical model of IPIA with a view to analysing differential patterns of consumption behaviour in order to measure significant changes in the well-being of participants is based on the Transcultural Health Care Utilisation Model, developed by Slikkerveer (1990) in an ethnomedical study among the Oromo population in the Horn of Africa. This model has also been used and adapted thereafter for similar ethnoscience-oriented research in Tanzania, Indonesia, Greece and a number of South-European countries (cf. Slikkerveer 1990; Agung 2005; Leurs 2010; Djen Ammar 2010; Ambaretnani 2012; Chirangi 2013; Aiglsperger 2014; Erwina 2019; Saefullah 2019; De Bekker in press).

As mentioned before, the related participatory research has successfully been conducted according to the 'Leiden Ethnosystems Approach' in emic-oriented research, in which the Historical Dimension (HD), the Field of Ethnological Study (FES) and the Participant's View (PV) are the key concepts in the fieldwork phase (cf. Slikkerveer 1991).

The construction of the analytical model and its components follows the objectives of the research, where the various categories of factors are structured along three levels of variables in the model with potential significant associations among the different variables: *independent, intervening* and *dependent* variables. Most variables are operational at the individual level of the respondent and are *categorical* as they can take on one of a limited, often fixed, number of possible values—usually five—assigning each individual or other unit of observation to a particular group of categories on the basis of a specific qualitative property. An example is the factor/variable 'blood type' of a respondent, where the answer categories include: 0—*don't know;* 1—*A;* 2—*B;* 3—*AB;* 4—*O*; 5 *other.*

Based on the findings and evidence of previous research, the categories in the present analytical model are in general encompassing well-defined types of factors, redefined as variables in the model, and include the following:

Independent variables:

- Predisposing variables: socio-demographic variables: 'household composition', 'household membership', 'gender', 'age', 'marital status', 'ethnicity', 'place of birth', 'education', 'religion', 'occupation';
- Predisposing variables: psycho-social variables: 'knowledge', 'opinion' and 'attitude' concerning the prevailing local and global phenomena, being the specific subject of the research,
- Household consumption variables: 'annual consumption', 'household consumption components', 'household consumption price indices', 'household goods', 'household services'
- Enabling variables: 'annual household income', 'annual household expenses', 'annual insurance expenses', 'annual subject-specific expenses', 'socio-economic status', annual subject-specific transport expenses;
- Perceived variables: 'perceived subject-specific need', 'perceived subject-specific demand';

- Institutional variables: 'accessibility of modern institutions', 'accessibility of indigenous institutions'.

Intervening variables:

- Intervening variables: 'impact of outside public forces/interventions', and 'impact of outside private forces/interventions'.

Dependent variables:

- Dependent variables: 'differential patterns of consumer behaviour' recorded over the selected period of time, and subdivided into variables of 'no change in consumer behaviour', 'positive change in consumer behaviour' or 'negative change in consumer behaviour'.

In general, the various categories of variables are brought together in the analytical model in order to facilitate the stepwise analysis for the explanation and prediction of measurable changes in consumer behaviour as an indication of poverty resulting from outside intervention. As mentioned above, the independent variables are categorised under the general heading of social variables of community characteristics, household characteristics, individual characteristics and institutional characteristics. Despite a general lack of research on the role of indigenous institutions in poverty-related behaviour, their significance is their organisational influence at the community level. Moreover, their traditional support for the poor and low-income families has continued over many generations. The intervening variables in the model refer to external interventions of policies, programmes and projects, and are categorised in public and private intervening variables. The study of patterns of consumption behaviour at the microlevel of household surveys renders it possible to study the differential individual behaviour of respondents from among different socio-economic statuses over time, including the poor and low-income members of the community (cf. Deaton and Grosh 2000).

Figure 7.4 shows a schematic representation of the analytical IPIA model, adapted for the stepwise analysis of the relationships and interactions among the various 'blocks' of independent and intervening variables in relation to the 'blocks' of consumption patterns of different SES groups as an indication of changes in their well-being, particularly of the poor in the target community.

The representation of the IPIA analytical model, shown in Fig. 7.4, encompasses the following 12 'blocks' of variables:

Independent variables:

Block 1 socio-demographic variables
Block 2 psych-social variables
Block 3 perceived need variables
Block 4 socio-economic status
Block 5 institutional variables
Block 6 community variables

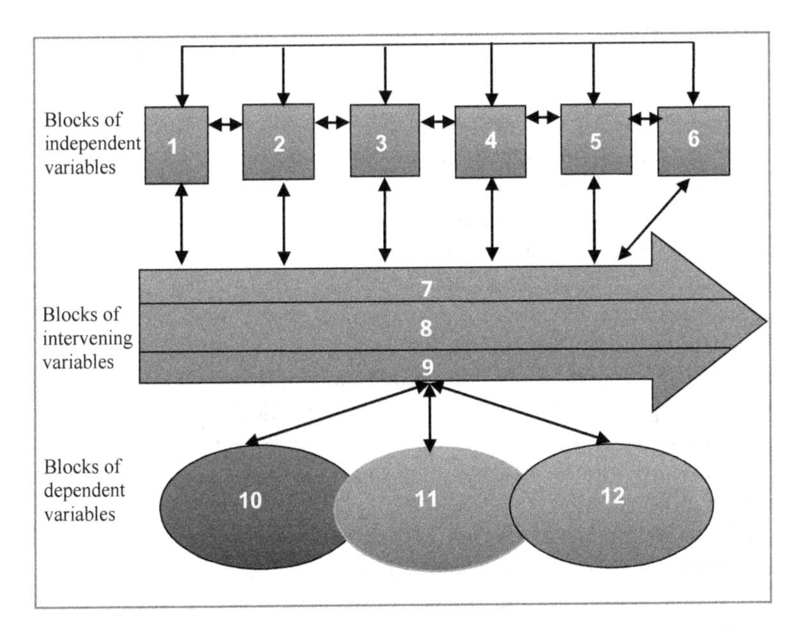

Fig. 7.4 Schematic representation of the IPIA analytical model, adapted for the evidence-based analysis of intervention impacts on consumption patterns of different SES groups as an indication of changes in their well-being, in particular of the poor members of the target community. *Source* Slikkerveer (1990)

Intervening variables

Block 7 past public project interventions
Block 8 past private project interventions
Block 9 present proposed project interventions

Dependent variables

Block 10 negative integrated impact of intervention on consumption patterns of poor, average and well-to-do participants in the community
Block 11 neutral integrated impact of intervention on consumption patterns of poor, average and well-to-do participants in the community
Block 12 positive integrated impact of intervention on consumption patterns of poor, average and well-to-do participants in the community

The next stages of the bivariate analysis (5), the mutual relation analysis (6), the multivariate analysis (7) and the multiple regression analysis (8) represent the execution of the stepwise analytical process, which is further elaborated in the next Paragraph.

7.6 Stepwise Analysis of Integrated Poverty Impact

The next four stages 5, 6, 7 and 8, of the evidence-based IPIA process are forming the core of the analysis of the relations and interactions among the variables of the datasets resulting as the outcome from the previous comprehensive research process, and imported into the SPSS format.

Basically, any statistical analysis is essential for the adequate progress, level and outcome of research because it facilitates not only reaching valid insights for complicated problems but also formulating sound recommendations for realistic policy planning and implementation. Presently, the advances in statistical methods and techniques of sequential activities, which start with the comprehensive research process including representative sampling, careful selection of the model, appropriate collection of qualitative and quantitative data, and construction of valid datasets, allow for the development of evidenced-based analytical description, explanation and prediction.

In order to determine the nature of possible statistical relationships between selected independent, intervening and dependent variables, and the categories in which they are classified, various statistical techniques are used to identify if these relationships, whether causal or not, refer to *dependence* or *association*. Although the term *correlation* is often used in linear relationships within the context of dependence, in statistical terms it does not imply a causal relationship between variables. Correlations, however, may indicate a predictive relationship between variables. The degree of such correlations can be measured with *correlation coefficients*, indicated as p or r. The most common of these is Pearson's correlation coefficient ($\chi 2$). Other correlation coefficients have been developed to give more weight to the correlation—that is, more sensitive to non-linear relationships, such as *Cramer's correlation coefficient*. The special data analysis process of the evidence-based IPIA model encompasses a stepwise analysis, built up of four stages: bivariate analysis, mutual relation analysis, multivariate analysis and multiple regression analysis, which are described as follows:

Stage 5: *Bivariate analysis*

Following the construction of the above-mentioned frequency tables and cross-tabulations of categorical variables in Excel, allowing for the first assessment of relationships among different variables, the actual analysis of the envisaged significant relationships among variables starts with the execution of the first step of the stepwise analysis of IPIA, being the bivariate correlation analysis of data in SPSS. As one of the easiest forms of quantitative analysis, it shows the empirical association between two variables, usually indicated as X and Y. In this context, association—or dependence—refers to any statistical relationship between two variables or bivariate data, albeit causal or not. The bivariate analysis does not only enable the examination of the relationship between two variables, but it also indicates the shape, direction and strength of that relationship (cf. Weinberg and Abramowitz 2002).

In development–oriented research, the analysis is generally focused on the degree to which the value of the dependent variable in the model is determined by the other independent variable. The degree of determination is shown in the calculated degree of significance, computed in the dataset in SPSS. The analysis should take into account that the variables are of the same type: categorical or ordinal, allowing for the use of multinomial regression analysis. However, in the case of the special data analysis for Integrated Microfinance Management (IMM) and Integrated Community-Management Development (ICMD), the dependent variables are continuous—either interval level or ratio level—allowing for a relatively simple analysis.

Since the type of data analysis in this Volume on process interventions of development and change is also interested in interaction rather than in associations among different variables, the appropriate analysis would eventually be non-linear, requiring the implementation of a non-linear correlation analysis (cf. de Leeuw and Meijer 2007). In practice, the bivariate analysis enables the examination of bivariate relationships between two variables in SPSS, where the significance can be determined by the application of significance tests. In a cross-tabulation in SPSS in which a contingency table is generated, different statistics can be used to test whether there is a statistically significant association between the row and column variables.

While Pearson's correlation coefficient (r) is used to determine if there is either a positive linear relationship, or a negative linear relationship, or no linear relationship between the variables, Pearson's chi-squared test (χ_2) is usually applied to different sets of categorical data with a view to determining if there exists a statistically significant association between two variables. The significance values assess the degree of such a probability, expressed as a p value. As Field (2009) observes, significance tests are based on a probability value of $p = 0.05$, which equals a confidence interval of 95%, indicating a probability of 5% that the association between the variables occurs by chance.

Pearson's chi-squared test (χ_2), however, does not only establish if the associations show significance, but is also able to measure the degree of significance (cf. Agung 2005; Leurs 2010; Djen Amar 2010). The degree is indicated below:

- $\chi^2 \geq 0.15 \rightarrow$ 'non-significant'
- $\chi^2 = 0.15–0.10 \rightarrow$ 'indication of significance'
- $\chi^2 = 0.10–0.05 \rightarrow$ 'weakly significant'
- $\chi^2 = 0.05–0.01 \rightarrow$ 'strongly significant';
- $\chi^2 = 0.01–0.001 \rightarrow$ 'very strongly significant'
- $\chi^2 \leq 0.001 \rightarrow$ 'most strongly significant'

An example of such a significant association is well documented in the bivariate analysis conducted in the OTC-SOCIOMED research project on the provision by general practitioners (GP) and the consumption by patients (PT) of 'over-the-counter' medicines in six Southern European countries. The variable of practical experience of the GPs as providers ('practgp') as an intervening variable shows a

Table 7.3 Significant intervening variable/determinant of provided medicines by 565 GPs of 6 European countries, distributed over prescribed and non-prescribed medicines (N = 1.664)

Variable	Non-prescribed medicines		Prescribed medicines		Total medicines	
	n	%	n	%	n	%
Practical experience of GP (‘*practgp*’)						
Less than 5 years	46	20.5	178	79.5	224	100.0
5–10 years	138	26.1	391	73.9	529	100.0
More than 5 years	772	36.6	139	63.4	811	100.0
Total	956	33.4	708	66.6	1.664	100.0
Pearson's Chi-square (Asymp. Sig. 2-sided) =0 .000						

Source Slikkerveer (2011)

‘most strongly significant’ association with the dependent variables, as indicated in Table 7.3 (cf. Slikkerveer 2011).

In this example, the provision of prescribed medicines is reported by GPs with less than 5 years’ experience to nearly two-thirds of the cases (79.5%), while GPs with more than 10 years’ experience report the highest prescription of non-prescribed medicines. The intervening variable of ‘practical experience of GP’ shows a X_2 value of 0.000 representing a ‘most strongly significant’ association between this intervening variable and the dependent variables of prescribing behaviour, of respectively ‘non-prescribed medicines’ versus ‘prescribed medicines’, rendering this variable a true determinant of the prescribing behaviour of GPs in the research area of six Southern European countries.

While the overall picture of the medicine-providing behaviour of the GPs (2864 in all 6 countries) is dominated by an interesting division in their prescribing behaviour of 956 (33.4%) non-prescribed medicines and 1908 (66.6%) prescribed medicines, a limited number of variables out of the total number of variables selected for the research show different degrees of significant relationships in the conventional list of all the variables of the research, each allowing for a descriptive interpretation (cf. Slikkerveer 2011).

In general, the statistical relationships emerging from the bivariate analysis are providing a basic insight into the overall process, and are indicating either dependence or association between two variables and tend to refer to a linear relationship with each other, which could be causal. Spearman’s Rank Correlation Coefficient (Rho) is a special formula to measure the direction and strength of the linear relationship, provided that the variables have the ordinal measurement level. In this way, Spearman’s Rank Correlation Coefficient indicates whether the variables are related or not to one another, and whether the relationship is positive or negative (cf. Field 2009).

For the implementation of the IPIA process, the bivariate analysis is useful to provide evidence that consumption behaviour is a more effective measurement than income to capture changes in poverty as a low level of well-being. Such a position is underscored by a recent study by Meyer and Sullivan (2002, 2012) who argue that consumption offers a more robust measurement of poverty than income.

Moreover, the bivariate analysis allows for a closer examination of the complex dimensions of poverty in terms of positive associations between, on the one hand, independent variables such as income, education, health, food security, housing, and socio-economic status, and, on the other hand, intervening variables of external policy and project interventions in relation to dependent variables of patterns of the consumption behaviour of specific poor members of the community. In the configuration of the variables in the analytical model, bivariate analysis has the capacity to show for each of the associations of the different variables the related level of significance, rendering the variable a strong determinant of consumption behaviour, or not. In this way, bivariate analysis is able to select, for each of the independent and intervening variables, the significant associations with the dependent variables.

Stage 6: *Mutual relation analysis*

Following the description and interpretation of the degree of significance/non-significance of associations among the single variables, i.e. independent versus dependent, and intervening versus dependent variables, Slikkerveer (2011) introduced a mutual relation analysis by clustering all those variables which are showing a differential degree of significance, ranging from 'indication of significance' to 'most strongly significant', into the pre-determined categories concerned, as represented as 'blocks' in the analytical model. In this way, the model provides a first insight into the role of these categories—or 'blocks' in the model—of variables as determinants with different values of significance of the related associations.

As an example, Fig. 7.5 shows the mutual relation analysis calculated from the data of the above-mentioned OTC-SOCIOMED study, projected in the analytical model of the provision behaviour of GPs. The analysis shows the results from the bivariate analysis of all significant variables measured in the above-mentioned study of the provision and consumption of prescribed and non-prescribed medicines in 6 Southern European countries (N = 2.864) (X_2 = 0.010–0.000). The mutual relation analysis clearly shows the dominance of two categories of, respectively, the psycho-social determinants and intervening determinants in the model of the medicine provision behaviour of GPs, with an indication of the level of significance of the association of each variable (cf. Slikkerveer 2011).

The advantage of the mutual relation analysis in this stage of the stepwise analysis is that it also paves the way for the final step of the later multiple regression analysis of the research data.

For the implementation of the IPIA process, the mutual relation analysis means that following the execution of the bivariate analysis, the selection of independent and intervening variables with associations with a certain level of significance with the dependent variables can be displayed in their position in the 'blocks' of the analytical model, giving a first indication of which 'blocks' are dominant in the overall process, such as for instance the psycho-social variables. In this way, the

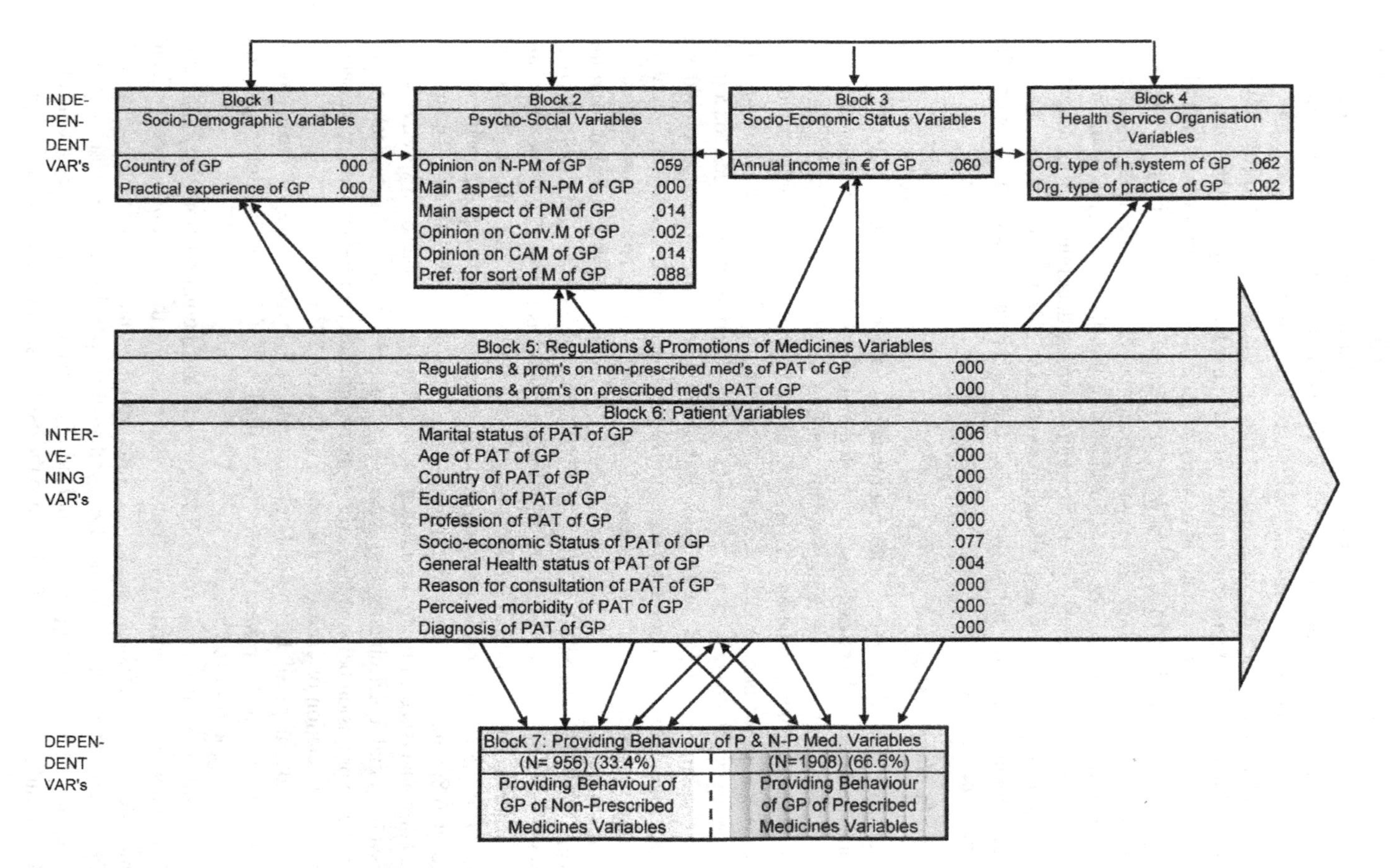

Fig. 7.5 Example of a mutual relation analysis calculated from the data of the OTC-SOCIOMED study, projected in the analytical model of the provision behaviour of GPs. *Source* Slikkerveer (2011)

mutual relation analysis contributes to the first level of explanation of the poverty impact analysis of the IPIA model.

Stage 7: *Multivariate analysis*

The third step in the stepwise analytical process of IPIA includes the execution of a multivariate analysis with the objective of identifying significant multivariate associations among multiple independent, intervening and dependent variables in the analytical model. After the classical 'univariate approach' to measure poverty through various ratios on the basis of a single dimension, generally income or consumption, as the only variable to measure poverty as an insufficient command over resources, multivariate approaches recently emerged to respond to the multi-dimensional conceptualisation of poverty with more comprehensive analyses and their relevance for policies (cf. Dagum and Costa 2004).

Multivariate analysis has emerged from the field of multivariate statistics, which is concerned with the observation and analysis of more than one statistical outcome variable simultaneously in the model. The multivariate analysis not only allows the measurement of the relationships between the variables themselves, but also the mutual interaction among all variables in the analytical model.

Although conventional linear multivariate analysis requires that each variable has a priori quantification and can be treated as numerical data, the analysis of the non-numerical variables in the data measured in the IPIA research cycle necessitates a non-linear multivariate analysis (cf. Van de Geer 1993a, b; De Leeuw and Meijer 2007; Leurs 2010).

The appropriate analytical programme, developed by the Department of Methodology and Statistics of the Faculty of Psychology of Leiden University are the CANALS and OVERALS multivariate statistical analyses, being non-linear generalised canonical analyses (cf. Van de Geer 1993a, b; De Leeuw and Meijer 2007; Van der Burg 1983, 1988). The advantage of the OVERALS analysis is that it allows to examine different sets of variables, and variables with different measurement levels, both nominal and ordinal. In addition to the SPSS output in a table of the calculated component loadings of the different sets of variables, the output can also graphically be made visible in the form of a projected plot of centroids in the canonical space, showing the solution in which two dimensions are presented, and where the component loadings serve as coordinates of the vectors. Such plots also indicate to what extent the labelled variables separate groups of objects. Figure 7.6 shows an example of the graphic representation of the component loadings of two sets of variables onto the canonical space of 30 variables in 2 dimensions of the MAC plants utilisation study in Crete, Greece (cf. Aiglsperger 2014).

The related factor analysis which enables the construction of an overview of all variables is known as the Principal Components Analysis (PRINCALS), which allows for the identification of dominant patterns in the data, including groups, trends and outliers, which can be shown in two projected plots. The Nonlinear Canonical Correlation Analysis (CANALS) refers to a procedure with several

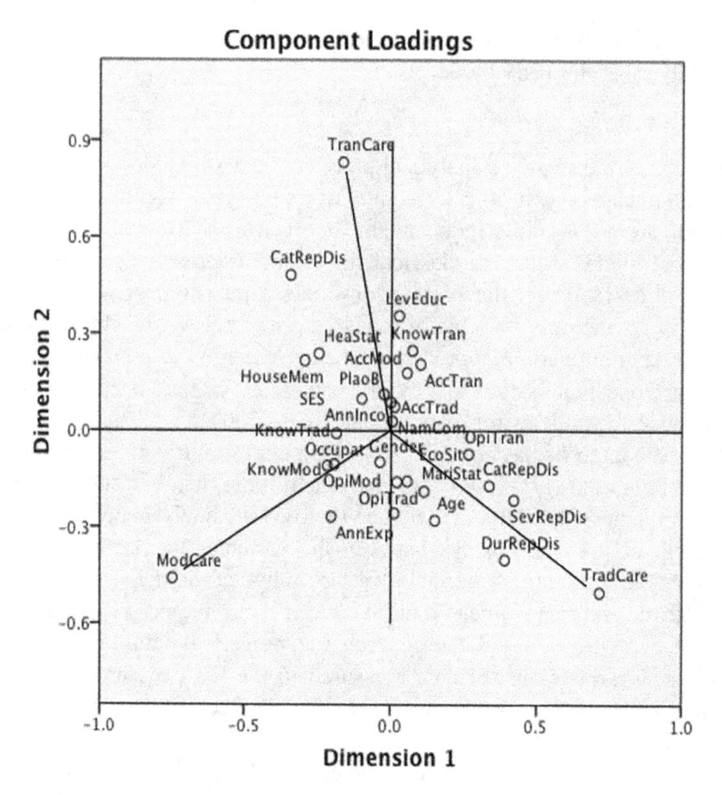

Fig. 7.6 Example of the graphic representation of the component loadings of two sets of variables onto the canonical space of 30 variables in 2 dimensions of the MAC Plants Utilisation Study in Crete, Greece. *Source* Aiglsperger (2014)

applications to analyse the relations and interactions between two or more sets of variables instead of between the variables themselves, as is the case in the Principal Components Analysis.

For the implementation of the IPIA process, the multivariate analysis means that a full explanation can be achieved of the relationships between the various variables and categories of variables at the three operational levels: independent, intervening and dependent variables. By introducing a comparison between a sample with and a sample without such external interventions, it becomes possible to provide an evidence-based explanation of the differential impact on the level of well-being of local participants in terms of consumption behaviour over time, representing the impact on the poor and low-income members of the community. This type of analysis is most appropriate for the IPIA process as it provides an explanation for the significant effects of all independent and intervening variables on the multiple dependent variables in the model.

Although the analysis of variables with multiple dimensions is rather complex and requires special software systems, the identification of dominant patterns in the data, including groups, trends and outliers, is important for an explanation of the effects of external impacts on the consumption behaviour of the poor respondents in the research.

Stage 8: *Multiple regression analysis*

The fourth and last step in the stepwise analytical process of IPIA includes the multiple regression analysis, which is an extension of the linear regression analysis, used to predict the value of a dependent variable on the basis of the value of two or more independent or intervening variables. The variable which will be predicted is known as the dependent outcome variable, while the variables which are used to predict the value of the dependent outcome variable are known as the independent predictor variables.

In order to determine the relative importance of the 'blocks' of independent and the 'blocks' of intervening variables in relationship with the 'blocks' of dependent variables, the above-mentioned OVERALS analysis programme is able to measure the association which can be expressed as functions of the 'eigenvalues' of the product matrix. In order to measure the associations, the multiple correlation coefficients (ρ) of the individual OVERALS analyses are used (cf. Van der Burg 1983; Leurs 2010). The related *Principal Component Analysis* (PCA) procedure allows not only for similar explanatory data analysis, but also for predictive data analysis. The most commonly used form of the multiple regression analysis is known as Multiple Linear Regression, and is used as a predictive analysis in order to explain the relationship between one continuous dependent variable and two or more independent variables, where the independent variables can be continuous or categorical.

In general, multiple linear regression analysis enables three analytical methods: (1) analysis of causal relationships; (2) prediction of an effect; and (3) prediction of a trend. The multiple linear regression analysis of a causal relationship between three or more categories of variables in the model is able to indicate directionality of effects, as it assumes that the independent variables have an effect on the dependent variable. The correlation among the variables in multiple regression analyses can be assessed with the coefficient of determination (R-squared or R2). The prediction of an effect of one or more independent or intervening variables on the dependent variables is possible in the multiple linear regression analysis by the prediction of values of the dependent variables. The estimation of the multiple linear regression analysis of the regression function of one variable can be used to predict the future values of the dependent variables. The prediction of a trend in the datasets is also possible in the multiple linear regression analysis by the calculation of data points in a scatter plot (cf. Statistics Solutions 2017).

For the implementation of the IPIA process, multiple linear regression analysis is most relevant as it provides the capacity to predict the changes in the patterns of the consumption behaviour of the poor as the result of the outside intervention of certain policies and projects. Also, the analysis provides further evidence that consumption is a more effective measurement than income to capture such changes in poverty as a strong indication of a low level of well-being. Such a position is underscored by the recent studies of Meyer and Sullivan (2002) and Galer (2012) who argue that consumption offers a more robust measurement of poverty than income.

7.7 The Building Blocks of the IPIA Process for the Final Report

As described in this Chapter, the major components of the *Integrated Poverty Impact Analysis* (IPIA) are linking up with the recent development of various policy impact assessment models, such as the *Environmental Impact Assessment* (EIA) and the *Social Impact Assessment* (SIA), and are founded on the analytical framework for IPIA.

In addition to the previous methods of the impact assessment of external interventions, largely based on descriptive assessments of various matrix-based overviews, the IPIA process is focused on the evidence-based explanation and prediction of policy and project impacts on the poor and low-income groups and individuals of the target population.

In order to reach such an extension of current analytical methods and techniques, a specific integrated analytical framework is constructed, in which various dimensions of the IPIA process are represented. Embarking on the three levels at which the integrated analytical activities are executed—the macro-, meso- and micro-level—the principles of a multidisciplinary approach, complementary research, evidence-based strategy, the integrated approach, the pro-poor orientation and the human-rights perspective all form the backbone of the IPIA process of the parallel execution of the project phases, the definition of stakeholders' positions and the IPIA stages.

As in all impact assessment models, the existing configurations and factors in the field are the starting points for the descriptions and explanations, while special attention is given to the validity and reliability of the complementary qualitative and quantitative research cycle, guided, monitored, controlled and evaluated in all its steps. The optimal execution of the subsequent steps in the cyclic research process is determining the empirical level of the analytical process, where the appropriate use of advanced parameters is crucial to safeguard the independent and objective reproducibility and replicability of the entire analytical process in order to achieve the status of an evidence-based analytical model.

The tasks of data collection, sorting and ordering in the Excel format, followed by importing these 'clean' data into the SPSS datasets, represents the groundwork for the construction of the analytical model and its components, based on the conceptual model underlying the previous execution of the household surveys. In the field of the objective measurement of poverty impacts over time—as is the case in intervention policies, projects and programmes—the research cycle can be extended by comparative research between a community which has experienced similar intervention and a community which has not experienced such an intervention. The construction of the analytical model adapted for the IPIA process is based on the behaviour models of Slikkerveer (1990), further developed by Agung (2005), Ibui (2007), Leurs (2010), Djen Amar (2010), Ambaretnani (2012), Aiglsperger (2014), Erwina (2019), Saefullah (2019) and De Bekker (in press).

The stepwise analysis specifically developed to complete the evidence-based IPIA process starts with the basic *bivariate analysis* of the associations between single variables categorised in the independent, intervening and dependent 'blocks' of the analytical model. As mentioned above, bivariate analysis provides a first general overview of statistically significant correlations among relevant variables, setting the tone for the following analytical steps to increase the understanding and prediction of policy impacts on the local communities.

The subsequent *mutual relation analysis* is based on the selection of variables which show a certain degree of significance in their associations with the dependent variables, categorised in the different 'blocks'. The schematic representation of the model has the advantage of presenting a first overview of the significant correlations between these 'blocks' in the model, which in a later analytical step can be further substantiated with calculated regression values.

The following step in the stepwise analytical process of IPIA includes the execution of a *multivariate analysis* to identify significant multivariate associations among and between all multiple independent, intervening and dependent variables in the analytical model. While multivariate approaches have shown to be most useful in the multidimensional conceptualisation of poverty, they have also shown to be rather important for the comprehensive analysis of associations and interactions among the variables involved in the process. As mentioned above, the appropriate statistical programme for the IPIA process is the OVERALS multivariate statistical analysis, being a nonlinear generalised canonical analysis, allowing the examination of different sets of variables, and variables with different measurement levels, scaled as multiple nominal, single nominal, ordinal, or numerical. The related factor analysis which enables the construction of an overview of all variables is known as the Principal Components Analysis (PRINCALS) which enables the identification of dominant patterns in the data, including groups, trends and outliers, which can be shown in two projected plots. The *Nonlinear Canonical Correlation Analysis* (CANALS) refers to a procedure with several applications to analyse the relations and interactions between two or more sets of variables instead of between the variables themselves.

The final step of the multiple regression analysis is an extension of the previous multivariate analysis as it enables the calculation of the correlation coefficients (r) between the categories of variables, represented as 'blocks' in the analytical model. By the measurement of the relationships and interactions between these 'blocks' of variables through the multiple regression analysis, the correlation can be calculated between observed values and predicted values in the model. In this way, specific categories of variables rather than individual variables are contributing to the prediction of differences in the dependent variables. In the case of the IPIA process, these differences are operationalised as changes in the patterns of consumption behaviour of groups of the target population including the poor, indicating a degree of impact on their well-being.

The stepwise analytical process of the evidence-based IPIA model enables in this way not only the *understanding* of the impact of outside interventions on specific local groups or individuals, identified in the bi- and multivariate analysis in terms of significant variables as *determinants*, but also the *prediction* of the expected impact of such interventions on specific local groups or individuals, identified in the multiple regression analysis in terms of *predictors*.

It is imperative to present not only the various considerations, but also the outcomes of the sequential phases, stages and steps which in a coherent way are leading up to the formulation of the final conclusions of the IPIA Report. The approach of the Integrated Poverty Impact Analysis (IPIA) to develop an evidence-based foundation for the design and implementation of a series of appropriate dimensions, phases, stages and steps for objective and independent decision-making processes for policy and project planning and implementation will eventually pave the way for the realisation of a positive effect on the position of the poor at the community level, and as such, contribute to the reduction of poverty and sustainable development around the globe.

References

Agung, A. A. G. (2005). *Bali endangered paradise? Tri Hita Karana and the conservation of the Island's biocultural diversity*. Ph.D. Dissertation. Leiden Ethnosystems and Development Programme. LEAD Studies No. 1. Leiden University. xxv + ill., p. 463.

Aiglsperger, J. (2014). *'Yiatrosofia yia ton Anthropo': Indigenous knowledge and utilisation of MAC Plants in Pirgos and Pretoria, Rural Crete: A community perspective on the plural medical system in Greece*. Ph.D. Dissertation. Leiden Ethnosystems and Development Programme. LEAD Studies No. 9. Leiden University, xxvi + 235, pp. ill.

Ambaretnani, P. (2012). *Paraji and Bidan in Rancaekek: Integrated medicine for advanced partnerships among traditional birth attendants and community midwives in the Sunda Area of West Java, Indonesia*. Ph.D. Dissertation. Leiden Ethnosystems and Development Programme. LEAD Studies No. 7. Leiden University. xx + ill, p. 265.

Carson, R. (1962). *Silent spring*. New York: Penguin Books.

Chambers, R., & Mayoux, L. (2005). Reversing the paradigm: Quantification and participatory methods. In *Paper for the EDIAIS Conference on 'New Directions in Impact Assessment for Development: Methods and Practice'*. Manchester: University of Manchester, UK.

Chirangi, M. M. (2013). *'Afya Jumuishi': Towards interprofessional collaboration between traditional & modern medical practitioners in the Mara Region of Tanzania*. Ph.D. Dissertation. Leiden Ethnosystems and Development Programme. LEAD Studies No. 8. Leiden University. xxvi + ill., p. 235.

Coudouel, A., Hentschel, J. S., & Wodon, Q. T. (2002). Poverty measurement and analysis. In Klugman, J. (Ed.), *A sourcebook for poverty reduction strategies* (Vols. 1 and 2). Washington, D.C.: World Bank.

Dagum C., & Costa M. (2004). Analysis and measurement of poverty: Univariate and multivariate approaches and their policy implications. A case study in Italy. In Dagum C., & Ferrari G. (Eds.), *Household behaviour, equivalence scales, welfare and poverty. contributions to statistics*. Heidelberg: Physica Verlag.

Deaton, A. (1997). *The analysis of household surveys: A micro-econometric approach to development policy*. Washington, D.C.: The World Bank.

Deaton, A., & Grosh, M. (2000). Consumption. In M. Grosh & P. Glewwe (Eds.), *Designing household survey questionnaires for developing countries: Lessons from 15 years of the living standards measurement study*. New York, NY: Oxford University Press for the World Bank.

de Bekker, J. C. M. (in press) *Transcultural health care utilisation in the mara region of Tanzania*. Ph.D. Dissertation. Leiden Ethnosystems and Development Programme. LEAD Studies. Leiden University.

Djen Amar, S. C. (2010). *Gunem Catur in the Sunda Region of West Java: Indigenous communication on the MAC plant knowledge and practice within the Arisan in Lembang*. Ph.D. Dissertation. Leiden Ethnosystems and Development Programme. LEAD Studies No. 6. Leiden University. xx + ill., pp. 218.

Erwina, W. (2019). *Iber Kasehatan in Sukamiskin: Utilisation of the Plural Health Information & Communication System in the Sunda Region of West Java, Indonesia*. Ph.D. Dissertation. Leiden Ethnosystems and Development Programme. LEAD Studies No. 10. Leiden University.

Field, A. (2009). *Discovering statistics using SPSS*. London: Sage.

Galer, S. (2012). *New study finds consumption measures poverty better than income. Uchicago News*. University of Chicago.

Ibui, A. K. (2007). *Indigenous knowledge, belief and practice of wild plants among the Meru of Kenya*. Ph.D. Dissertation. Leiden Ethnosystems and Development Programme. LEAD Studies No. 3. Leiden University. xxv + ill., pp. 327.

International Monetary Fund. (2009). *Poverty reduction strategy papers (PRSP)*. Washington, D.C.: IMF. http://www.imf.org/external/np/prsp/prsp.aspx

Interorganizational Committee on Principles and Guidelines for Social Impact Assessment (IOCGP). (2003). *Principles and guidelines for social impact assessment in the USA. Impact Assessment and Project Appraisal, 21*(3): 3:231–3:250.

Irizarry, R., Peng, R., & Leek, J. (2017). Reproducibility and replicability is a glossy science now so watch out for the hype. Simply Statistics. 2 March 2017. http://simplystatistics.org/2017/03/02.

Kanbur, R., et al. (Eds.). (2001). *Qualitative and quantitative poverty appraisal: Complementarities, tensions and the way forward. Contributions to a workshop*. Ithaca, NY: Cornell University.

Kent State University Libraries (KSUL). (2016). *Statistical consulting*. Kent, OH: Kent State University.

Klugman, J. (2002). *A sourcebook for poverty reduction strategies* (Vols. 1 and 2). Washington, D.C.: World Bank.

de Leeuw, J., & Meijer, E. (2007). *Handbook of multilevel analysis*. New York: Springer.

Leurs, L. N. (2010). *Medicinal, aromatic and cosmetic (MAC) plants for community health and bio-cultural diversity conservation in Bali, Indonesia*. Ph.D. Dissertation. Leiden Ethnosystems and Development Programme. LEAD Studies No. 5. Leiden University, xx + ill., pp. 343.

Meyer, B. D., & Sullivan, J. X. (2002). Measuring the well-being of the poor using income and consumption. In *Paper for the Joint IRP/ERS Conference on Income Volatility and Implications for Food Assistance*, Washington, D.C.

Meyer, B. D., & Sullivan, J. X. (2012). Identifying the disadvantaged: Official poverty, consumption poverty and the new supplemental poverty measure. *Journal of Economic Prospectives, 26*(3), 111–136.

Peng, R. D. (2009). Reproducible Research and Biostatistics. *Biostatistics, 10*(3), 405–408.

Practical Concepts Incorporated (PCI). (1979). *A manager's guide to a scientific approach to design & evaluation*. Washington, D.C.: PCI.

Saefullah, K. (2019). *Gintingan in Subang: An Indigenous Institution for Sustainable Community-Based Development in the Sunda Region of West Java, Indonesia.* Ph.D. Dissertation. Leiden Ethnosystems and Development Programme. LEAD Studies No. 11. Leiden University.

Slikkerveer, L. J. (1990). *Plural medical systems in the Horn of Africa: The legacy of 'Sheikh' hippocrates*. London: Kegan Paul International.

Slikkerveer, L. J. (1991). *Indigenous agricultural knowledge systems in Africa: The Ethnosystems' approach to sustainable development in Kenya.* SSCW/NWO Programme Report. Leiden: Leiden University. 86 pp.

Slikkerveer, L. J. (2011). *Assessment of the extent of the assessing the over-the-counter medication in primary care in southern Europe (OTC-SOCIOMED).* Leiden: LEAD Programme.

Slikkerveer, L. J. (Ed.). (2012). *Handbook for lecturers and tutors on integrated microfinance management for poverty reduction and sustainable development in Indonesia*. Bandung: LEAD-UL/UNPAD/MAICH/GEMA PKM.

Statistics Solutions. (2017). Statistics solutions: Advancement through Clarity. https://www.statisticssolutions.com.

Steinmetz, S. R. (1912–1913). Die Stellung der Soziographie in der Reihe der Geisteswissenschaften. *Archiv für Rechts- und Wirtschaftsphilosophie*, *6*(3) 492–501.

United Nations (UN). (2015). *Post-2015 agenda for sustainable development and sustainable development goals (MDGs).* New York: United Nations.

Vanclay, F. (2003). SIA principles: International principles for social impact assessment. *Impact Assessment and Project Appraisal, 21*(1), 5–11.

van der Burg, E. (1983). *Canals user's guide*. Leiden: Department of Datatheory, Leiden University.

van der Burg, E. (1988). *Nonlinear canonical correlation and some related techniques*. Leiden: Department of Datatheory, Leiden University.

van de Geer, J. P. (1993a). *Multivariate analysis of categorical data: Applications*. Newbury Park, CA: Sage.

van de Geer, J. P. (1993b). *Multivariate analysis of categorical data: Theory*. Newbury Park, CA: Sage.

Weinberg, S. L., & Abramowitz, S. K. (2002). *Data analysis for the behavioral sciences using SPSS*. Cambridge: Cambridge University.

World Bank (WB). (2003). *User's guide to poverty and social impact analysis*. Poverty Reduction Group (PRMPR) and Social Development Department (SDV). Washington, D.C.: World Bank.

World Bank (WB). (2009). *The living standards measurement study (LSMS)*. Washington, D.C.: World Bank.

World Bank (WB). (2017). *The living standards measurement study (LSMS)*. Washington, D.C.: World Bank.

L. Jan Slikkerveer is Professor of Applied Ethnoscience and Director of the Leiden Ethnosystems and Development Programme, Faculty of Science, Leiden University, Leiden, The Netherlands. He received his Ph.D. on his fieldwork in the Horn of Africa from Leiden University in 1983 and an Honorary Degree from the Faculty of Medicine of Universitas Padjadjaran in Bandung, Indonesia in 2005. He further extended advanced training and research in the newly-developing field of applied ethnoscience in several sub-disciplines, including ethno-economics, ethno-medicine, ethno-biology/botany, ethno-pharmacy, ethno-communication, ethno-agriculture and ethno-mathematics in combination with a focus on three target regions of South-East Asia, East-Africa and the Mediterranean Region. He is the supervisor of 25 Ph.D. students at Leiden

University and has published more than 100 books and 300 articles on the subject. While he has also received a number of substantial subsidies from the European Union in Brussels for international development programmes in South-East Asia, East-Africa and the Mediterranean Region, he also conceptualised both the IMM & ICMD approaches in Indonesia, and is Senior Lecturer in the International Master Course on Integrated Microfinance Management (IMM) at the Faculty of Economics and Business of Universitas Padjadjaran in Bandung, Indonesia.

Part III
Indonesia: Transitional Development Organisations and Institutions

Chapter 8
Governance, Policies, Rules and Regulations in Indonesia

J. C. M. (Hans) de Bekker and Kurniawan Saefullah

In democratic settings, government agencies and their officials in bureaucratic hierarchies are more ethical than self-interested individuals or firms in competitive markets.
H. George Frederickson (1999)

8.1 Introduction

The basic characteristics of a democratic society are built on the *Trias Politica, Constitutional Law, Equality and Freedom, Political Entities, Majority Rule, Proportional Representation* and a *Legal Framework for All levels of the Society* (UNDP 2005). The principals are the building blocks of a model adopted by the United Nations for worldwide promotion: *"Democracy is a universally recognised ideal and is one of the core values of the United Nations. Democracy provides an environment for the protection and effective realisation of human rights. These values are embodied in the Universal Declaration of Human Rights and further developed in the International Covenant on Civil and Political Rights, which enshrines a host of political rights and civil liberties underpinning meaningful democracies."* According to the *United Nations Development Programme Report* (UNDP) (2005), the characteristics of good governance are defined in societal terms as follows:

- *Participation*—All men and women should have a voice in decision making, either directly or through legitimate intermediate institutions that represent their interests. Such broad participation is built on the freedoms of association and speech, as well as the capacities to participate constructively.

J. C. M. (Hans) de Bekker (✉) · K. Saefullah
LEAD, Leiden University, Leiden, The Netherlands
e-mail: hansdebekker@hotmail.com

L. J. Slikkerveer et al. (eds.), *Integrated Community-Managed Development*, Cooperative Management, https://doi.org/10.1007/978-3-030-05423-6_8

- *Rule of law*—Legal frameworks should be fair and enforced impartially, particularly the laws on human rights.
- *Transparency*—Transparency is built on the free flow of information. Processes, institutions and information are directly accessible to those concerned with them, and enough information is provided to understand and monitor them.
- *Responsiveness*—Institutions and processes try to serve all stakeholders.
- *Consensus orientation*—Good governance mediates differing interests to reach a broad consensus on what is in the best interests of the group and, where possible, on policies and procedures.
- *Equity*—All men and women have opportunities to improve or maintain their well-being.
- *Effectiveness and efficiency*—Processes and institutions produce results that meet needs while making the best use of resources.
- *Accountability*—Decision-makers in government, the private sector and civil society organisations are accountable to the public, as well as to institutional stakeholders. The accountability differs depending on the organisation and whether the decision is internal or external to an organisation.
- *Strategic vision*—Leaders and the public have a broad and long-term perspective on good governance and human development, along with a sense of what is needed for such development. There is also an understanding of the historical, cultural and social complexities in which that perspective is grounded.

The *Human Factor*, as an explicit feature of effective governments in the traditional style of democracy, is valued by what is called 'the ability to sanction' and 'individual integrity', also known as 'responsible behaviour'(cf. Bovens 1990). It means that organisations which are available to the government to implement a certain ruling are actually capable of carrying out such a task without using coercive measures. Another feature is that officials of these governing bodies should be willing and able to detect any malfunctioning, and undertake action to change it, despite the threat to their professional position (cf. Kolthoff et al. 2007). Such a quality (or 'attribute') of personal integrity is often assumed to be present, but in reality many cases of ineffective government can be traced back to the choices made by individual people in crucial positions (cf. Khan 2006).

The purpose of government entails the provision of various public or collective goods and services, assuring a minimum level of access by individuals to goods and services (including housing, health, education, employment, etc.), resolution and adjustment of conflicts, providing an economic infrastructure, the maintenance of competition (a certain control over, and response to the market), the protection of natural resources, and the stabilisation of the economy. A cross-reference is made here to the 'Failed State Index', referring to a model for measuring the performance of a government in a global context. The index features twelve indicators which can classify a nation state as being able to perform its responsibilities (cf. Rotberg 2013).

The *Triangle Model* shows that social reality can be modelled by placing it in a framework of the state, including laws, regulations and public policies, the market including trade and business, and the civil society which is seen as organisations and groups with a social mission (cf. Gorzki 2007). Conceptualised in a triangle, these entities interact on the basis of their own power and mission. Each part is dependent on the other two parts in order to create social cohesion and a prosperous society. In the ideal situation, no part of the triangle dominates the other two parts. A 'democratic triangle' which has found the right balance between its three constituent elements can secure the quality of social existence. The second part in the triangle is the components of civil society i.e. charity NGO's, social (voluntary) movements, trade unions, professional, business, and religious associations, and the role of the media, with reference to the concept of *Public Private Partnership* (PPP). It means that an organisation founded on and run by citizens is responsible for a public service that may equally well be considered a government responsibility (cf. Schrijver and Weiss 2004).

An important question in this context is: what are the operational limits of the state? This covers the topics of the minimal state, or the so-called night-watchman state. Some scholars and politicians argue that the state should only protect its citizens and enforce the law. It is a concept where the government's responsibilities are so minimal that they cannot be reduced much further without becoming a form of anarchy, e.g. Tea Party ideologies in the U.S. Most political ideologies disagree with the notion of a minimal state. These ideologies introduce the concept of a welfare state in which the government also provides care and protection for disadvantaged or dependent people. It is a matter of politics and political preference as to how countries define the governmental radius of action.

The 'performance' of government is described in critical factors for successful reform in developing countries by Smith (2007): commitment from political leadership, pressure from civil society actors, available resources to implement reform, popular consensus on policy, and proper analysis of physical constraints. The relationship of these factors with other governance issues lies in the political emancipation of poor people, which is the target group actually participating in the decision-making process.

Democratic processes and development in Indonesia

Indonesia has been struggling with its own democratic development. Since its independence in 1945, the state has implemented different types of democratic systems. According to Azra (2000), the history of the development of democratic systems in Indonesia can be divided into four periods:

1. *Democracy during the revolutionary period (1945–1950)*

Since its independence in 1945 until circa 1950, Indonesia was still struggling with the Dutch who wanted to return to Indonesia. During this period, democracy was not really being implemented. At the beginning of independence, the political system was still centralised as it was governed by *Transition Law 1945, Article 4*. According to this law, although the country had a parliament and a court of justice, the president had absolute power, authorising and controlling the system.

2. *Democracy during the 'old order' (1945–1950)*

The second period of Indonesian democracy was the Old Order period which can be divided into two phases: the Liberal Democracy phase from 1950 to 1959, and the Guided Democracy from 1959 to 1966.

The first phase of liberal democracy started when the Government of Indonesia submitted the *Undang-undang Sementara* 1950 ('Temporary Law 1950') which was approved by the Parliament. According to this Law, Indonesia was practicing 'Pure Liberal Democracy' based on a Parliamentary system. The first democratic election was held in 1955 in which members of parliament were chosen, with 172 political parties participating in the election. However, the phase of pure democracy only survived for a short period of time due to conflict between the Parliament and the Cabinet. The conflict ended when the first President Soekarno released a Decree in 1959 which cancelled the 'Temporary Law 1950', returning to the Indonesian Law of 1945.

The second phase of this Old Order ran from this Presidential Decree in 1959 until 1965. During this phase, the country basically experienced a form of democracy which was based on the NASAKOM concept, a combination of nationalism, communism, and religion. The period showed a struggle between military and civil power on one side, and between different ideologies on the other. It led to one of the biggest tragedies in Indonesian history with the revolt of 1965 (*Gerakan 30 September*—G30S), where the Indonesian Communist Party was accused of being behind the killing of some high-ranking army generals. Considering that he didn't want the conflict to grow wider, President Soekarno did not take a firm stand towards the communist party although he was challenged to do so by the army and some groups of influential people. The conflict became a tragedy when almost half a million people judged as communists were killed by the military and civilian opponents in the name of saving the country. Roosa (2006) believes that President Suharto did a *coup d'état* backed by the CIA to eliminate President Soekarno. Roosa (2008) furthermore states that 1965 was the starting point for the regime of President Suharto, turning the Indonesian Democracy into an authoritarian style of leadership during his reign from 1966 to 1998.

3. *Democracy during the 'new order' (1965–1998)*

The period of the New Order ran from 1965 to 1998. Although democracy was still officially implemented, the participating political parties were under the control of Suharto's military. In general elections held as a symbol for practicing democracy, legislative representatives were selected, but in practice the bottom-up voices were silenced by his authoritarian leadership. The number of political parties that participated in the general elections decreased. From the 172 political parties that participated in 1955, only 10 participated in 1971 and only 3 contestants joined the general elections between 1971 and 1997. In addition to that, judiciary institutions did exist during this period of time; however their role was only an extension of the executive power of President Suharto. There were some civil uprisings during this period of time, such as in Aceh, Papua and East Timor (now an independent country known as 'Timor Leste') as a reaction to the authoritarian leadership established through military power.

Although people's freedom of expression and other liberties were neglected during this era, economic development and growth during the 80s during Suharto's presidency were quite substantial. Only after the Asian crisis started with the collapse of Thailand's Baht in June 1997 did the Indonesian economy suffer dramatically, with President Suharto finally stepping down in May 1998.

4. *Reform democracy (1998–present)*

The end of the authoritarian regime by President Suharto was marked by a succession of power when President Suharto handed over the presidency to the former Vice-President, B. J. Habibie. The transition of power also meant the start of democratic reform in Indonesia where people's voices and liberties were finally accommodated politically. There are some specific priorities on the reform agenda, such as increasing the role of the members of parliament, revising the state's code of conduct which should be free from corruption, and limiting the presidential reign to two periods or 10 years maximum. The period of President Habibie's tenure only lasted until 1999 when the Members of Parliament did not accept his Report submitted to the Special Assembly. Abdurrahman Wahid (known as Gus Dur) was elected as President, and Megawati Soekarnoputri, the daughter of the first President Soekarno, became Vice-President between 1999 and 2004.

During this period of time, some democratic reform was realised, such as regional autonomy for local government and legislation, and the approval of the direct election of a president and vice-president. President Abdurrahman Wahid did not survive his term to the end because of a political conflict and parliament again held a special assembly which resulted in Vice-President Megawati Soekarnoputri replacing him from 2002 until 2004 (cf. Budiarjo 1998; Azra 2000; Roosa 2008; Zonanesia 2016).

In the year 2004, the first election was held in Indonesia where people directly elected both positions of President and Vice-President: Dr. Susilo Bambang Yoedoyono and Mr. Jusuf Kalla were elected as the President and Vice-President, respectively, from 2004 until 2009. During the second period of Reform Democracy (2009–2014), President Yoedoyono was attended by Mr. Boediono as Vice-President, who was a former Governor of the Central Bank of Indonesia. The era witnessed more reforms such as freedom of the press, the end of military operations in Aceh, and the approval of the *Microfinance Law* which regulates formal and informal microfinance institutions. In 2014, Mr. Joko Widodo was elected as the President until 2019. One of the promises he made was to pay attention to indigenous people, bottom-up development, and economic inclusion.

One of the democratic reforms which started under his administration is the 'Village Law' of 2014 which promotes 'bottom-up' development initiatives. At the same time, the government allocated one billion rupiah to each of the 72,000 villages in Indonesia in order to support such initiatives (cf. Indonesian Law 6 2014). The President has assigned special staff to implement these local programmes. As a first step, the staff is currently in the process of documenting local traditions in all regions of Indonesia. The database will be used to include selected villages in the future development policies of Indonesia.

8.2 Spheres of Effective Government: Threats and Challenges

The assessment of the spheres in which the central government plays a role shows a wide variety in the existing literature, described in any type of assignment as a tool to start a general analysis of a specific region.

Geodemographic characteristics. The size of the country and the diversity of the population are factors which determine the type of infrastructure, such as geographical distance, means of communication, and the investments needed in terms of the distribution of services, i.c. personnel, public services and available resources. The contrast between urban and rural areas and the population density have become important parameters.

Economic conditions and natural resources. With an emphasis on rural areas, the scope of the 'local economy' is based on the livelihood of the majority of the local inhabitants. If it includes urban areas, then the state of the 'national economy' (in terms of scale) becomes increasingly important, and the relationship with established interests becomes an important factor. A problem may arise in the classification of the local population in terms of the definition of poverty.

Strength of official politics. The establishment of regional or district councils could lead to a difference in interest with regard to local policies with communities in which the traditional leadership is still very strong. In addition, ethnic, religious, and/or cultural diversity remain in the sub-regions.

Existing financial system. Regulation involves a direct or indirect relationship with the standards of commercial banking, and/or national financial policies. The definition of institutions refers to informal, semi-informal and formal institutions, depending on their relationship with the official financial system. Also, the number of lenders and the size of credit providers in terms of capital in one geographical area are important factors.

Inability of authoritative decision-making. Such a phenomenon is closely related to the individual credibility of people in office, the integrity of civil servants, and/or possibly the controversy brought about by unethical policies. Whatever the case, when authority loses its legitimacy, it is detrimental to the coherence of that society. The concept of the 'failed state' refers to the ultimate stage in which the central government ceases to be capable of implementing its rule and the state threatens to disintegrate, such as in Somalia, South Sudan, and Libya.

Inadequate provision of public services. It is primarily related to the availability of physical resources, although it may also have other dimensions such as unprofessional management, lack of consistent policies, or favouritism.

Lack of interaction with members of the international community. Part of the legitimacy also stems from external factors. When a government becomes isolated from international partners, it usually affects the economic trade, the exchange of knowledge or aid programmes, but it can also influence the social and ethical factors which need either the support or the sanctions of international organisations, particularly within the context of the process of globalisation.

Failure to collect revenues. It relates directly to the delivery of public services and involves the combination of legitimacy, sufficient staff and endorsement. Also, the topic cannot be separated from economic standards, considering that tax is also being collected from multinational companies or NGOs. The essence of this problem is in the invisible loop which it can create: insufficient taxes are leading to insufficient delivery of services, lack of legitimacy, corruption, and so forth.

Alternative economies. The effect of a large alternative economy is usually felt in two dimensions: the first refers to the diminishing resources for the government because of the absence of taxing possibilities, while the second concerns the fact that there is no ruling principle with regard to the market mechanism. The problem is that the imbalance may be inherent in terms of the (mis)use of resources, prices, accessibility, participation, distribution, illegal activities, or even waste disposal, which all are out of the reach of a formal regulatory organisation.

Widespread corruption and criminality. Another important aspect is the question of sequence. In societies with a colonial heritage, the position of the civil servants has often been closely related to the original social group, economic status, kinship, or to a dominant hierarchy within the bureaucratic system. The system leaves people in such a position rather vulnerable to the pressure of reciprocal activities, not defined by their office. The pretention often used is that corruption is a result of poverty. However, in many developed countries the misuse of means of power is equally widespread, albeit possibly on a different level of intensity or exposure. A related aspect is personal integrity, regardless of status. Criminality is often related to the level of development of the society, especially in terms of the accessibility of services, and the imbalance between the economic strata. As a consequence of the complementary effects in a situation with a poor system of bureaucracy, a low level of development and weak law enforcement, criminal behaviour will have an opportunity to flourish.

Involuntary movement of populations. The involuntary movement of populations is usually preceded by conflict, natural hazards, or economic decline, and in that sense the situation is the result of another process. A government confronted with such a situation must be able to find alternatives, while the support of international organisations will play an important role in solving this problem.

Considerable economic downfall. The consecutive effect of economic decline often shows an increasing lack of confidence in the central government, since it is expected to play a dominant role in the regulation of the economy. Lack of confidence may demonstrate itself in the growth of an alternative economy with black markets, and an increasing neglect of authority and legal structures.

Large-scale civil disobedience. The ultimate stage of civil disobedience refers to the situation in which the population no longer accepts their government, often preluding to a violent change. The condition can be the result of a combination of problems which are mentioned above, such as legitimacy, credibility, economic decline and corruption.

8.3 The Role of Public Policy

According to Shah (2006), public policy improvement as a means to reduce poverty refers to the five perspectives of: *Traditional Fiscal Federalism* (TF), *New Public Management* (NPM), *Public Choice* (PC), *New Institutional Economics* (NIE), and *Local Governance Networks* (LGN).

This Paragraph is focussing on the two perspectives of NPM and LGN, which have been introduced as a way to deal with bureaucracy. The first principle of NPM refers to managerialism, defined by Pollit (1991), as involving a continuous increase in efficiency, the use of sophisticated technologies, a labour force which is disciplined to productivity, a clear implementation of the professional management role, and finally, managers given the right to manage. According to Walsh (1995), the characteristics of the second principle of NPM are continuous improvements in quality, an emphasis on devolution and delegation, appropriate information systems, an emphasis on contracts and markets, the measurement of performance and increased audits and inspection (cf. Kolthoff et al. 2007). In this view, the local government requires a new 'Local Public Management Paradigm', in which the local government separates the advocacy of its policy from being the agent who does the actual implementation. The local government may have to outsource its services with higher provision costs, and subject the in-house providers to competitive pressure from outside providers in order to lower the transaction costs for the citizens. It means that a market principle is applied to the local organisation of the public services. There exists an interesting analogy in higher education, whereby the formal institutions have to prove their right of existence by delivering better results, even though they are simultaneously being subsidised by the same government. These institutions are also forced to secure part of their budget from private funds through the delivery of special services such as research assignments or the transfer of knowledge in training programmes.

Decentralisation as the other side of local governance refers to the reaction of the state to the local authority when it becomes successful. The central government may fear losing both its economic and political control, or even be confronted with excessive movements towards regional autonomy. The degree to which it affects the benefits of the central authority from that particular region will correspond to the support for decentralisation (cf. Green 2005). In the mechanism between the central and local government, one example which stands out is the degree of autonomy in terms of the collection of revenues and the public expenditure. It determines the extent to which a local authority has the flexibility of allocating available resources according to the local needs. The same applies to the status of 'civil servant' for the employees working in the local government, which is well protected, whereby the employees cannot be dismissed on an ad hoc basis. Such a situation does not provide a sound basis for local governments to be effective in terms of human resources management (cf. Ohmae 1996).

As Kamarck (2007) asserts, new types of government style are implemented, such as 'Government Reinvented' which refers to the introduction of a commercial managerial style in the execution of public policies, and 'Government by Network' which refers to the delegation of the execution of tasks with a contract to a specialised third party. Also, the style of 'Government by Market' makes use of a market mechanism as a regulation for achieving specific policy goals. The merits as well as the discrepancies of these styles should be compared in terms of not only the separation of the politician from the manager, but also the risk of purely efficiency-driven decisions—as opposed to socially desirable decisions—and the perils of conflicts of interest with commercial executors (cf. Kolthoff et al. 2007). An illustrative example of this phenomenon is presented in the introduction of a deposit on the return of aluminium cans, which proves to be more effective than banning the disposal of cans with regular trash, i.e. making use of a market mechanism as an economic incentive to implement special rules.

For a deeper insight in the direction of public policies in developing countries, the *Poverty Reduction Strategies* of the World Bank (2005) are functional as they show an inventory of various desired types of reform, in particular those relevant to the reduction of poverty, including the following (cf. Grindle 2004):

a. *Setting priorities, initiating political and judicial reforms*

The first lesson from this analysis deals with laying a basis for possible change through the setting of priorities in what to tackle first, and to begin by creating a formal basis in terms of law or legislation through which that change will be covered. The suggestion is that more often the identification of the problem is quite mechanical, and not associated with a possible solution. Most governments do not provide much detail on the proposed reforms and do not indicate the priorities, outline specific steps which are needed to be taken, or suggest how they would measure, monitor, and evaluate the tasks to which they have committed themselves.

b. *Measurable targets and timelines*

Following the first step of setting priorities, it becomes very important to reduce the scope of the problem by setting a target and a timeline which renders progress measurable. It might apply foremost in the sphere of budget and public expenditure, in order to develop a secure method of monitoring and accountability, which is visual to the community. The introduction of a specific time frame could add to the credibility.

c. *Awareness and number of incidents*

It is well indicated that in the process of identifying these reforms within the *Poverty Reduction Strategies*, a side effect emerged in rendering the officials more aware of the lack of the impact of their activities on poverty reduction, more so than the actual content of their policies. The reason may be that the inventory involved large-scale consultation, including the civil organisations. Another insight refers to the awareness that limited resources tend to prevent governments from engaging simultaneously in several reform programmes.

d. *Choice of effective assessment*

Another lesson brought forward is the relevance of the reform, or its effect in the long term, towards poverty reduction. It is illustrated by the example of popular emancipation, where Grindle (2004) argues that: '*the mobilisation of the poor into political parties, interest groups, unions, and NGOs may be a condition under which judicial reform, civil service reform, decentralisation, and other kinds of changes are most likely to have a significant impact on poverty or on the poor*'.

e. *Time factor and sequence: chicken and egg*

In an effort to assess the institutional reform agenda of international financial institutions, Chang (2000) finds that many factors currently considered preconditions for development are actually consequences of development. Indeed, Chang demonstrates that considerable economic development has occurred long before countries had fully institutionalised democracies, professional bureaucracies, rules for corporate governance, modern financial institutions, or extensive social welfare services. Similarly, Moore (1998) argues that as governments have become more proficient in tax collection, their overall organisational capacity tends to improve.

f. *Hidden dimension: political motives*

The hidden factor which is also described in the present analysis of several authors in the World Bank Workshop refers to the choice of priority and reform which is based on a political motive (cf. World Bank 2001). It is possible that a problem is tackled not because it really deserves that priority, but because it will deliver the short-term effect of consolidating the political basis of the government which is implementing the policy. In this way people do benefit, and it is argued that it might even be the right step to create sufficient support to tackle and implement another, maybe a less popular policy. The approach has also been termed 'political feasibility'. An example could be the policy to bring extra police on the streets at night in order to create a sense of security, instead of actually eliminating organised crime.

g. *Alternative providers*

Another interesting viewpoint refers to the argument for the idea that in line with decentralisation and local governance, public services could also be delegated to other actors apart from the government. In such a case, the definition should be wider, where the traditional authority can mediate in social conflict through arbitration, instead of through a local court. In Europe, there are interesting examples of the privatisation of public services as they are assumed to run more efficiently since they are based on market principles (cf. Kamarck 2007).

The above-mentioned considerations refer to the main characteristics which have been found absent in the *Poverty Reduction Strategies* (PRS). Similarly, their relationship with poverty reduction was also found to be absent in most of the policies described in this inventory. Moreover, there was rather a preoccupation with the improvement of the enabling system itself. The problems which are identified as the most urgent concern the reform to the judicial system, the public administration, anti-corruption, decentralisation and the management of public expenditure.

The role of NGOs as representatives of the poor in public policy reforms is also regarded in relation to their characteristics of reliance on outside resources, alliance with political partners, maintenance of a parallel agenda, rigid adherence to a mission statement, and insufficient consultation of target groups. The suggestion is that investments in the transition of poor people to the consolidated civil society are preferable to representation by proxy.

The level of governance and policy analysis in Indonesia

Table 8.1 shows the analytical assessment of governance in Indonesia, sub-divided into different approaches ranging from centralised to decentralised, in which a multi-level or hybrid approach is also used (cf. Fritzen and Brassard 2007).

As Table 8.1 also shows, in the case that development is implemented on the basis of centralisation, the 'objectivity' of the policy will not be questioned since all problems are generalised on the central level. However, the resource allocations determined by the bureaucracy of the central government are generally criticised because of the bias in the source of information, which is usually based on generalisations about the local situation.

The local people will have a chance to participate in the implementation of development policies since they are the people who know their problems, priorities and alternative solutions for local problems. However, this approach has also received criticism, not only because of its subjective conclusions about the local situation, but also because of the lack of ability of local people to implement appropriate solutions. In many cases, where the local people could not finance their development programmes, an appeal was made to the central government, so that decentralisation has not actually been achieved.

There is another alternative form of governance which is probably more suitable for the Indonesian context. It operates from a multi-level approach which synthesises between a centralised and a decentralised governance approach. The central government will negotiate with the local government based on a local development project proposal. The type of partnership between the parties of the government and the private people would be alternative. In theoretical terms, it could as such be a 'Public-Private Partnership' of which the *Program Nasional Pemberdayaan Masyarakat* (National Programme for Community Empowerment PNPM) could form an illustrative example (cf. Bappenas 2013). The development programme was initiated by the Ministry of Social Welfare and financed by the World Bank, whereby the Ministry delegates the staff from the central level to the village level. The staff act as facilitators to accommodate various development programmes at the local level, ranging from the educational sector through the infrastructure to health

Table 8.1 Policy analysis on different levels of governance

Approaches	Situation Analysis	Goal Setting	Program Planning	Implementation	Monitoring/Evaluation
Overly centralised (national level)	Relatively 'objective' based on large scale standardized surveys incl. mainly quantitative indicators	Aggregated goals: inherent biases due to sampling methodology and restricted choice of indicators	Output orientation: bureaucratic resource allocation	Sectoral: vertical, non-integrated lacking coordination	Focus on inputs: using additional and technical controls
Overly decentralised	Relatively 'subjective' based on participatory appraisals including mainly qualitative indicators	Few local goals: few specific measurable goals, only local goals	Donor-dependent lack of local financial resources and institutional weight	Bottom-up community participation, often weak links to technical agencies	Local project evaluation: assessment along multiple (often qualitative) indicators, local & people involvement
Towards a multi-level synthesis	Using mixed indicators: both qualitative and quantitative indicators are used	Outcome oriented local ownership and political will through goals 'negotiated'	Performance oriented: central transfer link to local accountability plus local mobilisation	Coordinated: partnership between state, private, civil society into each entity's strength (co-production)	Adaptive focus on: implementation gaps and needed adjustments in strategy to reach goals

Source Frietzen and Brassard (2007)

services. If a proposal for a local development programme were to be accepted by a lower level of the government—for instance at the municipality level—the proposal would be sent to the central government to be financed.

However, the financial support from the central government will only be implemented as an incentive and not in support of the whole project. It is expected that the incentive would motivate the people in the local community to support the project themselves. This type of programme nowadays represents the most well-known development programme in Indonesian local communities, using a multi-level or hybrid approach (cf. Asian Development Bank 2006; Bappenas 2013).

Public health policies

The Indonesian public health services are based on several regulations set by the central government. Indonesia has implemented the concept of the 'Indonesian health system' which is based on Law No 23/1992 about the Health Foundations, the Indonesian Legislative Judiciary No 7/2001 about the Vision of the Future of Indonesia, the Ministry Act No 99a/1982 about the Legal Status of the National Health System, Indonesian Law No 33 and 34/2004, as well as the above-mentioned Law of Local Governance.

The Indonesian Health System has as its main objective to achieve public health with quality services, healthy behaviour on the part of the people, and a healthy environment. Strategically, this main objective is to be reached by three strategic programmes: functionalise the center of health services, revitalise the center for family health, and implement the National Health Service programme, both for public and private services. The Indonesian Health System is implemented through the health-oriented development programme which consists of four strategic actions: (1) empower people to have healthy lives, (2) increase people's access to quality health services, (3) increase the surveillance system and health information monitoring, and (4) increase public health expenditure. The organisation of the Indonesian Health System is represented in Fig. 8.1.

There are no real achievements shown by the maternal health indicator with regard to child mortality. According to the *Report on the Indonesian Achievement of Millennium Development Goals* (Bappenas 2010) concerning Maternal and Child Health, Indonesia ranks lowest in the global achievements. The implementation of Indonesia's health system is supported by all public health institutions ranging from the central to the local level of the Indonesian structure of health services. The Maternal Mortality Rate (MMR) in Indonesia has gradually been reduced from 390 in 1991 to 228 per 100,000 live births in 2007. Extra efforts are still needed to achieve the aim of the *Sustainable Development Goals* by 2030 of the United Nations (UN 2015). The World Hall countries should reduce maternal mortality ratio (MMR) by at least two thirds of their 2010 baseline level. The average global target is a MMR of less than 70/100,000 live births by 2030. The supplementary national target is that no country should have a MMR greater than 140/100,000 live births, a number which is twice the global target by 2030 (cf. WHO 2015; Hort and Patcharanarumol 2017).

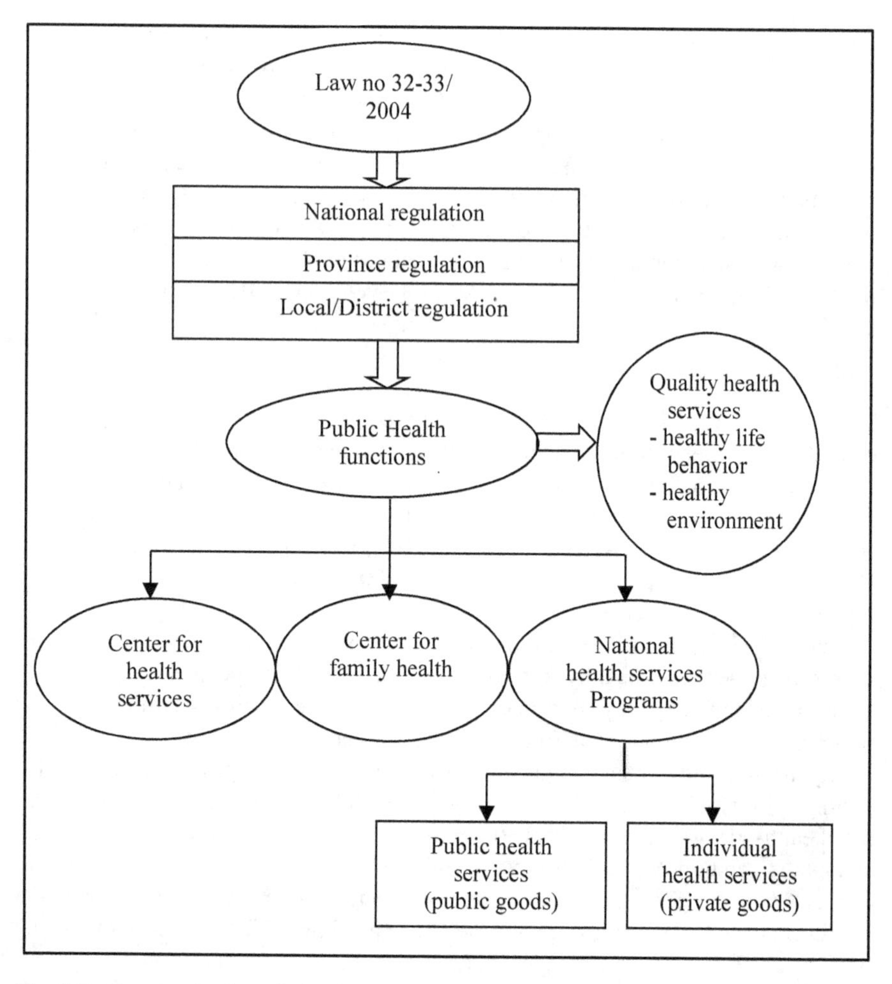

Fig. 8.1 The organisation of the public health system in Indonesia. *Source* Adapted from Indonesian Law No. 32-33 (2004)

Even though the rates for antenatal care and child births attended by skilled health personnel are relatively high, several factors, such as high risk pregnancy and abortion, are still playing a negative role and require special attention. Critical measures to reduce maternal mortality are underway, and measures are taken to improve contraceptive use, access to and quality of family planning and reproductive health services. In the future, maternal and child health will focus on expanding improved quality care and comprehensive obstetric care, improving family planning services and the provision of *Information, Education and Communication* (IEC) to the community.

In the area of the reduction of child mortality, Indonesia shows impressive figures over the past decades. The infant mortality rate has shown a significant decline from 68 in 1991 to 34 per 1000 live births in 2007. Such a decline promises the possibility to meet the target of the United Nations Sustainable Development Goal No. 3, to ensure healthy lives and promote well-being for all at all ages by 2030, and in particular, to end preventable deaths of newborns and children under 5 years of age, with all countries aiming to reduce neonatal mortality to at least as low as 12 per 1000 live births, and under-5 mortality to at least as low as 25 per 1000 live births, as well as to reduce the global maternal mortality ratio to less than 70 per 100,000 live births by 2030. Meanwhile, the performance of the health system in Indonesia is challenged by a changing environment, such as:

- the ongoing demographic and epidemiological incidents which are likely to increase the demand for health care and result in more costly and more diverse forms of health care services;
- additional pressure which will increase from emerging diseases and epidemics, such as HIV/AIDS, Avian Influenza (H5N1), Swine Influenza (H1N1), and related outbreaks;
- the implementation of Law No. 40/2004 on *Universal Health Insurance Coverage* (UHIC) which will further increase demand and utilisation;
- the national health infrastructure, although widely available for primary care, which will not have sufficient beds and health workers in order to respond to the increased health care needs;
- the pharmaceutical supplies, which are reasonable, but most Indonesians pay more than they need and most expenditure is paid with out of pocket money;
- the pressing need to address human resources distribution inequities and quality; and
- although the satisfaction levels are good overall, there is a high level of dissatisfaction with various specific aspects of health care, including pricing, staffing and quality.

In terms of the health budget, the new government is committed to implementing the reforms and securing access of all citizens to quality health services, and providing financial protection against the effects of unpredictable, excessively high costs of health care. In this type of health system and its associated policies, however, the accommodation and participation of local people, and the integration of indigenous traditions in health care and services are still missing. Although these often informal indigenous health practices are operational, the official support of their integration by the government is essential for improved health care for all Indonesians.

Public policies in education

Indonesia is putting education as one of the major priorities in national welfare. Education is regulated by the Basic Law of 1945 Chapter 31 Article 1-3, by which every civilian has the right of access to education and each civilian is obliged to enter basic education under government responsibility. Without considering its effectiveness, about 20% of the national budget is allocated for education on both a central and local level. The implementation of the education system in Indonesia is guided by Law No 14/2003 on National Education Systems. Basically, the services of the education systems are divided into three specific categories: *Academic-Oriented Education, Profession-Oriented Education,* and *Vocation-Oriented Education*. In terms of the operations, there are two ministries which co-ordinate the processes from the lower level to the higher level of education: the Ministry of National Education and Culture, and the Ministry of Religious Affairs. The length of study from the primary level to the highest level of education (doctorate) will take about 21 years. However Law No. 14/2003 only obliges people to follow school education up until the lower level of secondary education, i.e. 9 years of education.

According to the *Report on the Achievement of the MDGs in Indonesia* (Bappenas 2010), the government is on track to achieve the target for primary education and literacy. The government aims to continue its efforts by setting a new target for junior secondary education (SMP/*Madrasah tsanawiyah*-MTs, grades 7 to 9) in order to meet the universal basic education target. In 2008/2009 the *Gross Enrolment Rate* (GER) at the primary education level (SD/*Madrasah ibtidaiyah*/ Package A) was 116.77%, while the *Net Enrolment Rate* (NER) was 95.23%. At the primary education level, disparity in the participation in education among provinces has been significantly reduced with the *Net Enrolment Rate* (NER) above 90% in almost all provinces.

The main challenge in achieving the education target is improving equal access for children, girls and boys, to quality basic education. Government policies and programmes to address this challenge include: (i) extension of access to basic education particularly for the poor; (ii) improvement of the quality, efficiency, and effectiveness of education; and (iii) strengthening the governance and accountability of education services. The policy to allocate 20% of the government budget to the education sector will continue to accelerate the achievement of universal junior secondary education (cf. Bappenas 2010).

8.4 The Essence of Local Government

The importance of the assessment of the role of the local government for the subject of governance refers to the psychological as well as the physical distance between the local authority and the community. In terms of rural development, such a distance is important because it forms the challenge where integration starts to take place. Apart from the relationship with the central government and adherence to its policies, the roots of the local government are grounded in the local constituency. It relates to the relationship between the electorate and the civil service, and is not only dependent on the level on which it is based, such as municipality, district, province, region, etc., but also on the autonomy in terms of sanctioning the available capability. A further analysis is necessary to establish whether there is representation in the form of an assembly with elected members, or in the form of the relationship between the district assembly and the district council. These relationships may have their equivalent at any level. The identification of the underlying mechanism between the traditional authority and the influence which the local representation has on the policies can be determined locally as a basis of rural integration. Shah (2006) enumerates the concepts in order to explain the underlying principles of the transfer of power to the government at local level.

The principle of fiscal equivalency

Olson (1969) argues that if a political jurisdiction and the benefit area overlap, the problem of free riders is solved, and the marginal benefit equals the marginal cost of production, thereby ensuring an optimal provision of public services. Equating the political jurisdiction with the benefit area requires a separate jurisdiction for each public service, expressed in the motto: "*the money is spent where it is collected*".

The theorem of decentralisation

According to this theorem, Oates (1972) argues that: "*each public service should be provided by the jurisdiction having control over the minimum geographic area that would internalise the benefits and costs of such a provision*". His argument is based on the notion that local governments understand the concerns of local residents, and that local decision-making is responsive to the people for whom the services are intended. In this way, fiscal responsibility and efficiency will be encouraged, especially if the financing of services is also decentralised and unnecessary layers of jurisdiction are eliminated. At the same time, inter-jurisdictional competition and innovation will be enhanced.

The principle of subsidiarity

According to the principle of subsidiarity, taxing, spending and regulatory functions should be implemented by the lower levels of the government, unless a convincing case can be made to assign these tasks to higher levels of the government. In an attempt to further reduce the number of parameters with a view to analysing the status of the local government, the authors introduce the level involved, the legal status of that level, the population size of the area, the type of local public services to be delivered, the local revenues and budget, and the expenditure of the public sector. Together, these data enable a basic classification.

A significant characteristic of the local government is that the standard and performance of services is usually decided upon on a national level, while the sub-division of the district or municipal level is often tied to the elected local government. Also, the limited resources of the local government tend to increase the significance of the participation of the private sector. The services which are mostly delegated to special agencies include health, education, environmental planning, and the economic infrastructure. With regard to the administrative principles, a distinction is made between *economic efficiency* which determines that the means such as taxes, regulations and staff stay at the central level; *national equity* which regulates the distribution of these means and which determines the operational extent of the local government; *administrative feasibility* which assigns those taxes to the local government so that they can be assessed more appropriately at a local level; and *fiscal need* which decides that the tax levels as local revenue means should be determined by the local needs.

In his overview of the limitations and benefits of the local government, Green (2005) asserts that the transfer of staff is not only limiting the flexibility of human resources management, but also that the dependability on central transfers moves on average between 40 and 60% of the total budget. It means that in order to keep qualified staff, the more remote areas are always in dire need. Moreover, local governments are not locally accountable for the spending of funds for transfers, and have to find alternative ways for financing the local initiatives. In addition, there is a need to maintain a balance between elected and non-elected local officials. Especially in Indonesia, there is the influence of the local oligarchy in all spheres of the society.

Among the benefits of the principle of subsidiarity are the psychological and physical considerations of proximity to local people, a higher level of participation of the local people, more knowledge of the complexity of local problems, a smaller span of control, an existing administrative infrastructure already established by the central government, and the relative freedom to implement local rules and regulations.

Luebke (2009) introduces a model of relationships between the private sector and the local government, whereby he shows an increase of the involvement of economic actors in the local build-up of public-private partnerships as the government performance increases.

Motivations to engage in constructive dialogues with the government are primarily shaped by two counter-balancing 'cost functions'. On the one hand, economic actors are inclined to co-operate if they face fewer communication and co-ordination problems, i.e. 'transaction costs'. These 'transaction costs' decline exponentially with a rising economic concentration, i.e. a decreasing number of actors. On the other hand, economic concentration also affects the susceptibility to rent-seeking and corruption: as firms become more concentrated, they rapidly gain policy influence. Rising policy influence, combined with weak law enforcement, translates into rapidly growing 'opportunity costs' for efforts of productive reform, and a rising attractiveness of consumptive rent-seeking. By aggregating these two cost functions, the resulting 'disincentives' for cooperation with reforms tends to take a U-shape in the curve (cf. Fig. 8.2c).

Decentralisation and local governance

The new policy of decentralisation and regional autonomy in Indonesia was implemented by Law No. 22/1999 concerning 'Local Government' and Law No. 25/1999 concerning 'The Fiscal Balance Between the Central Government and the Regions'. The process gives the lower levels of the society, including local people, a chance to participate in the development process, and to decide what is best for their community. Policy in public sector management can thus be determined not only at the national level but also at the local level by a local

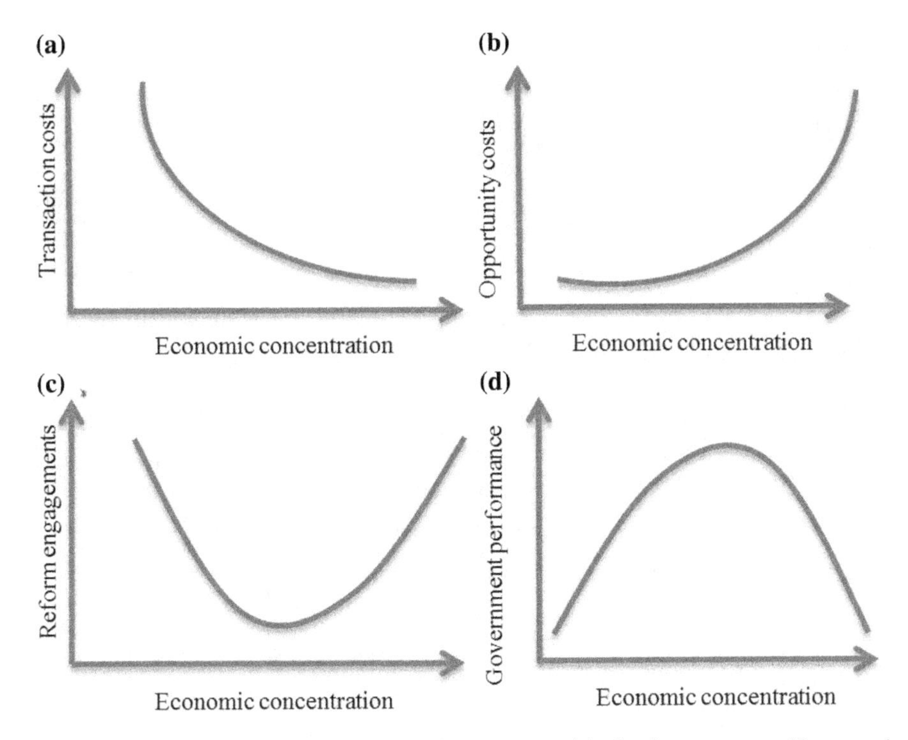

Fig. 8.2 Model of relationship between the private sector and the local government with cost and incentive functions of private sector groups. *Source* Luebke (2009)

government. Both above-mentioned laws are based on five principles: (1) democracy; (2) community participation and empowerment; (3) equity and justice; (4) recognition of the potential and diversity within regions; and (5) the need to strengthen the local legislation.

There are three key features in decentralisation and local governance which relate to poverty reduction programmes: *decentralisation, local governance and participation.* According to Rondinelli (1981), decentralisation in this context refers to the transfer of responsibility for planning, management and resource allocation from the central government to field units of the ministries of the central government, agencies, subordinate units or levels of the government, semi-autonomous public authorities or corporations, area-wide regional or functional authorities, and organisations of the private and voluntary sectors.

For the Indonesian situation, it is necessary to understand which approach of decentralisation is used for the various levels of structure in the society (cf. Fig. 8.3). Prior to the 'Reformation Era', Indonesia had not implemented any form of decentralisation, even though the structure of the government administration was divided from the national down to the municipal and even the village level. However, the plans and actions are controlled by the central government. In the

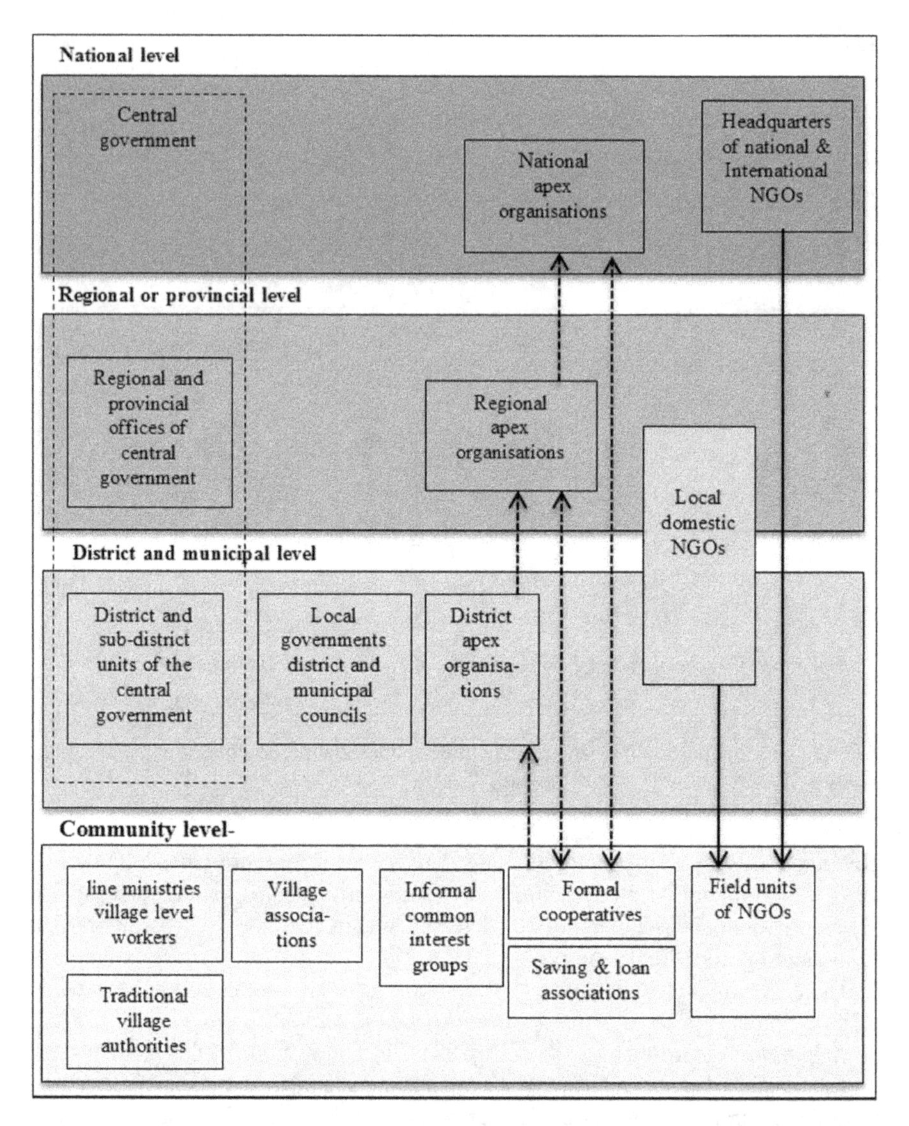

Fig. 8.3 Schematic representation of levels of decentralisation and governance. *Source* Antlöv (2002)

'Reformation Era' after 1998, participation at each level of the society had become more accommodated, particularly after Laws No. 22/1999 and 25/1999 had been applied.

Full autonomy—previously centralised—has been transferred to the *kabupaten* ('rural districts') and *kotamadya* ('urban municipalities') in a large number of sectors, including health care, education, public works, and natural resources management. Certain strategic policy areas, such as finances, the legal system, foreign affairs, defence and religion, have remained at the national level. Other sectors, such as the authority over roads and harbours, have been transferred to the provincial level, which acts essentially as an administrative branch of the central government (cf. Antlöv 2002).

Based on the framework shown in Fig. 8.3, the local governance refers to a mechanism at the local level to ensure that public services are implemented. Participation therefore refers to the involvement of any party at the local level, both municipality and community, in any development plan or implementation by a local government. The people at the municipal and community level who are urged to participate in development are usually defined as 'local people'. By using such a definition, local governance therefore can be regarded as a mechanism in which indigenous/local people are participating in the planning, execution, and control of local development, in order to achieve the local people's objectives. By consequence, the decentralisation process should be in line with this local governance mechanism. Indonesia has had particular experiences in this area, whether the participation of the local people was accommodated or not.

For instance, it would be interesting to analyse the cases of *Lembaga Perkreditan Desa* (LPD) ('Integrated Microfinance Institution') in the Province of Bali and *Lumbung Pitih Nagari* (LPN) ('Village Microfinance Institution') in West Sumatra. The first example refers to the success of how local government has taken the local culture into account, and accommodated the participation of the local people in the implementation of microfinance. The history of LPD documents that microfinance institutions worked well where they were integrated into the local Balinese culture, while LPN showed the opposite in West Sumatra, as they were eventually forced to follow the national banking system or 'external law'. LPD is owned, governed and financed by the customary village (*desa adat or desa pakraman*) and fully integrated into the Balinese culture. According to Siebel (2008), the functional integration is governed by the provincial law on Bali. The operation of LPD is similar to the way in which the banks are operating. The difference, however, becomes clear in the integration of livestock in the livelihood of the households engaged in micro-enterprise, the role of the customary village (*desa pakraman*) and the sustainability of the microfinance institution. The institution is owned by the customary village (*pakraman*), which is led by the Village Council (*prajuru desa*) and the Village Head (*bendesa*), appointed by the customary elections, while each *pakraman* consists of smaller communities (*banjar*). There are in Bali about 1433 customary villages (*pakraman*) and 3945 communities (*banjar*). The residents are divided between *krama ngarep* ('native/origin') and *krama tamiyu* ('guest'). The fact that LPD is owned and operated by the customary village means that all of its activities, including the transactions between the villages and the residents, are part of the communal practice of religious beliefs and conservation of their culture (cf. Agung 2005).

There has been a positive progress in the recent development of LPD on Bali. Their number shows a slight increase from 849 in 1995 to 1358 in 2008, covering about 95% of all customary villages, indicating that almost all people in Bali who live in customary villages are members of LPD. As regards its practices, LPD in Bali shows a certain similarity with LPN in West Sumatra, albeit with a stronger attachment to the customary/local traditions (*adat*). The *bendesa* acts both as the Village Head and on the Board of LPN. This means that in the case where the people in the *pakraman* are faced with socio-economic problems, LPN will take care of their needs.

The same principle applies to the problem of the repayment of loans. The *bendesa* is also involved in efforts to render LPD sustainable in the long term. In general, the local people feel ashamed if they cannot perform well with LPD. The local tradition has enabled LPD to assume the role of an inclusive microfinance institution by integrating the economic and the socio-cultural activities in the community. In contrast, in West Sumatra, where LPN declined dramatically because of its inability to cope with the regulations from the Central Bank of Indonesia, LPD could make a positive contribution to development by strengthening the influence of the local culture on the socio-economic situation.

LPN has been built and operated by the local people in order to develop the socio-economic situation in every village. The establishment itself was supported by Law No. 13/1983. The assets of LPN are collected from the people in the communities, and the accumulated funds are used for their socio-economic purposes. Additional capital comes from the local government through soft loans of 500,000 Rupiah for each institution, to be repaid at 475,000 Rupiah in 3 years (cf. Basa 2001). The implementation of LPN is supported by the Leader of a *nagari* —known as *kepala adat/pimpinan adat*—with the ability to become a *non-physical collateral/intangible collateral*. In some cases, the Leader of a *nagari* is also involved in the administration in order to ensure that the institution is running well for the benefit of the local people. The Leader can also be involved in determining which people in the community really need help and which do not. In the micro-banking system, this role is almost similar to that of a credit analyst. In the case where some clients are not able to pay back their loan, or delay repayment, the Leader will approach the clients individually and point out their own responsibility not only to LPN, but also to the community, making them aware of the social consequences in the local 'Minang' culture if they do not pay back their loan. In such a case, the Leader acts as a debt collector which is rather different from the situation in a conventional banking system, but shows to be more effective.

As reported by Basa (2001), the development of LPN had readily been accepted by the local people since it helped them in satisfying their socio-economic needs and in implementing their local initiatives. In 2001, there were about 521 LPN established in West Sumatra among 543 nagari. Their positive contribution, however, was challenged after the introduction of a new banking regulation of the government in 1998, which required compliance of any institution which collects money from people. The standard capital requirements were so high that they could not be met by the local initiatives of LPN. As a result, all the existing LPN, as well as other village banks, had to be transformed in order to obtain the status of a rural bank, known as *Bank Perkreditan Rakyat* (BPR), as a financial institution which accepts deposits only in the form of time deposits, savings deposits and other equivalent forms, and disbursed funds as a *Bank Perkreditan Rakyat* (BPR) business. The data of the LPN for 2001 show that only 71 out of 521 of these institutions have sustained their services since the introduction of the bank regulation of 1998, and that 51 out of 71 LPN with the status of a BPR, which have been able to survive, have faced many difficulties in complying with the new banking regulation. In 2014, a new ruling of the Indonesian Law No. 6/2014 on the Village (2014) was approved by the Parliament, giving more room for the local government at the community level to initiate 'bottom-up' development programmes. In addition, about 1 billion Rupiah was reserved by the central government for about 80,000 villages in Indonesia as support for the implementation of this new approach of development. However, the introduction of the new law has also raised many questions which need to be answered.

For instance, the operational guidelines and procedures of the new law have not yet been explicated, such as, for example, in what way the new 'bottom-up' initiatives are accommodated in the case of the submission of a wide range of initiatives. During a recent Workshop on 'New Law, New Villages? Changing Rural Indonesia', held in Leiden by KITLV, such questions also include what the impact of Indonesia's 'Village Law' will be on the character of villages and their role in Indonesia's economic and political development, and how this change can be studied across Indonesia in a comparative manner (cf. Vel et al. 2016). Similarly, an analysis of what lessons there are to be learned from the history of village governance in the context of change processes taking place in rural areas needs to be conducted.

During the Workshop, the Director General of Village Administration of the Indonesian Ministry of Home Affairs made a plea for further cooperation to deepen the insight into the process and find answers to some critical questions raised, such as:

- enhancing capacity for the villages;
- apparatus and village institutions;
- facilitation of how to manage the budgets and assets of the villages;
- structuring villages' territory;
- developing the villages' ICT; and
- other matters which are relevant to village management and development (cf. Vel et al. 2016).

The relevance of these points for the present Chapter is the endeavour to be undertaken, in order to study and analyse the potential and the impact of the new law for the integration of the central policies and local cultures with a view to increasing prosperity through sustainable development and to reduce poverty in the rural areas of Indonesia.

8.5 Policies Within the Local Socio-Economic Context

Gulli (1998) introduces a model in which the factors of the socio-economic context at the household level and the transformation from the macro- to the meso-level are forming the basis of assessment. The model shows the influence of all relevant economic, political and socio-cultural factors on the welfare of the subjects, apart from the institution-related factors. On the meso-level, the inventory starts with the living conditions in the broadest meaning of the word, including household composition, economic activity and the social and cultural environment, including social networks. A distinction is made between means and opportunities: unskilled labour, the availability of land, both as a tenant or in ownership, produce, poultry or livestock, trading commodities or craftsmanship, and group membership. The opportunities are qualified as the existence of a local market, the possibility to collaborate with other existing enterprises, the option of cooperative activities, individual retails, or irregular demands for unskilled labour.

In addition, there are some identifiable constraints. These can be the physical well-being, household relationships in terms of male dominance or care for children, and moral obligations to the family or group members. There is also the absence of relatives to rely upon or sufficient assets. Apart from the opportunities, the factors could be serious impediments to starting a micro-enterprise. External factors which influence the setup of a micro-enterprise could include unsuitable financial assistance, political or regulatory challenges, environmental hazards or ultimately infrastructural limitations of proximity, transport, etc. (cf. Brett 2006).

As already indicated earlier, the trigger to the search for financial assistance can be a combination of working capital, consumption, or prior debt relief. It means that further identification could indicate that there are hidden motives such as loan diversions as a result of social obligations, either to family or to group members, and the psychological pressure of group repayment obligations. It is important to consider that most micro-starters have poverty-driven motives, and are not necessarily entrepreneurs (cf. Mayoux 2005). Simultaneously, the local market situation must allow for a business process which makes enough profit to support the household, pay off debts, keep group obligations and make provisions for social services. Thus, a feasible opportunity requires the availability of suitable financial support, a social network of relationships, a proximate market opportunity, and a household situation which is manageable. Moreover, health care and education provisions should be included in the assistance, or they should be part of a credit-saving scheme, which is government or NGO-based.

Sustainable development policies therefore have to include an analysis of all aspects of the socio-economic situation of the potential borrower. The calculation of financial support itself is going to prove insufficient if no attention is paid to the potential turnover, the time factor and the provision of social services, regardless of local embedding in an existing executive body of the community. In this context, the influence of the indigenous culture on governance is significant and deserves special attention. A significant contribution to the further understanding of the meaning of the indigenous culture within the context of governance for the envisaged integration of the indigenous institutions to attain sustainable community development and poverty reduction in Indonesia is provided by the *United Nations Declaration on the Rights of Indigenous Peoples* (UNDRIP 2007).

As an elementary aspect of indigenous culture, common recognition is needed of the historical dimension of the country, the language, practices, symbols, religion and traditions, as an expression of the cultural heritage. The indigenous culture provides social coherence through kinship and functional economic relationships. Also, the traditional authority and leadership are directly related to the execution of local governance. All these important factors combined are leading up to an accumulation of 'Indigenous Knowledge', which, however, may not always be available to all community members at all times (cf. Warren et al. 1995).

For the purpose of the analysis of the influence of the indigenous culture on governance, the communities should be identified as centered around some type of economic activity, i.e. rural agricultural communities, coastal fishing communities, local mining communities, including transitional communities under the impact of urbanisation. Here, careful consideration is required concerning the role of traditional leadership and communal assembly, the consultation of elders, the associations established on the basis of craft or interest, religious associations, the function of conflict arbitration, and the phenomenon of communal labour. All these characteristics are in some form or other often still functioning within indigenous communities (cf. Posey 1999). Other characteristics include territorial boundaries, the oscillating effect of larger communities splitting up into smaller units, introspective tendencies within the community such as a focus on interpersonal relationships, and the survival of traditional communal values through time. Functional relationships may determine status to a large extent. A strong cohesion remains as the result of social and economic interdependence, as defined by reciprocity. Among the counterbalancing influences to the homogeneity and preservation of the indigenous communities are increased population pressure, forced transition to economic activities, migration, and the increasing influence of modernisation which may lead to the alienation of some members from their ancestry. There exists a dual relationship between indigenous communities and their environment, as on the one hand, there is a strong connection in terms of knowledge and respect, while on the other hand, there may be many opportunities of exploitation without intent, basically because of the total economic dependence, such as in over-fishing, firewood collection, single crop soil exhaustion, and water pollution. In indigenous communities, there is generally limited vertical mobility, which is sometimes positively interpreted as an 'egalitarian' class structure, which may result in the exodus of ambitious young people. As mentioned before, recent experience shows that external aid can distort social cohesion when there is no integration with the existing indigenous institutions.

Development needs a multidisciplinary input of sciences addressing issues of concern to the developing countries, with the general aim of achieving well-being for all citizens through sustainable community development. Previous theories on development have largely focused on the measurement of achieved quantitative economic data of income, while more recent theories embark on a more comprehensive assessment of the well-being of the entire population. The concept of sustainable development implies that development should be viewed as a dynamic process of change, in which the exploitation of resources, the direction of investments, the orientation of technological development and institutional changes are made consistent with both current as well as future needs.

Rogers et al. (2008) assert that any development project should maintain the balance of fulfilment on three dimensions of sustainability: maintaining a balance in the interaction of economic development, environmental quality, and social quality. Maintaining the economic dimension means maximising income while maintaining a constant or increasing stock of capital. Maintaining the environmental quality refers to the preservation of the resilience and robustness of biological systems, while maintaining the socio-cultural quality means supporting the stability of the social and cultural systems.

Furthermore, sustainable development consists of at least three criteria, formulated by Swisher et al. (2009) as: *'environmentally sound, economically productive, and socially just.'* Reijntjes et al. (1994) elaborated the criteria for sustainable agricultural development in such a way that they should be: (1) environmentally friendly, (2) humane, (3) just, (4) economically viable within the community, and (5) adaptable. In this respect, the perspectives of the indigenous peoples themselves as the 'participant's view' further substantiate the *emic* approaches of sustainable development as a counterbalance to the *etic* perspectives and suggestions of outsiders on development (cf. Warren et al. 1995).

It is remarkable that the definition of poverty used by the *Commission on Legal Empowerment of the Poor* (UNDP 2008) and the *Sustainable Development Goals* (SDG) of the United Nations (2015) are not properly reflecting or representing the realities and priorities of the indigenous peoples. The development planners often do not consider the cultural aspect of indigenous people. Both definitions tend to ignore the important cultural aspects in terms of the indigenous peoples' knowledge, beliefs, traditions, ceremonies, rituals and cosmologies. The pioneering work by Warren et al. (1995) is among the early documentations of the significance of the various systems of knowledge, beliefs and practices of indigenous peoples in development. Later, Woodley et al. (2006) and Loeffelman (2010) confirmed the importance of these cultural aspects in development. Marsden (1994) stated that: "*It is now commonplace to hear that there are many paths to development, each built on a different cultural base, and using different tools, techniques and organisations. The assumptions underlying the view that it would be sufficient to transfer western technology and expertise no longer hold. It is not the 'native' who is backward, nor is it a failure to incorporate the 'human factor' which is at fault, but the essential inappropriateness of the western package that was on offer*".

Presently, the challenge is to design a locally suitable programme which includes co-ordination with other public services, preparation of an operational structure for funding, target group identification and an assessment of individual eligibility. This challenge leaves the option of whether integration with other services is executed as a part of the programme, i.e. as a participation requirement, or whether the integration with other services is co-ordinated at the departmental level, i.e. the task is offered to eligible individuals. It simultaneously implies compliance with not only the actual rules and regulations within the policies of the local government, but also with an operational structure ensuring the involvement of community members in the decision-making process and external support, preventing 'donors' from setting the requirements. While the assessment of applications should include the socio-economic factors as mentioned above, the provision of other public services implies a funding structure which ideally encompasses three sources: the members' input, NGOs and the government.

In this context, a comparison with the functioning of *Self Help Groups* (SHGs) is relevant. These groups consist of people who work together for financial services, often beginning with periodic, compulsory savings and then mainly loans, and sometimes even social services. SHGs are usually managed by their members, with varying degrees of external support. SHGs may at some point in the future evolve to a commercial enterprise at a higher level. SHGs have been operating rather successfully in India, and both their growth and the call for more creditability clearly define a need for the establishment of a legal framework for these groups. Some examples of organising and arranging the local configurations include the following:

- The local configuration represents an informal collaboration based on local practices and traditions and it will stay that way. It is the competence of the development agent to deal with these informalities and to help deliver social services and encourage small enterprises;
- The local configuration will be structured as a project. A project is a temporary activity under the guidance and responsibility of a certain authority with a starting date, specific goals and conditions, defined responsibilities, a budget, a plan, a fixed end date and multiple parties involved. After the end date, the activities have to be incorporated into some other entity. The development agent can play a central role as the project manager, who will be working towards the realisation of the defined goals;
- Part of the local configuration will be structured as a non-profit organisation with full legal status. A non-profit organisation is a corporation or an association which conducts business for the benefit of the general public without shareholders and without a profit motive. Non-profit organisations are created according to the state law. They must file a statement of corporate purpose—statutes or constitution—with the state, conduct regular meetings, and fulfil other obligations, usually accountability of expenditure, to achieve and maintain corporate status;
- Part of the local configuration is transformed into an entity with full legal status, for example a co-operative or a private company. A cooperative is characterised by open membership, with members involved in governance, meeting the criteria to become a healthy cooperative in terms of transparency, accountability, etc. Co-operatives may be for-profit or non-profit. A private company is a legal entity which is set up to run a for-profit business. It has a commercial mission,

but even social services can be delivered on a commercial basis given the right circumstances;

- The local configuration is transformed into a public-private partnership. A public-private partnership is a partnership arrangement in the form of a long-term performance-based contract between the public sector, i.e. any level of government, and the private sector, usually a team of private sector companies working together to deliver a public infrastructure for citizens or to improve the systems for delivering public services. The development agent should operate as an intermediate between public and private entities.

The modelling of the local configuration requires that the delivery of certain social services, e.g. education and health care, could be taken over from the government by other organisations in order to reach out to the very poor, but it is still likely that the delivery of vital social services remains a task of the local government and thus the local configuration has to be built on that condition. Integrated sustainable community development is a concept which focuses on the integration of social services and the emerging people on the community level, and is therefore built on an approach in which public and private modes of operation are combined. Local values and methodologies are leading elements while respecting binding policies and regulations of a higher order. They encourage social cohesion by strengthening both horizontal and vertical relationships. The integration of social and commercial missions can range from an informal organisation or a combination of functions to a formal or legal structure. From the angle of the sustainable delivery of services to the very poor, it is also important to assess the solidity of emerging local configurations. Community-based configurations are often managed without any solid legal status, while the legal status of a group or organisation determines who has the ownership and who has the decision-making power. As it seems meaningful to explore whether any formalisation is useful, the related legal status will be useful if it is based on a local governance framework with the full participation of the local people.

References

Agung, A. A. G. (2005). *Bali Endangered Paradise? Tri Hita Karana and the Conservation of the Island's Biocultural Diversity*. Ph.D. Dissertation. Leiden Ethnosystems and Development Programme. LEAD Studies No. 1. Leiden University. xxv + ill., p. 463.

Antlöv, H. (2002). The making of democratic local governance in Indonesia. Logo Link International Workshop on Participatory Planning: Approaches for Local Governance. 20–27 January 2002. Bandung, Indonesia.

Asian Development Bank (ADB). (2006). *A review of community-driven development and its application to the Asian Development Bank*. September 2006.

Azra, A. (2000). *Demokrasi, Hak Asasi Manusia, dan Masyarakat Madani*. Jakarta: Kencana Media Group.

Bappenas. (2010). *Report on the achievement of the millenium development goals in Indonesia*. Jakarta: Ministry of National Development Planning/National Development Planning Agency.

Bappenas. (2013). *Evaluasi PNPM Mandiri*. Jakarta: Direktorat Evaluasi Kinerja Pembangunan Sektoral Kementrian Perencanaan Pembangunan Nasional/Bappenas.

Basa, Z. (2001, December). Revitalisasi Lumbung Pitih Nagari 'Skenario' Menjemput Modal Sosial untuk Kembali Ke Sistem Nagari di Masa Depan. *Jurnal Analisis Sosial: Lembaga Keuangan Mikro dalam Wacana dan Fakta: Perlukah Pengaturan, 6*(3). Bandung: Akatiga.

Brett, J. A. (2006). We sacrifice and eat less: The structural complexities of microfinance participation. *Human Organization, 65*(1), 8–19.

Bovens, M. A. P. (1990). *Organisatie en Verantwoordelijkheid*. Leiden: Universiteit Leiden.

Budiarjo, M. (1998). *Demokrasi di Indonesia: Demokrasi Pancasila*. Jakarta: Gramedia.

Chang, H-J. (2000). Institutional development in developing countries in a historical perspective: Lessons from developed countries in earlier times. Unpublished paper, University of Cambridge.

Frederickson, H. G. (1999). Ethics and the new managerialism. *Public Administration Management, 4*(2), 299–324.

Fritzen, S. A., & Brassard, C. (2007). Multi-level assessments for better targeting of the poor: A conceptual framework. *Progress in Development Studies, 7*(2), 99–113.

Gorzki, E. (2007). Civil society, pluralism and universalism. *Polish Philosophical Studies,* VIII, Volume 34. Cultural Heritage and Contemporary Change. Washington DC: Council for Research in Values and Philosophy.

Green, K. (2005). Decentralisation and good governance: The case of Indonesia. Personal RePEc Archive Paper No. 18097. Munich: MPRA.

Grindle, M. S. (2004). Good enough governance: Poverty reduction & reform in developing countries. *Governance: An International Journal of Policy, Administration, and Institution*, 17 (4): 525–548, October.

Gulli, H. (1998). *Microfinance and poverty: Questioning the conventional wisdom.* Microenterprise Unit-Sustainable Development Department. Washington D.C.: Inter-American Development Bank (IADB).

Hort, K., Patcharanarumol, W. (2017). The Indonesia health system review. *Health System in Transition*, Vol. 7 No.1. Asia-Pacific Observatory on Health System and Policies – World Health Organization.

Indonesian Law No. 6/2014. (2014). *Undang-Undang Republik Indonesia No. 6 tahun 2014 tentang Desa* (Indonesian Law No. 6/2014 on the Village). Jakarta: SEKNEG.

Indonesian Law No. 32/2004. (2004). *Undang-Undang tentang Pemerintahan Daerah tahun* 2004 (Indonesian Law on Regional Administration). Jakarta: SEKNEG.

Indonesia Law No. 33/2004. (2004). *Undang-Undang Republik Indonesia tahun* 2004. Jakarta: SEKNEG.

Indonesia Law No. 34/2004. (2004). *Undang-Undang Republik Indonesia tahun* 2004. Jakarta: SEKNEG.

Kamarck, E. C. (2007). *The end of government as we know it: Making public policy work.* Boulder (Colorado): Lynne Rienner Publishers.

Khan, M. H. (2006). *Governance and anti-corruption reforms in developing countries.* Washington DC: The World Bank.

Kolthoff, E., Huberts, L., & Van den Heuvel, H. (2007). The ethics of new public management: Is integrity at stake? *Public Administration Quarterly, 30*(4), 399–439.

Loeffelman, C. (2010). Gender and development through western eyes: An analysis of microfinance as the west's solution to third world women, poverty, and neoliberalism. Thesis. Washington D.C.: George Washington University.

Marsden, D. (1994). Indigenous management: Introduction. In S. Wright (Ed.), *Anthropology of organizations* (pp. 35–40). London: Routledge.

Mayoux, L. (2005). *Women's empowerment through sustainable microfinance.* Organisational Gender Training, Quetta, Pakistan: Taraqee Foundation. Gender and micro-finance website: http://www.genfinance.net and http://lindaswebs.org.uk.

Moore, M. (1998). Death without taxes: Democracy, state capacity, and aid dependence in the fourth world. In M. Robinson & G. White (Eds.), *The democratic developmental state* (pp. 84–123). Oxford: Oxford University Press.

Oates, W. (1972). *Fiscal federalism*. New York: Harcourt-Brace.

Ohmae, K. (1996). *The end of the nation state: The rise of regional economies*. New York: Harper Collins.

Olson, M. (1969). The principle of 'fiscal equivalence': The division of responsibilities among different levels of government. *American Economic Review, 59*(2), 479–487.

Pollit, C. (1991). *Managerialism and the public services: The Anglo-American experience*. Oxford: Blackwell Publishing Ltd.

Posey, D. A. (1999). *Cultural and spiritual values of biodiversity: A complementary contribution to the global biodiversity assessment*. Intermediate Technology Publication – UNEP. London/Nairobi: Intermediate Technology Press/UNDP.

Reijntjes, C., Haverkort, B., & Waters-Bayer, A. (1994). *Farming for the future: An introduction to low-external-input and sustainable agriculture*. Leusden/London: ILEIA/Macmillan Press.

Rogers, P., Jalal, K. F., & Boyd, J. A. (2008). *An introduction to sustainable development*. London: Earthscan.

Rondinelli, D. A. (1981). Government decentralization in comparative perspective: Theory and practice in developing countries. *International Review of Administrative Sciences, 47,* 133–145.

Roosa, J. (2008). Pretext for mass murder: The September 30th Movement and Suharto's Coup d'État in Indonesia. New Perspectives in South East Asian Studies. Madison, WI.: The University of Wisconsin Press.

Rotberg, R. (2013). Failed and weak states defined. *The Oxford companion to comparative politics* (pp. 383–389). New York: Oxford University Press. https://robertrotberg.wordpress.com/2013/02/11/failed-and-weak-states-defined/.

Schrijver, N. J., & Weiss, F. (2004). *International law and sustainable development: Principles and practice*. Leiden: Brill.

Shah, A. (2006). *The new vision of local governance: Evolving roles of local governments*. Washington, DC: World Bank.

Siebel, H. D. (2008). Desa Pakraman dan Lembaga Perkreditan Desa: A study of the relationship between customary governance, customary village development, economic development, and LPD development. ProFi Working Paper Series.

Smith, B. C. (2007). *Good governance and development*. New York: Palgrave McMillan.

Swisher, M. E., Rezola, S., & Stern, J. (2009). *Sustainable community development*. IFAS Extension. Gainesville FL.: University of Florida.

United Nations Declaration on the Rights of Indigenous Peoples (UNDRIP). (2007). *Adoption of the United Nations declaration on rights of indigenous peoples*. New York: United Nations News Centre.

United Nations Development Programme (UNDP). (2005). *A guide to civil society organizations working on democratic governance*. New York: UNDP.

United Nations Development Programme (UNDP). (2008). *Making the law work for everyone*. Commission on Legal Empowerment of the Poor. New York: United Nations.

United Nations. (UN). (2015). *Transforming our world: The 2030 agenda for sustainable development*. New York: United Nations.

Vel, J. A. C., Berenschot, W., & Daro Minarchek, R. (2016). *Report of the workshop 'new law, new villages? Changing rural Indonesia'*. Leiden: KITLV.

Von Luebke, C. (2009). Striking the right balance: Economic concentration & local government performance in Indonesia and the Philippines. In *Annual Meeting of the American Political Science Association (APSA)*. Toronto, September 3–6, 2009.

Walsh, K. (1995). *Public services and market mechanism: Competition, contracting and the new public management*. London: Macmillan.

Warren, D. M., Slikkerveer, L. J., & Brokensha, D. (1995). *The cultural dimension of development: Indigenous knowledge systems. IT studies on indigenous knowledge and development*. London: Intermediate Technology Publications Ltd.

Woodley, E., Crowley, E., Pryck, J. D., & Carmen, A. (2006). Cultural indicators of indigenous peoples' food and agro-ecological systems. (SARD) Initiatives. Paper on the 2nd Global Consultation on the Right to Food and Food Security for Indigenous Peoples. 7–9 September 2006, Nicaragua.

World Bank (WB). (2001). Global public policies and programs: implications for financing and evaluation proceedings from a world bank workshop. In Gerrard, C.D., Ferroni, M., Mody, A. (Eds.), Washington, D.C.: World Bank.

World Bank (WB). (2005). *Poverty reduction strategies—Operational approaches and tools in support of the country-driven model*. Washington, D.C.: World Bank.

World Health Organisation (WHO). (2015). *Strategies toward ending preventable maternal mortality* (EPMM). Department of Reproductive Health and Research. Geneva: World Health Organization. www.who.int/reproductivehealth.

Zonanesia. (2016). *Periodisasi Sistem Pemerintahan Indonesia*. http://www.zonanesia.net/2014/10/periodisasi-sistem-pemerintahan.html.

J. C. M. (Hans) de Bekker is the Coordinator of Market Research & Analysis at Inholland University of Professional Education since 1992. He graduated from Leiden University in Medical Anthropology and is currently finalising his Ph.D. research on traditional, transitional and modern health care utilisation in Tanzania. He was Secretary of *Horizon Holland Foundation* until 2011 and is also Senior Lecturer in the Course on *Integrated Microfinance Management* (IMM), Bandung.

Kurniawan Saefullah is Lecturer of the Faculty of Economics and Business, Universitas Padjadjaran, Bandung, Indonesia since 1998. He graduated from Universitas Padjadjaran and the International Islamic University, Malaysia. He has published several books and articles with his colleagues, including Introduction to Management and Introduction to Business. In 2009, he joined the International Project on Integrated Microfinance Management (IMM) for Poverty Reduction and Empowerment in Indonesia. As a Ph.D. Researcher at the Leiden Ethnosystems and Development Programme (LEAD), Faculty of Science of Leiden University, The Netherlands, he is finalising his Ph.D. Dissertation on 'Gintingan in Subang: An Indigenous Institution for Sustainable Community-Based Development in the Sunda Region of West Java, Indonesia' under the supervision of Prof. Dr. L. J. Slikkerveer. The focus of his research is on the important role of the local cosmology of Tri Tangtu in the process of sustainable community development, while he also developed a model of integration of community-support institutions on the basis of the Indigenous Knowledge Systems (IKS) in combination with the IMM and ICMD approaches, such as financial, medical, educational, communication and social services through the indigenous institutions of the local communities in the Sunda Region of West Java.

Chapter 9
Bank Rakyat Indonesia: The First Village Bank System in Indonesia

Kurniawan Saefullah and Asep Mulyana

> *I initially thought that personalities were the explanation for the successful reforms; some good people in BRI, talented technical assistance, and a. supportive and strong Minister of Finance. ... It may be hard to believe, but the Indonesians were doing microfinance before Grameen Bank.*
>
> (Klaas Kuiper 2004)

9.1 The Era After Independence

Indonesia has a unique rural financial sector whereby an indigenous or traditional institution and a non-indigenous or modern institution—also known as a formal and an informal institution, respectively—are co-existing throughout the rural areas. Different categories of thousands of micro-banking units, hundreds of thousands of semi-formal financial units, and millions of savings and credit associations based on the traditional approach have been operating harmoniously in the country (Seibel 2005). During the regime of President Soekarno, the availability of vast land and fertile soils in Indonesia inspired the Government of Indonesia to target agriculture as the primary sector for development.

Soon after independence in 1945, Mr. I. G. Kasimo, Minister of Welfare during the era of President Soekarno, introduced a development package known as the *Kasimo Plan* which aimed at increasing production in order to achieve self-sustained food security. The programme was adapted from a similar policy, named *Olie Vlek*, which had been implemented during the colonial administration

K. Saefullah (✉)
FEB, Universitas Padjadjaran, Bandung, Indonesia
e-mail: kurniawan.saefullah@unpad.ac.id

A. Mulyana
Department of Management of the Faculty of Economics and Business, Universitas Padjadjaran, Bandung, Indonesia

L. J. Slikkerveer et al. (eds.), *Integrated Community-Managed Development*, Cooperative Management, https://doi.org/10.1007/978-3-030-05423-6_9

of the Dutch. Unfortunately, the plan failed due to several reasons such as political instability, and lack of facilities, knowledge and technology. Although President Soekarno had adapted the initiative in the 1960s by launching a programme of replacement projects, *i.e. Bimbingan Masyarakat* ('Advocacy and Training of the People') (BIMAS), the programmes were also unsuccessful as the result of a lack of funding and capital support. The failure of these projects also contributed to the economic recession in 1963, whereby the government ordered the people to consume corn as a replacement to rice (cf. Booth and McCawley 1990).

The programme of development packages introduced during the regime of President Soekarno ended soon after his succession by President Soeharto in 1967. Under the 'New Order', the programme was replaced by the *Pembangunan Lima Tahun* (*Pelita*) ('Five-Year Development Plan') and the *Rencana Pembangunan Lima Tahun* (*Repelita*) ('Five-Year Plan'). Agriculture became the first priority to realise the objectives of the first Pelita. In this new plan, the development of agricultural technology was still targeted to support the agriculture sector. A specific programme to reach self-sustained food security was implemented, known as the *Revolusi Hijau* ('Green Revolution'), to support the agriculture sector, particularly the growth of domestic rice production. The *Revolusi Hijau* was successful in securing food security through the production of rice in 1984. Soon thereafter, Indonesia was awarded with a Prize by the Food and Agricultural Organisation (FAO) in 1986 for its achievement in self-sustained food security through the production of rice.

One of the factors that supported both the green revolution and the five-year development plan which aimed at achieving self-sustained food security was the involvement of the banking sector in the development programme, particularly in the rural area where farmers cultivated their crops and plants. The *Bank Rakyat Indonesia* (BRI) was a rural bank which was promoted by the government in its support for the Green Revolution. The subsidy scheme of the government enabled BRI to establish the *Unit Desa* ('Village Unit') and the *Badan Kredit Desa* ('Village Credit Unit') in every village in the rural areas. Both programmes provided credit for farmers in any form of agriculture support including purchasing seeds and agricultural implements, as well as transportation to sell the harvest. Although the achievements in self-sustained food security and the roles of *Unit Desa* and *Badan Kredit Desa* of BRI were never questioned, the programme of the village banks discontinued after the 1990s, particularly after the financial crisis of 1997–1998, which struck Indonesia hard. In order to keep its banking activities running, it transformed its organisation into a commercial banking system, which enabled its activities to expand beyond the provision of microcredit. However, the main business of BRI, which is to provide services to micro- and small enterprises, still continues. In 2016, BRI was awarded with the title of the best bank in Indonesia according to *Global Finance* (cf. Berita Satu 2016).

9.2 Brief History of *Bank Rakyat Indonesia* (BRI)

Bank Rakyat Indonesia (BRI) was established in December 1895 in Purwokerto in the Province of Central Java. In the beginning, BRI was initially operating with a view to providing small loans to poor people, enabling them to avoid the local money lenders and loan sharks. The history of the bank's establishment began with the initiative of *Raden Aria Wirjaatmadja* in Banyumas, who sought to find a way for people to avoid loans from moneylenders with high interest rates by initiating small loans with lower interest rates to assist the poor people in the society (cf. Schmit 1994; Rachmadi 2002). Although his successful initiative became widely known among the local people, his initiative was prohibited by his superior *E. Siedburgh,* Assistent-Resident of Banyumas. Notwithstanding, *Raden Aria Wirjaatmadja* pursued his initiative to support poor people and he finally established Indonesia's first Rural Bank with the official support of the Residency of Banyumas. Originally, its name was *Hulp en Spaarbank der Inlandsche Bestuurs Ambtenaren* ('Aid and Savings Bank of the Local Civil Servants'). Throughout its early history, the village bank changed its name, and progressed from *Algemene Volkscredietbank* ('General Peoples' Credit Bank'), through *Syomin Ginko* during the Japanese occupation, into *Bank Rakyat Indonesia* (BRI) after independence in 1945. Since 2003, it became known as *PT Bank Rakyat Indonesia (Persero) Tbk*; the bank made a *Public Offering* after which the bank was listed in the Indonesian Stock Exchange.

Since its establishment, BRI was known as a bank which provides financial support to people in the lower economic strata, particularly in the rural areas. Recently, BRI has become known as a commercial bank which provides services mainly for small and medium enterprises, as stated in the bank's corporate mission (cf. Ismawan 2002). Since the 1960s, BRI has been cooperating with the government of Indonesia, acting as a 'development agent', by implementing various government programmes to support the poor people of the country, particularly in the rural areas.

As mentioned above, BRI had extended its services through its units *Badan Kredit Desa* (BKD) and *Unit Desa* (UD). Both institutions were established by the BRI in order to provide financial and non-financial support to the people in rural areas through microcredit and financial services, mainly to farmers and fishermen. BRI was supported by the Government of Indonesia in the 'Green Revolution' programme which was aimed at achieving self-sustained food security.

Although this programme had contributed to the position of self-sustained rice security for the country in 1984, it was discontinued in 1990, after which the *Pelita* ('Five-Year Development Plan') also stopped. The credit portfolio of BRI for the micro-enterprises is about four-fifths of the bank's lending allocation. Throughout its history, BRI has focused its services on the lower level of the economy and consistently supported small and medium enterprises, both in rural and urban areas (cf. Universitas Indonesia 2008).

9.3 *Unit Desa:* The Village Unit of *Bank Rakyat Indonesia*

The role of Unit Desa in achieving self-sustained food security

One of the well-known units which provides services to the micro-enterprises, particularly in rural areas, is the *Unit Desa* (UD), initially created in the 1970s and formalised in 1984. It was established in cooperation with the Government of Indonesia in order to support the *Revolusi Hijau* ('Green Revolution') programme to achieve self-sustained food security through increased agricultural production.

Unit Desa (UD) was initiated as a savings and borrowing institution at the village level, providing micro-loans to the people in the village, mainly farmers, with the aim to support their economic activities through increased rice production. The farmers used the loan mainly to buy seeds, cows, as well as other agricultural tools. *Unit Desa* (UD) is cooperating with a local institution, established to support the activities, known as *Koperasi Unit Desa* ('Village Unit Cooperative Institution') (KUD), in which some representatives of the farmers are assigned to be managers of the institution.

In order to avoid the misuse of loans—such as their use for consumer goods—the government extended its funding in cooperation with BRI with a training programme called *Bimbingan Masyarakat* (BIMAS). The training and promotion was conducted by farmer groups in every village. This programme supported the farmers not only with know-how of farming methods and techniques, but also with skills to utilise the funding support by the government through *Unit Desa* (UD).

Between 1967 and 1973, the government revised the BIMAS scheme and the programme was also allocated to farmers who already had a high production level. Soon, the BIMAS programme was extended with the *Intensifikasi Masal* (INMAS) in order to widen the target of support. Between 1973 and 1987, further enhancement of *Unit Desa* (UD) was achieved by *Intensifikasi Umum* (INMUM) and *Intensifikasi Khusus* (INSUS) in order to focus on support for targeted groups. In 1984, when Indonesia successfully reached the state of self-sustained food security, the rural financial sector increased rapidly. As indicated by Seibel (2005), the sector comprises about six thousand formal microfinance institutions and more than forty thousand semi-formal institutions, serving more than forty million deposit accounts and thirty million loan accounts. About four-fifths of the institutions, particularly *Unit Desa* (UD), continue to operate under the structure of BRI.

The reformation of Unit Desa (UD) *and Badan Kredit Desa* (BKD)

One of the factors which have contributed to the success of *Unit Desa* (UD) and *Badan Kredit Desa* (BKD) of BRI is the regulation of the interest rates by the government. As the government-owned financial institutions dominated the financial sector, the proposal for the establishment of a new commercial bank has been restricted, particularly in the rural areas. Furthermore, the government was also providing subsidies to the farmers through *Unit Desa* (UD), rendering the programme rather successful. However, after oil prices dropped in the early 1980s, the decline of the GDP of Indonesia—together with the heavy losses from the BIMAS programme caused by a repayment rate of only 50% and the high transaction costs

—brought *Unit Desa* (UD) into a difficult situation. As a result, *Unit Desa* (UD) was challenged by policy makers to be either closed down or reformed (cf. Feekes 1993; Seibel 2005).

After the implementation of the deregulation policy of the financial sector by the Government of Indonesia in 1983, *Unit Desa* (UD) was transformed into a commercial unit, enabling it to become a self-sustaining profit center. The staff of *Unit Desa* (UD) received a profit-sharing incentive, in addition to the policy of penalties for arrears which exceeded 5%. Following the transformation of *Unit Desa* (UD), two commercial products were introduced: *Simpanan Pedesaan* (SIMPEDES) and *Kredit Umum Pedesaan* (KUPEDES). Unlike the initial scheme of the *Bimbingan Masyarakat* (BIMAS) programme where *Unit Desa* (UD) mainly provided credit in order to support farmers with their production of crops and cultivations, the new scheme provided all kinds of rural credit or rural savings for any purpose, for production or even for consumption. Seibel (2005) provides an overview of the achievement by *Unit Desa* (UD) from the time of its transformation from a supporting institution for a self-sustained food security programme to a self-sustaining profit center, documenting an increase from 2655 accounts with an amount of $39,300,000 in 1964 to 29,869,197 accounts with an amount of $3,527,300,000 in 2003.

The development of *Unit Desa* (UD) demonstrates that a financial institution can effectively support development programmes in rural areas. Its integration into government programmes which support the agricultural sector has shown that such collaborative actions in rural areas are important for reaching sustainable development goals. However, several considerations need to be taken into account in order to sustain these programmes, such as high transaction costs and low repayment rates, while training and promotion of new cadres of community-based managers are needed to support innovative approaches such as the *Integrated Community-Managed Development* (ICMD) approach for achieving sustainable community development in the rural areas.

Presently, however, BRI is no longer acknowledged as a bank specializing in services for micro-enterprises, particularly in the rural areas. Although *Unit Desa* is still the largest provider of rural financial services, BRI has become a global bank which now provides the same variety of services as other commercial banks in the financial market. While other global financial institutions are struggling for survival in the uncertain circumstances of the world market, BRI has shown its competitiveness at the global level. With assets of 846 trillion Rupiah (circa $65 billion), an annual capital growth of 16.7% and an average annual net profit of 25.2 trillion Rupiah (circa $1.9 billion), BRI has underscored its financial strength and sustainability. Experience shows that a financial institution which focuses on micro-enterprise and the empowerment of local communities can operate in a sustainable way, compared to those institutions which mainly provide services to large projects and private institutions.

References

Feekes, F. (1993). Extending small credits profitability in Indonesia. *Small Enterprise Development, 4*(2), 33–38.

Berita Satu. (2016). BRI Jadi Bank terbaik di Indonesia, *Berita Satu Daily Online* (Jakarta), April 4, 2016. http://www.beritasatu.com/bank-dan-pembiayaan/358169-2016-bri-jadi-bank-terbaik-di-indonesia.html.

Booth, A., McCawley, P. (eds) (1990). Ekonomi orde baru. *Lembaga Penelitian Pendidikan Penerapan Ekonomi dan Sosial* (LP3ES). Jakarta.

Ismawan, B. (2002). *Membangun Indonesia dari desa melalui keuangan mikro*. Jakarta: Gema PKM.

Kuiper, K. (2004). ACT OR ACCIDENT? The birth of the village units of bank rakyat Indonesia. Eschborn: Deutsche Gesellschaft für Technische Zusammenarbeit (GTZ) GmbH.

Rachmadi, A. (2002). *Bank rakyat Indonesia pelopor: Microfinance dunia, melayani seluruh lapisan masyarakat Indonesia: Membangun Indonesia dari desa melalui keuangan mikro*. Jakarta: Gema PKM.

Schmit, L. (1994). *The history of the "volkscredietwezen" (popular credit system) in Indonesia 1895-1935*. Hague: Ministry of Foreign Affairs.

Seibel, H. D. (2005). The microbanking division of bank rakyat Indonesia: A flagship of rural microfinance in Asia. In M. Harper & S. S. Arora (Eds.), *Small customers, big market: Commercial banks in microfinance*. Rugby: Intermediate Technology Press.

Universitas Indonesia. (2008). Cases in management: Indonesia's business challenges. Seri 2. *Case Center- Departemen Manajemen, Fakultas Ekonomi Universitas Indonesia*. Jakarta: Salemba Empat.

Kurniawan Saefullah is Lecturer of the Faculty of Economics and Business, Universitas Padjadjaran, Bandung, Indonesia since 1998. He graduated from Universitas Padjadjaran and the International Islamic University, Malaysia. He has published several books and articles with his colleagues, including Introduction to Management and Introduction to Business. In 2009, he joined the International Project on Integrated Microfinance Management (IMM) for Poverty Reduction and Empowerment in Indonesia. As a Ph.D. Researcher at the Leiden Ethnosystems and Development Programme (LEAD), Faculty of Science of Leiden University, The Netherlands, he is finalising his Ph.D. Dissertation on 'Gintingan in Subang: An Indigenous Institution for Sustainable Community-Based Development in the Sunda Region of West Java, Indonesia' under the supervision of Prof. Dr. L. J. Slikkerveer. The focus of his research is on the important role of the local cosmology of Tri Tangtu in the process of sustainable community development, while he also developed a model of integration of community-support institutions on the basis of the Indigenous Knowledge Systems (IKS) in combination with the IMM and ICMD approaches, such as financial, medical, educational, communication and social services through the indigenous institutions of the local communities in the Sunda Region of West Java.

Asep Mulyana is Lecturer at the Department of Management and Business of the Faculty of Economics and Business of Universitas Padjadjaran, Bandung, Indonesia. He received his Ph.D. Degree from Bogor Agricultural University (IPB), Indonesia, and is a specialist in the field of the Management of Cooperatives & Small-Medium Enterprises. He was the Coordinator of the Course on *Integrated Microfinance Management* (IMM) in Bandung from 2012–2015.

Chapter 10
Bina Swadaya: A Community-Based Organisation

Bambang Ismawan

> *Bina Swadaya has trained over 10,000 community leaders, spawned the creation of more than 12,000 grassroots self-help groups serving 3.5 million people, and launched 650,000 microfinance institutions with 13.5 million members.*
>
> Ashoka (2008)

10.1 The Foundations of *Badan Pengembangan Swadaya Masyarakat*

Bina Swadaya is an acronym of *Badan Pengembangan Swadaya Masyarakat* referring to a *Community-based Mutual Aid Organisation*, and is one of the largest Non-Government Organisations (NGO) in Indonesia. As an organisation which supports the empowerment of communities, the main purpose of its establishment has been the initiation of a 'bottom-up' development approach, which provides various services to many self-reliant communities, particularly farmer communities in the rural areas.

History, philosophical foundations, vision, and mission

Bina Swadaya was established in Jakarta on the 24th of May, 1967 by *Ikatan Petani Pancasila* ('Association of Pancasila Farmers') (IPP) with the initial name of *Yayasan Sosial Tani Membangun* ('Social Foundation of Farmers' Initiatives') (YSTM). It was established as a 'bottom-up' development project by IPP in order to provide farmers with various forms of support. Considering the political situation in Indonesia, IPP, together with another fourteen farmer organisations, merged in 1973 into a single farmers' association, named *Himpunan Kerukunan Tani Indonesia* ('Indonesian Farmers' Association') (HKTI).

B. Ismawan (✉)
Bina Swadaya Indonesia, Jakarta, Indonesia
e-mail: b.ismawan@gmail.com

L. J. Slikkerveer et al. (eds.), *Integrated Community-Managed Development*, Cooperative Management, https://doi.org/10.1007/978-3-030-05423-6_10

Since its establishment, YSTM provided training and promotion as well as financial support to the agricultural sector until 1985. Through its medium of publication, YSTM is publishing the *Trubus* magazine which has been distributed regularly to the farmers. The magazine contains various significant articles which relate to agriculture, from planting methods, harvesting, and funding support, to the marketing of agricultural products. However, in 1985, the Government of Indonesia issued a new regulation which restricted the press and publishers from undertaking activities which go beyond their main objectives. Subsequently, some members of the Management Board of YSTM established the *Yayasan Bina Swadaya* ('Bina Swadaya Foundation') with the same Management Board as the YSTM. The new foundation managed programmes which were distinct from the publication of the *Trubus* magazine. Later onwards, all activities were consolidated under the management of *Bina Swadaya*, which has evolved into three phases along the changes of socio-political development of Indonesia:

1. *the Pancasila Social Movement Era* (1954–1974)
   In this period of time, *Bina Swadaya* became a mass organisation, focusing on community empowerment, particularly in the agricultural sector in rural areas;
2. *the Socio-Economic Development Institution Era* (1974–1999)
   In this era, *Bina Swadaya* assumed its role as a Socio-Economic Development Institution, profiling itself as a social institution for people participating in community development. *Bina Swadaya* engaged in many cooperation programmes with various institutions, implementing programmes both locally and globally; and
3. *the Social Entrepreneurship Institution Era* (*since* 1999)
   During this period of time, *Bina Swadaya* became a social entrepreneurship institution with a focus on community empowerment by implementing the approach of development 'from the bottom'.

The related activities are in compliance with the new concept of *Integrated Microfinance Management* (IMM) which has been introduced by the Leiden Ethnosystems and Development (LEAD) Programme of the Faculty of Science of Leiden University in The Netherlands. The current endeavours of *Bina Swadaya* are also rather diverse in terms of sectors and services. It serves mainly the public sectors in agriculture, forestry and fisheries, while its activities are integrating various institutions, both vertically and horizontally, ranging from public to private sectors (cf. Slikkerveer 2007; Ismawan 2012). The activities of *Bina Swadaya* are based on several philosophical principles, motives and goals, the formulation of its *Philosophy*, *Vision*, *Mission* and *Strategy:*

Philosophy: In serving the community to eradicate poverty, *Bina Swadaya* is based on the following philosophical foundations:

- serving other people is a noble vocation;
- the people are able to help themselves;
- the best results can only be reached through sincere cooperation and willingness to grow and flourish together;

- social entrepreneurship is an effective vehicle for boosting the capacity and power of the community; and
- good intentions, if carried out ethically, earnestly and consistently, will certainly yield good results.

Vision: The philosophical foundations of *Bina Swadaya* are translated into the institution's *Vision* towards the society. The vision is to become an institute which is pioneering and superior in raising the people's competence and power through social entrepreneurship.

Mission: Bina Swadaya has its own *Mission*, which embarks on the motto to awaken and develop the competence and power of the poor and the marginalised communities in terms of the economic aspects through the facilitation of capacity improvement, institutional community development and access to resources, with a view to:

- exert influence on development policies to be more supportive to the common and marginalised people;
- develop innovations which provide advantages and benefits to the poor and marginalised people;
- forge partnerships with many parties in order to promote the capacity of providing services for the public; and
- safeguard and sustain the life of the institute (cf. Ismawan 2012).

10.2 The Operation of *Bina Swadaya*

From mutual aid to a socio-entrepreneurship organisation

In the implementation of its programmes, *Bina Swadaya* is attempting to contribute in all its activities to one of the Sustainable Development Goals (SDGs) of the United Nations (2015) in terms of strengthening strategic partnership for sustainable community development. As mentioned above, *Bina Swadaya* started its operations by empowering the communities with an approach towards the peoples' local economy to reach a more equitable process of growth. After some years of its engagement with the communities, *Bina Swadaya* supported the organisation of farmers into affinity groups, called *Self-Help Groups,* with a view to facilitating the delivery of technical assistance. *Bina Swadaya* applied the principle of mutual aid in those situations where any community development programme was implemented on the basis of local initiatives from the communities, providing full advisory and training support. The implementation of this principle has evidently shown that both parties can have a mutual benefit, where the one empowers the other.

Although *Bina Swadaya* evolved during the era of the 'New Order' in which a 'top-down' approach in development was dominantly implemented, it did not refrain from its intention to empower the people at the community level. The successful integration of the *Bina Swadaya* programmes into the development plans of the government, such as the *Pembangunan Lima Tahun* ('Five-Year Development Plan') (*Pelita*) which focused on the agricultural sector up until the 1980s, has rendered wide acceptance to the organisation, both by the government and by the communities. *Bina Swadaya* also cooperated with the Government Institution for National Planning (Bappenas), particularly in the implementation of *Inpres Desa Tertinggal* (IDT), a government programme that intended to raise the development of local people and communities in remote areas.

The cooperation between *Bina Swadaya*, the Ministry of Home Affairs and the Ministry of Education and Culture in the distribution of various books for schools has also further strengthened the position of *Bina Swadaya*'s *Trubus* publisher, later renamed *Penebar Swadaya*. This successful cooperation has also extended the role of *Bina Swadaya* beyond the agricultural sector in the rural areas to profile itself as a socio-economic development organisation, particularly from 1974 to 1999, a year after the end of the era of the 'New Order'.

With the beginning of the era of 'Reformation' in 1999, the progress of the democratisation process also created new opportunities for the implementation of regional autonomy, which have strengthened the position and activities of *Bina Swadaya*. It has transformed from a socio-economic organisation into a socio-entrepreneurship organisation which implements the 'bottom-up' approach in combination with the additional motive of 'self-sufficiency' in order to ensure that its activities will retain its sustainability.

The Bina Swadaya Network

Bina Swadaya has created a large network of partners working in the implementation of its programmes, both nationally and globally. At the national level, *Bina Swadaya* has been active in building up cooperation networks with various institutions, including the Secretariat of *Bina Desa*; *Asosiasi LSM Mitra Lembaga Keuangan* ('Association of Financial Institute Partners NGOs') (ALTRABAKU); *Forum Komunikasi Kehutanan Masyarakat* ('Community Forestry Communication Forum') (FKKM); *Forum Kerjasama Pengembangan Koperasi Indonesia* ('Indonesian Cooperative Development Forum') (FORMASI); *Gerakan Bersama Pengembangan Keuangan Mikro Indonesia* ('Indonesian Movement for Micro Financial Development') (Gema PKM); *Jejaring Kerja Pemberdayaan Masyarakat* ('Community Empowerment Network') (JKPM); the *Partnership in Development Forum* (PDF); and many other organisations and institutions.

In terms of its international networks, *Bina Swadaya* has strong cooperation relations with various institutions all over the world. In Asia, they include, *i.a.*: the *Japan Partnership Network for Poverty Reduction* (AJPN, Tokyo); the *Asian NGO Coalition for Agrarian Reform and Rural Development* (ANGOC, Manila); the *Asia Pacific Banking with The Poor Network* (Brisbane, Australia); the *Asia Pacific Rural and Agriculture Credit Association* (APRACA, Bangkok); and the *INASIA Network* (Colombo). It also cooperates with: the *International Council on Social Welfare* (Geneva); the *International Leaders Forum on Development Finance* (New York); the *Microcredit Summit Global Campaign* (Washington); and many other networks (cf. Ismawan 2002).

Training Programmes

One of *Bina Swadaya's* core competences is to provide training courses for community empowerment. Some of its most popular courses include the following:

- Training for *Community Self-Help Units Facilitators* (*Tenaga Pengembangan Kelompok Masyarakat Swadaya*—TPKS);
- Training on *Community Self-Help Development Management* (*Pelatihan Manajemen Pengembangan Swadaya Masyarakat*);
- Training on *Participatory Rural Appraisal*;
- Training on *Collective Savings-Loans Business/Credit Unions* (*Pelatihan Usaha Bersama Simpan Pinjam*);
- Training for *Small-Scale Entrepreneurs* (*Pelatihan Wirausaha Kecil*); and
- Training on *Household Economic Management* (*Pelatihan Pengelolaan Ekonomi Rumah Tangga*).

At present, *Bina Swadaya* has a large variety of training programmes, and a number of training courses are still in preparation. They include training courses on *Micro-Financial Institution Management* and *Small-Scale Industry/Business Technical Skills*, as well as various training courses on *Disaster Risk Mitigation and Disaster Handling*, *Farming Technical Skills* and so forth.

The number of *Bina Swadaya's* alumni from training programmes since 1980 has reached about 30,000 persons (cf. Budianta 2007; Ismawan 2012).

Supporting Community Development

Although training programmes would eventually need proper venues with decent facilities, at the early stage however such expectations may need to be postponed. Often, urgent training programmes have to be launched soon with whatever facilities are available. In the early days, the *Community Self-Help Units Facilitators* (*Tenaga Pengembangan Kelompok Masyarakat Swadaya*—TPKS) training courses were held in a poultry slaughter house converted into a class room, where the participants were placed in local peoples' houses as their internship accommodations near the slaughter house. These modest 'facilities' might have served as an opportunity for the trainees/participants to experience the real life of the rural villages where they would be likely to live in the future. Later, with the support from international donor organisations, *Bina Swadaya* built a complete training center with its own cottages and bedrooms for 48 persons. On the 18th of August 1982, the center was inaugurated as the *Bina Swadaya Education and Training Campus*, better known today as *Wisma Hijau* ('Green House').

The facilities of *Wisma Hijau* presently include 200 beds and a number of large and small meeting rooms, facilitating several training courses, workshops and seminars with programmes and itineraries running in parallel sessions. The users of these facilities are usually government agencies, NGOs, religious institutions, universities and private organisations. The annual occupancy rate of *Wisma Hijau* is about 70%, rendering the Education and Training Campus an excellent facility for community development and income-generating enterprises.

Policy dialogue and consultancy

The experience of *Bina Swadaya* shows that only about 3000 *Community Self-Help Units* (KSMs) have received the services of *Bina Swadaya*. But when the community empowerment programmes were conducted through dialogues and cooperations with a number of government and corporate agencies, the number of KSMs which received the assistance of *Bina Swadaya* reached about 1 million, with about 25 million families or 100 million people as its members. Several of these programmes have been carried out, including the following:

- Cooperation with the *Badan Koordinasi Keluarga Berencana Nasional* ('National Family Planning Coordinating Board') (BKKBN) and the *United Nations Population Fund* (UNFPA) with financial support from the *Ford Foundation* in the *Women Participation in Development Programme.* Between 1983 and 1989, the programme was successfully set up and it gave assistance to 650,000 units of the *Usaha Peningkatan Pendapatan Keluarga Sejahtera* ('Improvement of Marginal Families' Earnings Scheme') (UPPKS) with about 13.5 million families as members.
- Cooperation with *Perum Perhutani* ('Perhutani State Corporation') with support from the *Ford Foundation* for the *Social Forestation Programme on Java Island*, involving about 50 NGOs and a number of Universities in the establishment and provision of advocacies to 9000 *Kelompok Tani Hutan* ('Forest Farmers'Groups') between 1986 and 1998.
- Cooperation with *Bank Indonesia* and *Bank Rakyat Indonesia*, with support from the Deutsche Gesellschaft für Technische Zusammenarbeit ('German Technical Cooperation' (GTZ)) for the Bank and KSM Relations programme in the establishment and provision of advocacies to 34,227 Community Self-Help Units/KSM, with 1.026.810 families obtaining financial services at over 1000 bank offices between 1987 and 1999. Policy dialog activities have also been carried out by *Bina Swadaya* as a Member of the *Micro-Financial Draft Law Formulating Team* along with *Bank Indonesia* in 2001, and with the *Regional Representatives' Council* (DPD) in 2008. However, both efforts failed to support the establishment of a Microfinance Law in Indonesia. Between 2010 to 2011, *Bina Swadaya* joined the team which was established by the Ministry of Home Affairs to draft the Indonesian Law on Civil Society Organisation.

Supporting Integrated Microfinance Management (IMM)

In its efforts to improve the capacity of the poor, particularly in rural communities, *Bina Swadaya* is convinced that poor people are not the same as the people called the 'have nots'. They are just the people who have very little. The potency of the local people, whatever small, if well-managed by offering appropriate education in advanced skills, developed in a manner of collaboration and solidarity, and provided with advocacies in a proper manner, could enable them to overcome their problems by themselves. In the application of this idea for community empowerment, *Bina Swadaya* has developed three strategies:

1. encouraging the establishment and development of a solidarity institution called *Kelompok Swadaya Masyarakat* ('Community Self-Help Units' or 'Self Help Groups');
2. promoting the production and marketing of the products by publishing the '*Trubus*' Magazine; and
3. developing and providing services for capital demands through integrated micro-financial services.

Since 2009, *Bina Swadaya*, under the *Gerakan Masyarakat untuk Pemberdayaan Keuangan Mikro* (GEMA PKM), joined the *International Consortium on Integrated Microfinance Management* (ICIMM), together with the *Leiden Ethnosystems and Development Programme* (LEAD) of Leiden University in the Netherlands; the *Faculty of Economics and Business* (FEB) of Universitas Padjadjaran, Bandung in Indonesia; and the *Mediterranean Agronomic Institute of Chania* (CIHEAM-MAICh) in Crete, Greece. In 2012, the ICIMM Consortium successfully established the new Master Course on *Integrated Microfinance Management* for Poverty Reduction and Sustainable Development in Indonesia (M.IMM) with support from the Ministry of Economic Affairs of The Netherlands ((EVD/INDF). The development of the new Master Course, hosted by the Faculty of Economics and Business (FEB) of Universitas Padjadjaran in Bandung, Indonesia, is further documented in Chap. 12.

Over the years, *Bina Swadaya* has demonstrated that a 'bottom-up' approach can indeed be successful in scaling up its activities from a community-based organisation to eventually become a local institution of the indigenous peoples in the rural areas by focusing on sustainable community development. In addition, the thriving integration of *Bina Swadaya* with various institutions, horizontally and vertically, at local and global levels across different sectors, has shown that an integrated approach for sustainable community development through a 'bottom-up' approach is crucial in order to realise the Sustainable Development Goals in Indonesia by the year 2030 (cf. UN 2015).

References

Ashoka. (2008). Bambang Ismawan-The New Idea. URL: https://www.ashoka.org/ en/fellow/bambang-ismawan.

Budianta, E. (2007). *Menuju Indonesia Swadaya*. Jakarta: Bina Swadaya.

Ismawan, B. (2002). *Membangun Indonesia dari desa melalui keuangan mikro*. Jakarta: Gema PKM.

Ismawan, B. (2012). *Bina Swadaya–45 Years (1967-2012)*. Jakarta: Bina Swadaya. http://binaswadaya.org/bs3/history/.

Slikkerveer, L. J. (2007). *Integrated microfinance management and health communication in Indonesia. Cleveringa Lecture*. Jakarta: Trisakti School of Management.
United Nations (UN). (2015). *Post-2015 Agenda for Sustainable Development and Sustainable Development Goals (MDGs)*. New York: United Nations.

Bambang Ismawan is the Founder and Chairman of the *Bina Swadaya* Foundation and General Secretary of *Gema PKM Indonesia*. He holds two Master Degrees in Agriculture Economics from Gadjah Mada University and in Rural Sociology from Bogor Agricultural University, and is currently finalising his Ph.D. research at Leiden University. He is also Senior Lecturer in the Course on *Integrated Microfinance Management* (IMM), Bandung.

Chapter 11
Recent Government Policies of Poverty Reduction: KDP, UPP and PNPM

L. Jan Slikkerveer and Kurniawan Saefullah

> *PNPM represents to a large extent a "laboratory" that offers lessons and innovative practices on community planning, capacity building, and targeting more and marginalised groups*
> World Bank (2011)

11.1 Indonesia's Poverty Reduction Policies

Following the global financial crises of 1997–1998 which particularly affected the poor and low-income families in the developing countries, many development planners and experts have directed their attention to the subject of poverty reduction on a worldwide scale. Much interest has been shown in the development of new approaches of assessing the impact of socio-economic policy interventions (cf. Chap. 7). The failure of the previous 'top-down' approaches to achieve poverty reduction has also inspired many scientists to pay more attention to the empowerment of local peoples and communities to participate in development. The impact of the above-mentioned financial crises made it clear that after a prolonged experience with overall disappointing community development programmes, largely implemented through 'top-down' strategies, alternative approaches were needed to achieve a true reduction of poverty among the poor in an effort to attain genuine, sustainable development in the near future.

Linking up with the experience of various forms of community development which have been initiated over the past decades, such as *Integrated Rural*

L. J. Slikkerveer (✉)
LEAD, Leiden University, Leiden, The Netherlands
e-mail: l.j.slikkerveer@gmail.com

K. Saefullah
FEB, Universitas Padjadjaran, Bandung, Indonesia
e-mail: kurniawan.saefullah@unpad.ac.id

L. J. Slikkerveer et al. (eds.), *Integrated Community-Managed Development*,
Cooperative Management, https://doi.org/10.1007/978-3-030-05423-6_11

Development (IRD), *Community Learning and Development* (CLD) and *Community Capacity Building* (CCB), new approaches of *Community-Driven Community Development* (CDCD) have been developed to give more power to local decision-making processes and control over local resources management to indigenous peoples and communities with a view to increasing peoples' participation in sustainable development. As mentioned before, neo-ethnoscientists, including Warren et al. (1995) have amply illustrated that embarking on the integration of indigenous knowledge systems as the expression of the cultural dimension of development in various settings promotes local participation as a major prerequisite for attaining sustainable development, which, in turn, is intimately interrelated with community-based poverty reduction (cf. Warren et al. 1995; Alkire et al. 2001; Platteau 2004; Asian Development Bank 2006).

The following paradigm shift not only evoked a growing interest in indigenous knowledge systems within the context of development and change, but also provided the basis for the implementation of related community-level decision-making processes; follow-up research has later been conducted to include the closely related indigenous institutions, and the role these could play in 'bottom-up' sustainable development policies and programmes. By consequence, the new community-based approaches have further been elaborated in advanced training in newly-developing fields of *Integrated Microfinance Management* (IMM) and *Integrated Community-Managed Development* (ICMD).

The impact of the severe monetary crisis in 1997–1998 in Indonesia has also reinforced the notion that the 'top-down' approach of development has failed to reach equal distribution of goods and services as well as well-being for the entire population (cf. Palumbo et al. 1984; Matland 1995).

Although the overall economic growth in Indonesia showed positive figures in the decades before the crisis, the country's poverty rates however tend to remain rather high, posing a continuing challenge to the central government. By using the Indonesian measurement of the poverty line, which was calculated in 2013 at 211,726 Rupiah per month—equal to $16 per month or $0.45 per day, which is far below the World Bank's recent indicator of the poverty line of $1.90 per day—the number of people who are still suffering from poverty is reaching 30 million inhabitants or 13.3% of the population in 2014. The poverty rate initially declined by 1% annually from 2007 to 2011 and since 2012, poverty has declined by an average of only 0.3% points per year. The World Bank (2015) asserts that: *"Approximately 40% of the entire population remain vulnerable to falling into poverty, as their income hovers marginally above the national poverty line"* (cf. Neraca 2011; Bappenas 2013; World Bank 2015).

In addition to the measurement of the low income of the poor, the disparities between the rich and the poor of the country are also considered rather high after the crisis of 1997–1998. According to the World Bank (2015), Indonesia's Gini coefficient—a measure of inequality—of 0.40 is higher than any of its neighbouring countries, as about 1% of the wealthy people own about 28% of the total Indonesian wealth, leaving behind the remaining 72% of the population, *i.e.* approximately 185.5 million people.

As in the case of poverty reduction, the 'top-down' approach has also failed to decrease the disparities between the rich and the poor in Indonesia, rendering the need for an alternative approach in this area as similarly urgent. As a result of these negative aspects of the socio-economic situation, the Government of Indonesia eventually also decided to change its policies with the 'bottom-up' approach which incorporates community participation in development (cf. Davies 2008; Winters 2011; Budiantoro 2011; Latifah 2011). The timely adoption of the 'Bali Promise' of the *World Culture Forum* (2013) calls for a measurable and effective role, as well as the integration of culture in development at all levels in the Post-2015 Development Agenda of the United Nations to contribute to the improvement of the situation of the poor in Indonesia.

Thus, after the global financial crisis, during the era of President Gusdur and Vice-President Megawati, the government made an effort to empower rural communities in order to tackle the problem of poverty in an integrated and sustainable manner by the introduction in 1998/1999 of a national programme implemented at the community level.

As a modification of past poverty reduction programmes, such as the *Underdeveloped Village Inpres* (IDT) and the *Disadvantaged Village Support Infrastructure Development Assistance* (P3DT), the new programme encompassed the *Program Pengembangan Kecamatan* ('Kecematan Development Programme') (KDP) focusing on community development in the rural areas, and the *Program Penanggulangan Kemiskinan di Perkotaan* ('Urban Poverty Programme') (UPP) focusing on community development in the urban areas. Both programmes, funded by the World Bank (2015) in the form of a soft loan to the Government of Indonesia, have been designed to alleviate poverty by empowering the community and promoting participation in development.

11.2 *Kecamatan* as the Administrative Unit for Community Development

Indonesia is a large country in which the administrative organisation is divided into 34 provinces, 508 districts, 6694 sub-districts, and 77,465 villages. During the 'New Order', when the number of provinces was only 27, development was implemented from the top by the central government in the capital of Jakarta. Provinces (*Propinsi*), Districts (*Kota/Kabupaten*), Sub-Districts (*Kecamatan*), and Villages (*Kelurahan/Desa/Kampung*) were basically subjected to the plans and policies of the central government. The administrative structure had been designed to ensure that the government policies would be executed throughout the country in an effective manner.

During the period of time of the 'New Order', the implementation of Law No. 5/1979, known as the *Village Government Law,* required the nearly 67,000 *Desa* ('Villages') to adopt a uniform system of government. The *Kepala Desa*

('Village Head') became a civil servant in the national government, assisted by the *Lurah* ('Village Secretary') and some *Kepala Urusan* ('Section Chiefs'). The law also assigned the establishment of the *Lembaga Musyawarah Desa* ('Village Consultative Council') (LMD) and the *Lembaga Ketahanan Masyarakat Desa* ('Village Community Resilience Council') (LKMD), whose tasks were later in 1981 and 1984 extended with a mission to activate community participation to carry out development in a coordinated way.

Soon, however, domestic and international NGOs started to criticise Indonesia's *Village Government Law* as a dangerous homogenising policy which undermined the traditional institutions of the indigenous socio-cultural groups throughout the country, and in particular their indigenous resources management systems which had developed over many generations on the basis of unique traditional environmental knowledge (TEK). As Lindsey (2008) asserts: *"The 'one-size-fits-all' Village Government Law shunted aside these time-tested structures and institutions, replacing them with a new culture of government that was accountable not to local communities and livelihoods, but to political rulers in far-off Jakarta"*. Soon, it became clear that most *adat* functionaries did not meet the legal criteria to serve as *Kepala Adat* in terms of age, education, literacy, knowledge of the national law etc., allowing only some of them to take up such a position temporarily. Notwithstanding, the position of *Kepala Desa* became much desired, as it enabled diverting village development funds for personal and family gain, and monopolising and expropriating village-owned territory and resources (cf. Lindsey 2008).

Figure 11.1 shows the organisational structure in Indonesia, in which the *Kecamatan* represents the local government in the implementation of the national community development programmes (cf. Indonesian Law No. 21/2001; No. 6/2014; No. 23/2014; No. 47/2016). In the Indonesian development policy, the concept of a 'local community' refers to the components of the administrative structure which encompasses the level of the *Kota* ('Municipality')/*Kabupaten* ('District') down to the lowest level of the *Rukun Tetangga* ('Household Group'). In this structure, the *Kecamatan* is the sub-district which coordinates the development of the villages and is responsible for reporting to the level of the *Kota/Kabupaten* of government. So when the government started to launch the community development policy, its approach to community empowerment was based on the local administrative level which had direct connections with the central government.

The renewed development approach, however, proved rather insufficient, largely because the indigenous concept of local communities was not taken properly into account in the related development projects and programmes. In particular, the *Lembaga Adat* ('Traditional Institution') and its indigenous structures prevailing among most ethno-cultural groups were not completely accommodated in the community approach of the central government.

There are some areas in Indonesia which have been granted the privilege to become a special area with freedom to implement their own culture, such as Aceh and Yogyakarta, which received the status of *Daerah Istimewa* ('Special Area').

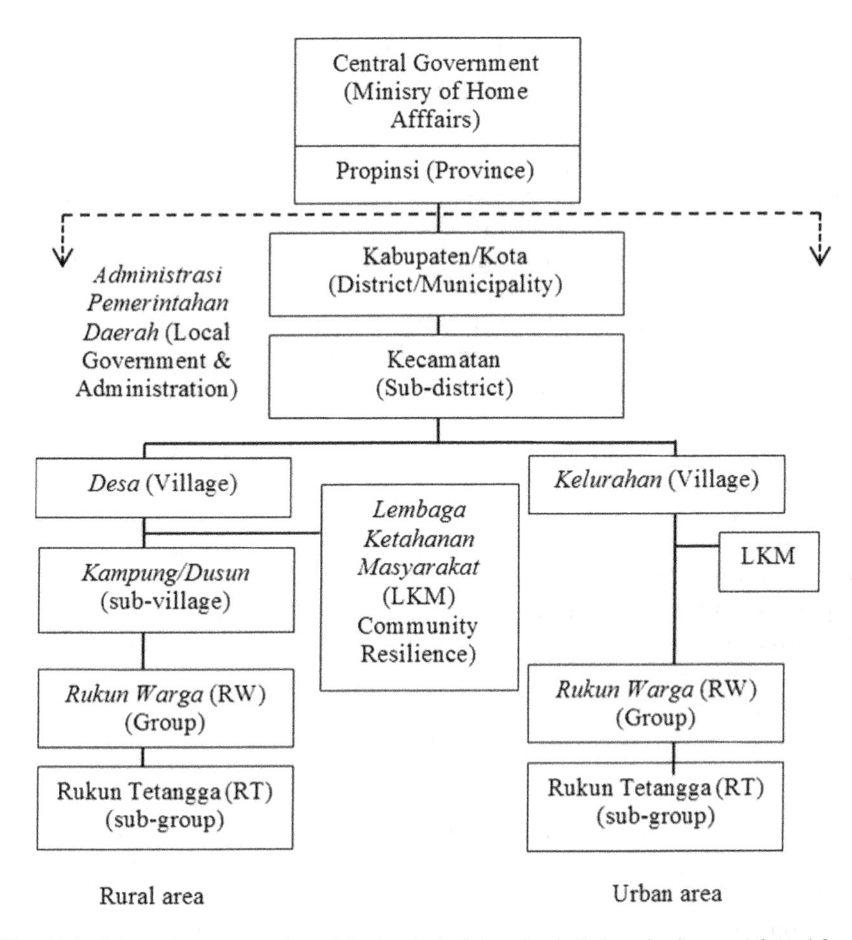

Fig. 11.1 Schematic representation of the local administration in Indonesia. *Source* Adapted from Indonesian Law No. 21/2001, Law No. 6/2014, Law No. 23/2014, and Law No. 47/2016

They could develop their area by integrating their cultural views with their development policies. However, most geographic areas in Indonesia do not enjoy such a special privilege. In this context, the cultural diversities have not yet been accommodated in the development plans and policies of Indonesia. However, the implementation of the local administrative structure according to Fang (2006) happened for three reasons: (1) earlier research by the *Program Pengembangan Kecamatan* (PPK) ('Kecematan Development Programme' (KDP)) had revealed that poverty tends to be homogenous at the sub-district level; (2) early successful Village Infrastructure Projects (VIP) suggest a similar approach; and (3) *Kecamatan* showed good coordination with the villages and the bureaucracy was considered smooth (cf. Sirait 2000; Fang 2006; Mulyono 2014).

11.3 Program Pengembangan Kecamatan (Kecamatan Development Programme)

The *Program Pengembangan Kecamatan* (PPK) ('Kecamatan Development Programme') (KDP) refers to a development programme by the Government of Indonesia to alleviate poverty after the monetary crisis of 1997, particularly in the rural areas of the country. The programme had been designed to empower communities to participate in the development process, which had not been realised in the 'New Order' when President Soeharto was in power. While during the 'New Order', all development policies were decided by the central government, PPK which had started the first phase as a three-year programme in August 1998 began to accomodate the 'bottom-up' approach of development. It stimulated the people in the communities to identify their own development problems and to come up with a proposal for improvement, using the local government structure as shown in Fig. 11.1. PPK was targeted at the poorest *Kecamatan* ('sub-district') in Indonesia. It aimed to foster a 'bottom-up' approach of development and participatory forms of local governance by strengthening the *Kecamatan*, including village capacities and improving community participation in development. Local communities at the village level can prepare any development proposal which is then submitted to the *Kecamatan*. All the proposals from the *Kecamatan* level will then be submitted to the central government, under the coordination of the district and provincial levels of administration. The Government of Indonesia will support *Kecamatan* projects ranging from 500 million to 1 billion Rupiah, which is about $43,000–$125,000 per year, depending on the type of programme as well as on the size of the *Kecamatan*. All proposed PPK projects were aimed at empowering the villagers to make their own choices about development projects which they need and want (cf. PSF 2002).

The PPK has been implemented under the Directorate of Community Development of the Ministry of Home Affairs. The Directorate coordinated the PPK together with the teams of facilitators and consultants from the village level to the national levels, and provided some technical support and training. At the village level, a 'Village Facilitator' was selected in a public forum by the local community to coordinate the programme. Project decisions were made locally and the 'Village Committees' were responsible for procurement, financial management, project implementation and oversight. The projects on infrastructure used local building materials, suppliers and labourers. Indonesian civil society organisations such as the 'Association of Journalists' and NGOs based in the provinces have been providing independent monitoring for the PPK, which had started in 1998 during a period of great political and economic turmoil. After three years of its operation, the PPK had reached a coverage of 4048 *Kecamatan* and 69,168 *Desa* (Table 11.1).

From all the sectors' coverage, around 73% was allocated to infrastructure, including roads, bridges, water, sanitation facilities, *i.a.*, while about 27% was allocated to the economic sector and other sectors, including agriculture and fisheries (cf. Table 11.2). Over 7000 roads were built, which covered 10,800 km. About 2170 bridges, 2370 water supply and sanitation units and 3200 irrigation

Table 11.1 Coverage of the *Program Pengembangan Kecamatan* (Kecamatan Development Programme)

Year one of KDP (1998–1999)	Year Two (1999–2000)	Year Three (2001–2002)	Total in Indonesia
20 provinces	20 provinces	22 provinces	32 provinces
105 districts	116 districts	130 districts	341 districts
501 *kecamatan*	727 *kecamatan*	984 *kecamatan*	4048 *kecamatan*
3542 *desa*	1325 *desa*	15,481 *desa*	69,168 *desa*

Source Wong (2003)

Table 11.2 Sectors Allocated by the *Program Pengembangan Kecamatan* (Kecamatan Development Programme) (Years 1 & 2)

Sectors allocated by KDP	Percentage of allocation (%)	Type of sector
Roads	43.0	Infrastructure
Bridges	6.8	Infrastructure
Water	8.2	Infrastructure
Irrigation/drainage	7.3	Infrastructure
Sanitation facilities	1.4	Infrastructure
Markets	1.4	Economy
Trading	5.0	Economy
Animal husbandry	6.1	Agriculture
Loans & savings	11.5	Economy
Plantation	2.9	Agriculture
Fisheries	1.6	Fisheries
Other infrastructure	4.8	Infrastructure
Total allocation	100.0	

Source Adapted from the PNPM Support Facility (2002)

schemes were also executed by the PPK (cf. PNPM Support Facility 2002). Although the first phase (1998–2001) of the PPK was evaluated as not successful in terms of empowering local people to participate in development policies and programmes, the successful coverage of the programme encouraged the government to continue it.

Second Phase of the Kecamatan Development Programme (PKK) (2002–2005)

The second phase of the PKK was implemented for four years from 2002 to 2005. The World Bank provided more soft loans to the Government of Indonesia with an additional USD 320.8 million to implement the second phase of the PKK whose objective was to accelerate poverty alleviation in order for communities to reach self-independence by improving the capacity building of the local people and strengthening local institutions, both within the villages as well as between villages. Furthermore, the programme supported a broad construction programme of social and economic infrastructure in accordance with the development needs of the villagers.

The specific objectives of the second phase of the PKK (KDP-2) (2002–2005) are as follows:

1. to improve community participation in the development of the decision-making process for planning, implementation, monitoring and sustainability;
2. to improve the role of women in the development of the decision-making process;
3. to make efficient use of local resources and create potential for development;
4. to support participatory planning and development management in the villages; and
5. to provide support in economic, education and/or health infrastucture based upon community self-identified needs.

Evaluation

While the initial motive was to increase equality in development, the PKK can be considered as rather successful in reaching its target. However, it has also been reported that the successful coverage was largely determined by a form of one-way communication and a 'top-down' approach. The local participation which was meant to be reached by the PKK had failed to be realised. The implementation of a 'bottom-up' approach had also failed. Many cases have been reported in which the proposal for a development project had not been designed by the local people at the village level, but rather by the coordinator at the *kecamatan* level, where in many cases even the decisions on the development sector had been taken.

Almost a thousand complaints were made on the violation of the procedure of the PKK, while more than 700 cases concerned the misuse of funds, all complaints totalling to almost 2000 during the first phase of the operations of the PKK. As shown in Fig. 11.2, the implementation of the PKK was only successful in 1999

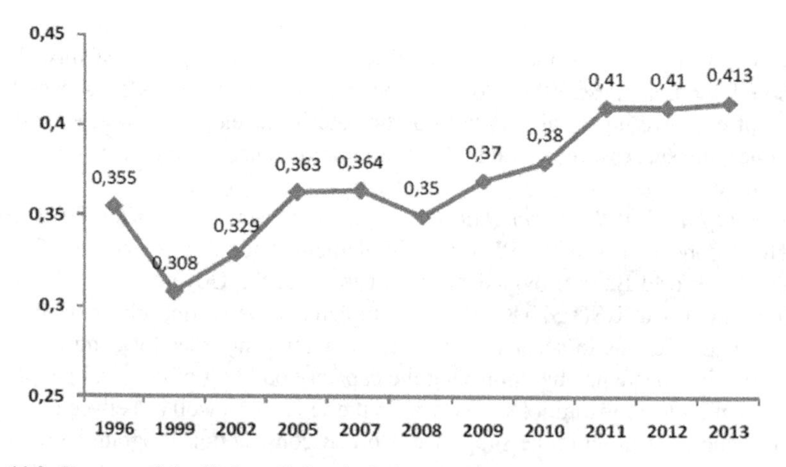

Fig. 11.2 Progress of the Gini coefficient in Indonesia 1996–2013. *Source* Bappenas (2013)

with a decrease of the Gini coefficient to 0.308. However, the Gini coefficient increased again to 0.329 in 2002 when the first phase of the PKK ended.

The second phase of both the PKK (KDP-2) and the *Program Penangggulangan Kemiskinan di Perkotaan* (P2KP) ('Urban Poverty Programme') were implemented from 2002 to 2005. However, as indicated in Fig. 11.2, the disparities increased, as illustrated by the increased Gini coefficients, *i.e.* from 0.329 through 0.363 to 0.364.

Although the second KDP-2 and the P2KP were not as yet successful in decreasing the disparities, they continued under Pesident Susilo Bambang Yudhoyono, albeit with an adaptation in the name. Then, the government launched a similar empowerment programme, named *Program Nasional Pemberdayaan Masyarakat (PNPM) Mandiri* or *PNPM Mandiri* ('National Programme for Community Empowerment') which focused on the implementation of the 'bottom-up' approach, but still with 'top-down' support by the government. The programme adapted the approaches of the PKK and the P2KP, and used the administrative structure of the *Kecamatan* ('Sub-District') and *Kelurahan/Desa* ('Village') as the basis for the community development approach. The programme which was also funded by the World Bank sought to involve people at these two levels to accommodate both the 'bottom-up' approach and participation in development.

Although the decrease in the above-mentioned disparity was a major target of the PKK, Fig. 11.2 shows—as mentioned above—that the Gini coefficient increased again, highlighting that the programme remained unsuccessful in terms of realising its objective to decrease the income gap of the people (cf. Wong 2003; Bappenas 2013).

11.4 *Program Penangggulangan Kemiskinan di Perkotaan* (P2KP) ('Urban Poverty Programme') (UPP)

The *Program Penangggulangan Kemiskinan di Perkotaan* (P2KP) ('Urban Poverty Programme', UPP) is a similar programme to the PKK, which focused on the urban areas for its implementation. Unlike the PKK which fell under the responsibility of the Ministry of Home Affairs, the P2KP fell under the management of the Ministry of General Affairs (*Kementrian* PU). It was based on a soft loan from the World Bank where USD 112 million was made available in the first phase of the programme and USD 126 million for the second phase. As indicated by the World Bank (2005), the initial development objective of the P2KP was described as follows:

> Through a bottom-up and transparent approach, the project seeks to improve basic infrastructure in poor urban neighborhoods and to promote sustainable income generation for its poor urban residents who are mostly long-term poor, have incomes eroded by high inflation, or lost sources of income in the economic downturn. Also, the project seeks to strengthen the capability of local agencies to assist poor communities.

As regards its coverage, the P2KP has covered 2621 *Kelurahan* ('Villages') from 2800 project targets in five provinces of Northern Java. These *kelurahan* covered about 7.8 million households or 31.2 million people. About 3.9 million households benefitted directly and indirectly from the P2KP. In the first phase, the project covered about 55% of poor beneficiaries (defined by a household earning less than 250,000 rupiah per month). In the second phase of the project, about 80% of poor beneficiaries could be covered. The project involved 12,000 facilitators where 9271 volunteers were trained to implement the project. In terms of gender equality, about 38% of beneficiaries were women. In terms of the project's allocation, in general, the P2KP has distributed funds for micro- and SME-enterprises in the form of microcredit (30%), other social services (18%), sub-national government administration (18%), the general industry and trade sector (17%) and housing construction (17%).

In terms of reaching the poorest of the poor, the P2KP seems unsuccessful in reaching this group in the community. According to the Report, the volunteers and consultants preferred to lend the money to the economically active poor and micro-enterprises rather than the poorest of the poor, due to the risks that the project had to face. Only in the second phase did the P2KP allocate funds to support the poorest of the poor in order to enable them to meet their basic needs. The Report by Bappenas (2008) reveals that in terms of the physical infrastructure, the project can be considered successful in the implementation of the P2KP. However, in terms of the budget allocation to the socio-economic infrastructure, several critical remarks can be made on its implementation:

- In terms of economic support, the P2KP has supported the communities by providing them with credit to manage their micro-enterprises. However, there emerged problems with the repayment, where many clients could not repay their loans, largely because of the lack of social capital for the people in the communities;
- In terms of social and educational support, the allocated budget supported the accessibility of people to education in terms of buying uniforms, books, and other products; and
- Most of the people still need to be guided and facilitated by the consultants and facilitators. The idea of a 'bottom-up' approach was not completely implemented as the involvement of the local people remained insufficient (cf. World Bank 2005; Bappenas 2008).

11.5 PNPM-Mandiri: The Shift from Community

Economic Empowerment

Following the Community-Driven Development (CDD) programmes of the PKK and the P2KP, the Government of Indonesia decided to continue both programmes with an adaptation of their names (Fig. 11.3).

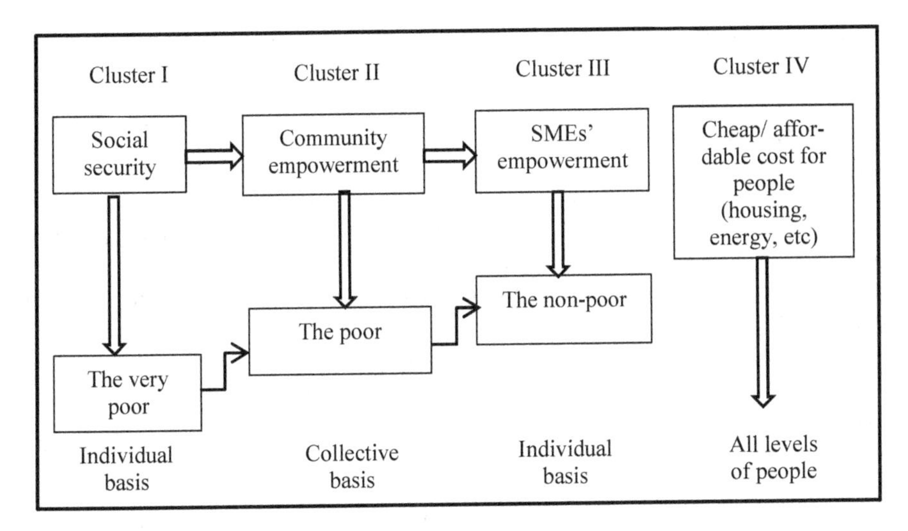

Fig. 11.3 Four clusters of the national programme for poverty alleviation ('PNPM Mandiri') *Source* Adapted from Agus (2013)

In April 2007 the *Program Nasional Pemberdayaan Masyarakat* ('PNPM Mandiri') was launched with two main programmes: *PNPM Perkotaan* ('Urban PNPM') and *PNPM Perdesaan* ('Rural PNPM'). In this way, *PNPM Mandiri* evolved from its establishment in 2007 to its implementation in the following years, where initially, two programmes were implemented: *PNPM Perdesaan* ('Rural PNPM Mandiri') for rural areas and *PNPM Perkotaan* ('PNPM Mandiri') for urban areas.

Since 2008, both programmes expanded to focus on a number of specific target sectors and groups of people including the PNPM-*Peduli* ('Care-PNPM'), to reach the vulnerable and marginalised groups of the population, including the victims of trafficking, prostitution, orphanages, HIV/AIDS, etc. (cf. Bappenas 2013).

11.5.1 The Operation of PNPM-Mandiri

The new *Program Nasional Pemberdayaan Masyarakat* adapted the Community-Driven Development approach with the support of the existing administrative structure of local governance in its implementation, as previously described. PNPM Mandiri became a part of the *National Programme for Poverty Alleviation*. The main objective of this national programme was to improve the well-being of the poor and provide job opportunities for them. Specifically, the programme aimed at: (1) increasing community participation in development, especially by the poor people; (2) increasing the capacity building of the people and the local government to improve their ability to put poverty alleviation into practice; (3) building a synergic network in poverty alleviation; and (4) strengthening social capital innovation for community empowerment and poverty alleviation (cf. Bappenas 2013).

Any PNPM Mandiri project, implemented at the community level, is based on a proposal submitted by the *Kelurahan/Desa* ('Local Community') to the *Kecamatan* ('Sub-District'). In the villages, there is one trained facilitator whose task is to coordinate and integrate local people and institutions in order to prepare the local initiatives and projects, based on the needs of the local people, varying from economic needs to infrastructure. The local initiatives include not only the kind of development project which will be proposed, but also the way in which the project will be executed, and who will provide the resources for the project. In most cases, the resources are a combination of local sources in the form of a donation from the people themselves and a subsidy from the central government. The submitted proposals are brought to the district and provincial levels where approval will be made. The central government, supported by a specific investment loan from the World Bank, allocates specific funds every year for each province on the basis of the priorities and submitted project proposals. As an implementation of the *Community-Driven Development* approach, two main factors are taken into consideration for approval: the local participation in the proposal and the level of priority of the project which relates to the local development programme as planned by the local government (cf. World Bank 2011).

11.5.2 Coverage of the Programme

Some observations can be made about the implementation of the PNPM Mandiri programme. In general, *PNPM Perdesaan* ('Rural PNPM') has increased people's consumption by 9.1%, while about 300,000 people who were unemployed could get employment during the implementation of the programme, particularly in the infrastructure sectors.

The accessibility of the people to basic health care services also increased. According to Bappenas (2008, 2013), Bali reached 100%, Java 98%, Sumatra 95%, and Sulawesi 94%. However, in some areas such as Maluku, accessibility still remained at 83%, West-Papua 74% and Papua the lowest by 59%.

Access to education amounted to about 98% in urban areas for 9 years of compulsory education, while in rural areas the number was lower at 71%. From 1998 to 2009, more than 9.9 million people could be accommodated by the projects implemented by the PNPM Mandiri programme. From 2007 to 2014, the Government of Indonesia allocated about 72.56 billion Rupiah. The funds have supported the *PNPM Mandiri Perdesaan* ('Rural PNPM') in the rural areas in building more than 520,878 road projects with a completed distance of more than 650,978 km, 314,315 bridges, 319,463 irrigation systems, 350,000 water sanitation projects, and 2733 traditional markets, as well as providing electricity to more than 3600 villages, and renovating more than 80,000 schools all over Indonesia.

The *PNPM Perkotaan* ('Urban PNPM') in the urban areas has covered 125,988 project units, including road projects with a completed distance of more than 470,000 km, 77,555 drainage units, almost 8000 bridges, 300,000 microenterprises, more than 14,000 houses, 393 schools and 1200 health care services (cf. Bappenas 2008, 2013; Setkab 2014). Although the programme has made satisfactory achievements in development, there is some criticism on the implementation of the *Community-Driven Development* approach through the PNPM Mandiri programme. Several disadvantages can be identified on the implementation of the programme, which are listed as follows:

- The infrastructure is still the dominant sector to which the programme funds are being allocated;
- The facilitators in the *kecamatan* and *kelurahan* cannot fully execute their roles of responsibility in the community;
- In West Java, a case was reported where the implemented budget was below the allocated budget as a result of the inconsistency between the databases at the central government and the lower level of government; and
- The level of local participation still tends to remain low, as reported for Jogja and Bali, rendering the aim of achieving increased participation as not fullfilled (cf. Bappenas 2008, 2013).

Although in some cases, the participation of the local communities has been achieved, it turned out that certain decisions were already made beforehand by the facilitators or by the elites in the local government. Manusri and Rao (2004) criticise the domination of the elites in the implementation of the community-driven development projects, particularly concerning the inadequate results of the causal relations between the project outcomes and the participation of the communities. Although the programme had initially been planned to empower the community, it was found that without an adequate involvement of the local people to take their own decisions, the programme simply became an adjusted development approach *from the top* with a justification *from the bottom.* In this way, the implementation of the programme remained limited to an economic empowerment model without the participation of the community. Unfortunately, factors of failing leadership and elite capture at the local level have largely contributed to the decline in the achievement of the community-based development approach throughout the country (cf. Manusri and Rao 2004; Balint and Mashinya 2005; Dutta 2009).

References

Agus, E. N. (2013). *Program Pengentasan Kemiskinan Berbasis Partisipasi Masyarakat: Tantangan Dan Kendala Program PNPM Mandiri.* Makalah dipaparkan pada seminar Pengayaan Evaluasi PNPM Mandiri, Jakarta: PNPM.

Alkire, S., Bebbington, A., Esmail, T., Ostrom, E., Polski, M., Ryan, A., et al. (2001). *Community-driven development.* Washington, D.C (Draft).: World Bank.

Asian Development Bank. (2006). *A review of community-driven development and its application to the Asian Development Bank.* Manila: ADB.

Balint, P. J., & Mashinya, J. (2005). The decline of a model community-based conservation project: governance, capacity and devolution in Mahenye, Zimbabwe. *Geoforum, 37,* 805–815.

Bappenas. (2008). Kajian Evaluasi Program Penanggulangan Kemiskinan di Perkotaan, Laporan Akhir, Staf Ahli Menteri Negara PPN/Kepala Bappenas. Bidang Pemberdayaan Masyarakat dan Penanggulangan Kemiskinan. Jakarta: Bappenas.

Bappenas. (2013). Evaluasi PNPM Mandiri. Direktorat Evaluasi Kinerja Pembangunan Sektoral Kementrian Perencanaan Pembangunan Nasional/Bappenas. Jakarta: Bappenas.

Budiantoro, S. (2011*). Kemiskinan Melonjak, Jurang Kesenjangan Melebar Kekayaan 40 Orang Terkaya, Setara Kekayaan 60 Juta Penduduk*. Prakarsa Policy Review, Nov 1–4.

Davies, B. J. (2008). *Personal wealth from a global perspective*. Oxford: Oxford University Press.

Dutta, D. (2009). Elite Capture and Corruption: Concepts and Definitions Bibliography with an Overview of the Suggested Literature. Paper of the National Council of Applied Economic Research, New Delhi: NCAER.

Fang, K. (2006). Designing and implementing a community-driven development programme in Indonesia. *Development in Practice, 16*(1), 74–79.

Indonesian Law No. 21/2001. (2001). Undang-Undang Republik Indonesia No. 21 tahun 2001 tentang Otonomi Khusus bagi Propinsi Papua (Indonesian Law No. 21/2001 on Special Authonomy for Papua Province). Jakarta: SEKNEG.

Indonesian Law No. 6/2014. (2014). Undang-Undang Republik Indonesia No. 6 tahun 2014 tentang Desa (Indonesian Law No. 6/2014 on the Village). Jakarta: SEKNEG.

Indonesian Law No. 23/2014. (2014). Undang-Undang Republik Indonesia No. 23 tahun 2014 tentang Pemerintahan Daerah (Indonesian Law No. 23/2014 on Local Governance. Jakarta: SEKNEG.

Indonesian Law No. 47/2016. (2016). Peraturan Menteri Dalam Negeri No. 47 tahun 2016 tentang Administrasi Pemerintahan Desa (Ministrial Law No. 47/2016 on Village Government Administration). Jakarta: SEKNEG.

Latifah, E. (2011). Harmonisasi Kebijakan Pengentasan Kemiskinan di Indonesia yang Berorientasi pada: MDGs. *Jurnal Dinamika Hukum, 11*(3), 402–413.

Lindsey, T. (2008). *Indonesia: Law and society*. Annandale: Federation Press.

Mansuri, G., & Rao, V. (2004). Community-based and—Driven development: A critical view. *The World Bank Research Observer, 19*(1).

Matland, R. E. (1995). Synthesizing the implementation literature: The ambiguity-conflict model of policy implementation. *Journal of Public Administration Research and Theory, 5*(2), 145–174.

Mulyono, S. P. (2014). *Kebijakan Sinoptik Penerapan Hukum Adat dalam Penyelenggaraan Pemerintahan Desa*. Yustisia Edisi 89: May–August.

NERACA. (2011). *Jumlah Penduduk Miskin Masih Tinggi*. NERACA Harian online. Edisi 2011. http://www.neraca.co.id/article/7969/jumlah-penduduk-miskin-masih-tinggi.

Palumbo, D. I., Maynard-Moody, S., & Wright, P. (1984). Measuring degrees of successful implementation. *Evaluation Review*, 45–74.

Platteau, J. P. (2004). Monitoring elite capture in community-driven development. *Development and Change, 35*(2), 223–246.

PNPM Support Facility (PSF). (2002). *Kecamatan development program: Brief overview*. Jakarta: PNPM Support Facility (PSF) Library. http://psflibrary.org/catalog/repository/Brief%20Overview%20of%20KDP%20(English%20version).pdf.

Setkab, R. I. (2014). PNPM-Mandiri Membantu Membangun Infrastruktur. http://setkab.go.id/pnpm-mandiri-membantu-membanguninfrastruktur-perdesaan/.

Sirait, M., Fay, C., & Kusworo, A. (2000). *Bagaimana Hak-hak Masyarakat Hukum Adat dalam Mengelola Sumber Daya Alam Diatur*. Southeast Asia Policy Research Working Paper. No. 24. Bogor: ICRAF.

Warren, D. M., Slikkerveer, L. J., & Brokensha, D. (Eds.). (1995). *The cultural dimension of development: Indigenous knowledge systems. IT studies on indigenous knowledge and development*. London: Intermediate Technology Press.

Winters, A. J. (2011). *Ancaman Oligarki dan Masa Depan Politik Indonesia. Public Lecture & Discussion. Perhimpunan Pendidikan Demokrasi (PPD)*—Solidaritas Masyarakat Indonesia untuk Keadilan. 3 Juni 2011, Jakarta: Rumah Integritas.

Wong, S. (2003). *Indonesia Kecamatan development program: Building a monitoring and evaluation system for a large-scale community driven development program*. Region discussion paper. Environment and Social Development Unit of East Asia and Pacific, Washington, D.C.: World Bank.
World Bank. (2005). *Indonesia implementation completion report* (IDA-32100). Project information document. Washington, D.C.: World Bank.
World Bank. (2011). *Indonesia: The second urban poverty project* (PNPM Urban, Article 5, April 2011). Washington, D.C.: World Bank. http://www.worldbank.org/en/results/2011/04/05/indonesia-second-urban-poverty-project.
World Bank (WB). (2014). *Community driven development in Indonesia*. August 2014. Washington, D.C.: World Bank. URL: http://www.worldbank.org/en/country/indonesia/brief/community-driven-development-in-indonesia.
World Bank. (2015). Overview Indonesia. Washington, D.C.: World Bank.
World Culture Forum (WCF). (2013). *The Bali Promise of the World Culture Forum*. Paris: UNESCO.

L. Jan Slikkerveer is Professor of Applied Ethnoscience and Director of the Leiden Ethnosystems and Development Programme, Faculty of Science, Leiden University, Leiden, The Netherlands. He received his Ph.D. on his fieldwork in the Horn of Africa from Leiden University in 1983 and an Honorary Degree from the Faculty of Medicine of Universitas Padjadjaran in Bandung, Indonesia in 2005. He further extended advanced training and research in the newly-developing field of applied ethnoscience in several sub-disciplines, including ethno-economics, ethno-medicine, ethno-biology/botany, ethno-pharmacy, ethno-communication, ethno-agriculture and ethno-mathematics in combination with a focus on three target regions of South-East Asia, East-Africa and the Mediterranean Region. He is the supervisor of 25 Ph.D. students at Leiden University and has published more than 100 books and 300 articles on the subject. While he has also received a number of substantial subsidies from the European Union in Brussels for international development programmes in South-East Asia, East-Africa and the Mediterranean Region, he also conceptualised both the IMM & ICMD approaches in Indonesia, and is Senior Lecturer in the International Master Course on Integrated Microfinance Management (IMM) at the Faculty of Economics and Business of Universitas Padjadjaran in Bandung, Indonesia.

Kurniawan Saefullah is Lecturer of the Faculty of Economics and Business, Universitas Padjadjaran, Bandung, Indonesia since 1998. He graduated from Universitas Padjadjaran and the International Islamic University, Malaysia. He has published several books and articles with his colleagues, including Introduction to Management and Introduction to Business. In 2009, he joined the International Project on Integrated Microfinance Management (IMM) for Poverty Reduction and Empowerment in Indonesia. As a Ph.D. Researcher at the Leiden Ethnosystems and Development Programme (LEAD), Faculty of Science of Leiden University, The Netherlands, he is finalising his Ph.D. Dissertation on 'Gintingan in Subang: An Indigenous Institution for Sustainable Community-Based Development in the Sunda Region of West Java, Indonesia' under the supervision of Prof. Dr. L. J. Slikkerveer. The focus of his research is on the important role of the local cosmology of Tri Tangtu in the process of sustainable community development, while he also developed a model of integration of community-support institutions on the basis of the Indigenous Knowledge Systems (IKS) in combination with the IMM and ICMD approaches, such as financial, medical, educational, communication and social services through the indigenous institutions of the local communities in the Sunda Region of West Java.

Chapter 12
'*Satu Desa, Satu Milyar*': Village Law No. 6/2014 as a Rural Financial Development Programme

Benito Lopulalan

> *Working toward inclusive government and society remains a challenge in Indonesia's villages. One of the most effective ways to do this is through strengthening the implementation of the Village Law by ensuring that the distinct and varied voices of those most excluded from society are given a platform to speak and make change.*
>
> (Anggriani 2016)

12.1 Introduction

In 2014, after years of public discussions and debates, and enriched by demonstrations of village leaders nationwide, the Indonesian Law No. 6/2014 (2014), popularly known as *Undang-undang Desa* ('Village Law') was officially ratified by the Parliament. It highlights a new era of sharing of power, governance and decentralisation at the village level in Indonesia. The law is based on the notion of the village as a self-governing community with local-level government, which recognises and respects the origin of the village and its indigenous rights. The introduction of the principle of subsidiarity acknowledges not only the initiatives of governance at the village level, but also the practice of consensus (cf. Indonesian Law No. 6 2014).

Before the implementation of the 'Village Law', especially before the era of decentralisation and *Reformasi*, a village was regarded as the lowest level in the government hierarchy, embedded in the obligation mainly to implement the agenda of central government. The implementation of 'Village Law' had several consequences for the process of community development at the village level. A new aspect of development emerged in which the state, by law, should allocate 10% of

B. Lopulalan (✉)
Sinergi Indonesia, Depok, Indonesia
e-mail: benito.lopulalan@gmail.com

L. J. Slikkerveer et al. (eds.), *Integrated Community-Managed Development*,
Cooperative Management, https://doi.org/10.1007/978-3-030-05423-6_12

the annual national budget of the state to the villages under the new account of *Dana Desa* ('Village Fund'). The total budget for the villages in 2016 amounted to IDR 104.6 trillion, or about IDR 1.4 billion per village per annum (cf. Rofiq et al. 2016). In August 2016, the actual projection of the national budget for 2017 has been estimated above the amount of IDR 1 billion per village (cf. Kompas 2016).

The 'Village Law' also takes the socio-economic development of the villages into consideration by the introduction of an economic entity known as *Badan Usaha Milik Desa*, or *Bumdesa* ('Village-owned Enterprise'), which is an important instrument for socio-economic community development. A *bumdesa* is established by village consensus and village regulation, and the enterprise may engage in any business implemented for the benefit of the village. Today, the income of the villages is derived from various sources, such as village business, asset development, self-funded sources, the national budget, 10% of regional tax income, regional funds etc. (cf. Rofiq et al. 2016).

12.2 Historical Context of *Undang-Undang Desa* ('Village Law')

The context in which the 'Village Law' has been created cannot be separated from the processes of decentralisation and democratisation of the country which followed the first five decades of the country's political and economic centralisation. According to Sarosa (2006), the process of centralisation during the decades of national government under the subsequent leaderships of President Soekarno and President Soeharto has been based on the notion that the diverse country needed a strong central government leadership to unify the nation. Under the regime of President Soeharto, the implementation of the Indonesian Law No. 5/1979 (1979) on Village Government and similar regulations were based on the notion of a strong government with a centralised political structure in which the ruling party of Golkar and the army were embedded in a government hierarchy with control and power over the rural people, their labour and their land, while the priorities of the 'New Order' were promoted (cf. Sarosa 2006; Antlov and Eko 2012).

The processes of decentralisation and democratisation were implemented during the era of the *Reformasi*, which took place after the drastic changes followed by the downfall of the regime of the 'New Order' of President Soeharto in 1998. As Sarosa (2006) argues, the political notion of *decentralisation* in Indonesia is basically similar to the notion of *centralisation*, in order to unify the nation or to avoid separatism as the result of regional unrest. The process of the regulation of decentralisation was initiated through the enactment of the Indonesian Law 22/1999 (1999) on Regional Administration, later amended by the Indonesian Law 32/2004 2004. The law on Local Governance, popularly known as the 'Decentralisation Law', has transferred the obligations and functions of the central government to the local government. Exceptions are made for the central functions, including the national defense and security, foreign affairs, the judicial system, fiscal and monetary affairs, macro-economic planning, standardisation and a few other functions such as the prerogative of the national government (cf. Sarosa 2006). In the financial system, the aspirations of the *Reformasi* era have also created a balance in the national and local government, as reflected in Law 25/1999 on the fiscal balance between central and local governments, which was later amended by the Indonesian Law 33/2004 (2004) (cf. Turner et al. 2003; Sarosa 2006).

12.3 Village and Inequality

The comparison of the population in Indonesia between the people living in urban and rural areas showed a percentage of people living in the rural areas of around 80% in 1950. By 2010, the statistical data indicates that 51.2% of the population were rural dwellers, while in 2012, estimations show that people living in urban areas have exceeded the rural population by around 8%, as it is estimated that 54% of Indonesians are living in urban areas. The swift changes during the past decade are largely the result of the rapid economic development and the improvement of the infrastructure, which have turned some areas from rural to urban. Together with regional development, the disparity between rural and urban infrastructure, especially in health, education and job creation, has encouraged the migration of people from rural to urban areas (cf. Kompas 2012). The statistical data indicate that 52% of the rural population have to live under circumstances of poor sanitation, illustrated by the fact that every two hours a pregnant mother in Indonesia dies because of bad health care (cf. Ismawan 2013). The disparity between the rural and urban socio-economic situation has encouraged the emigration of people with better skills or better education, known as 'brain-drain'.

This process has been defined by Dodani and LaPorte (2005) as the migration of skilled or educated personnel in search of a better standard of living and quality of life, and improved access to advanced technology and more stable political conditions in different places. In Indonesia, such migration for better opportunities has taken place more frequently in some areas outside the island of Java, such as in Borneo, East Nusa Tenggara and Papua, where many young, well-educated people have left their village or town, to migrate to other parts of the country or even abroad. As a result, the rural areas are suffering from a lack of educated people.

In addition, the reality of the villages shows a process in which the mobilisation of village savings is not followed by an intention from the financial banks to provide loans to the villagers (cf. Ismawan 2003). Most of the money which has been mobilised from rural savings has been dispatched to financing corporate businesses in the urban areas, considered more profitable for the commercial banks. This situation can indeed be called 'capital drain' from the rural to the urban economies.

The above-mentioned phenomena of 'brain drain' and 'capital drain' have determined the conditions of village development in Indonesia. The flight of human capital in comparison to the phenomenon of 'brain drain' tends to influence the social infrastructure of the village, whereby 'capital drain' influences the fabric of the financial infrastructure of the village.

12.4 The Practice of Village Development

For many people, the process and the results of the implementation of the Indonesian Law No. 6/2014 on the Village (2014) have given new hope and a promise that 'Village Law' opens up the opportunity to free the villages from the chain of poverty, feudalism, backwardness and exploitation of their natural resources: *Villagers will become subjects of development, no longer objects.* However, there is a chronic problem of the paradigm in rural development in Indonesia, which has also contributed to inequality at the village level. Since the colonial era, the practices of rural development have largely been following the paradigm of 'betting on the strong'. This classical paradigm embarks on a development agenda which is set to be harmonised with the capabilities of more advanced villagers, and a more educated, mostly male, village elite. As Wertheim (1964) notes: "*they show a greater responsiveness to all kinds of innovations and technical improvements and are much easier to approach, they will set an example for the backward ones, who are expected to follow the model.*" Such a notion has encouraged the colonial government policies, as Booth (1988) describes: "*to put less emphasis on improving the general level of welfare and more on encouraging the emergence of outstanding individuals with genuine entrepreneurial ability.*" The classical approach is also called 'betting on the rich' (cf. Ludden 2011).

Notwithstanding, there has been a wide range of programmes in Indonesia in the past which implemented this approach of 'betting on the strong'. For instance, one of the famous agricultural programmes, known as *Bimbingan Massal* (BIMAS) ('Massive Intensification'), was based on the provision of credit for inputs in advanced technical improvements and agriculture innovation, eventually requiring cash repayment. During its first period from 1966 until 1983, the BIMAS strategy, including its credit programme, was betting on the strong by creating agricultural 'demonstration plots' at the village level. BIMAS worked largely with landowners-farmers who managed to control the local workforce and to finance and supply them with pesticides and hybrid seeds in order to deploy intensive cultivation of their land. During BIMAS' second period from 1984 until 1989, its credit programme was abolished, and by 1989, the whole BIMAS programme had been put aside (cf. Resosudarmo and Yamazaki 2011). As Jackson and Pye (1980) document, the decline of the BIMAS programme was largely the result of corruption, selling BIMAS fertilisers on the black market and non-payment of BIMAS credits, together with maladministration and peasant discontent. In addition, the massive influx of foreign technology and capital ignoring the indigneous farmers' knowledge and practices had certainly increased the negative attitude towards the BIMAS programme handed down from the central government.

Opposed to the strategy of 'betting on the strong' is the approach of 'betting on the many', in which the development agenda takes the ideas of the majority of the people in the village into account, and does not rely only on the strong, but also on the weak, including women, and encourages them to join the process of development. The approach of 'betting on the many' is similar to the concept of 'inclusive development' which ensures that *all marginalised and excluded groups are stakeholders in development processes regardless of their gender, ethnicity, age, sexual orientation, disability or poverty* (cf. UNDP 2016).

The approach of 'betting on the many' relies mainly on the development of social capital and on building collective actions and commitments throughout the process of institutionalisation, in which every villager is taken into account, and where all have the opportunity to grow together regardless of their poor or rich, educated or uneducated, and skilled or unskilled status. As Rauniyar and Kanbur (2009) assert, social capital development refers to: "*increasing the opportunity of the poor to participate in decision-making and self-managed community services, such as creating community-based groups in microfinance, health, and natural resources management*" Also, 'betting on the many' is an important process to avoid the process of marginalisation of the poor or less educated members of the community, since 'betting on the strong' can make "*the rich grow richer and the poor grow poorer*" (cf. Wertheim 1968).

The related process of developing inequality among the population has also been documented for Indonesia, where the Gini coefficient has increased from 0.34 in 2007 to 0.41 in 2014, although in the last two years the inequality has slightly declined in the Gini coefficient to 0.39 (cf. Indonesia-Investments 2016).

12.5 Redefining the Concept of Infrastructure

In common understanding, particularly in Indonesia, the concept of 'infrastructure' refers to physical infrastructure such as roads, bridges, buildings and irrigation systems. In the *Masterplan of Development Indonesia* (2011), the infrastructure of connectivity is defined as: "*the construction of transportation routes, information and communication technology (ICT), and all regulations associated with them.*" However, the concept of infrastructure goes far beyond its physical meaning. Maurer and Mainwaring (2012), in their view about the anthropology of business, point to infrastructure of money, a payment infrastructure, while Yaron (1994) underscores the importance of the infrastructure of human development. Sartori (1968) reviewed the classical conception held by de Tocqueville (1856), followed by Bellah (1959) who reviewed 'Durkheim and History', elaborating on the backbone of democracy in the form of an 'intermediate structure' of independent groups and voluntary associations. According to Sartori (1968), the intermediate structure may shape a vital and active 'infrastructure'.

At the village level, where a preference for the approach of 'betting on the many' prevails, Ismawan (2016) defines three basic forms of infrastructure which are important for building the pillars of inclusive development. The first is the 'Social Infrastructure' of the village, the second is the 'Financial Infrastructure', and the third is the 'Production or Supply-Chain Infrastructure'. These three pillars of development are connected and integrated in order to build a strong fabric of development at the village level. Such an approach to the concept of infrastructure pays attention to the postulate of Slikkerveer (1990) on the 'dual use' phenomenon involving transcultural utilisation of indigenous and modern medical systems, in his study of the utilisation of traditional, transitional and modern medical systems by the local population in the Horn of Africa. A similar coexistence of traditional, transitional and modern medical systems has also been documented in Crete, Greece by Aiglsperger (2014).

In line with the organisational structure of the new Master Course on *Integrated Microfinance Management* (IMM), conceptualised by Slikkerveer (2012), the infrastructure of the integrated development approach, based on the understanding of Integrated Microfinance Management (IMM), is represented in Fig. 12.1.

The social infrastructure of the village is specifically developed through the integration of the indigenous institutions which are rooted in the cultural foundations of the local communities over many generations, well illustrated in local knowledge and practices in education, health and natural resources management.

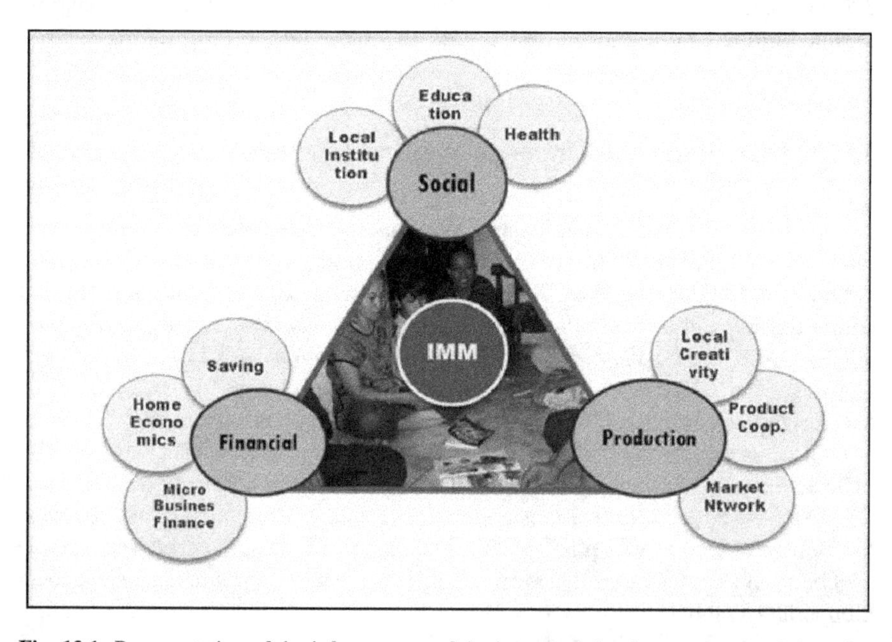

Fig. 12.1 Representation of the infrastructure of the integrated development approach based on the understanding of Integrated Microfinance Management (IMM). *Source* Adapted from Slikkerveer (2012)

In general however, local institutions or organisations may refer to modern socio-economic and financial institutions and agencies imported from outside along the process of extension by the government or by private donors, but most important for achieving sustainable community development are the indigenous institutions at the community level. These indigenous institutions have successfully implemented various social strategies of mutual aid, self-help and cultural activities for many generations, and are crucial for the creation and maintenance of the social fabric of the villagers.

Education as an important form of development infrastructure not only refers to formal education, but more importantly to indigenous education, applying a kind of social bonding in order to create a learning community among the villagers, among whom skills and knowledge are systematically distributed. Eventually, the integration of traditional and modern education systems has brought up a community of modern Indonesian citizens who have a special respect and unique feeling for their ethno-cultutral background. Similarly, health care as another form of development infrastructure not only refers to modern health care, but more importantly to traditional medicine; this is especially important in rural areas, where indigenous knowledge and practices of MAC (Medical, Aromatic and Cosmetic) plants are used for a variety of illnesses, such as knowledge and practices of *jamu* ('herbal medicine') in Java (cf. Slikkerveer 2006; Slikkerveer and Slikkerveer 1995).

The financial infrastructure has largely been developed since the second half of the 20th century through the introduction of modern practices of savings, deposits, improvements in home economics and capacity building, and micro-business finances. Savings can be made in local saving clubs, cooperatives or local banks, while the improvement of home economics is important in developing households into strong economic entities. Micro-business finance in the rural economy is important since most villagers are micro-entrepreneurs, who need micro-business services and capacity building. Consequently, the financial infrastructure is not an independent sector, but it is integrated with the social infrastructure as well as the third kind of infrastructure, that of production.

In general, the infrastructure of production comprises local creativity, production cooperatives and market networking. The villagers in Indonesia have been involved in the infrastructure of production, contributing and gaining not only from modern agriculture, but also from various creative activities derived from the legacy of indigenous knowledge and skills, which have evolved into a wide range of well-adapted indigenous methods and techniques. Indigenous food processing methods, including the preparation of local products such as *terasi*, *krupuk*, *kripik*, *sambal*, etc., have become a major source of knowledge in further developing creativity, the organisation of production and a supply-chain mechanism for market networks.

In order to build such an integrated infrastructure, the development of social capital is a key issue. The villagers can be organised into *community-based organisations*, being associations of people who unite themselves to undertake special socio-economic activities to develop social capital in terms of promoting the welfare of all members of the community. Usually, these *community-based organisations* encompass solidarity organisations comprising 15–50 participants, thereby forming a vehicle of mutual learning to identify problems, make appropriate decisions, mobilise resources and communicate with other parties. In this way, the community-based approach will enable the local villagers not only to organise their activities collectively, but also to seek the benefit of a set of collective goals or interests (cf. Grootaert 1999).

An important question concerning the process of development, including the kind of development proposed in the Indonesian Law No. 6/2014 on the Village (2014), relates to the way in which the local people will be included to participate in the programme. Hence, the issue is how to organise the people in the process of development at the village level, and put the considerations beyond the interests of the village elites. The failure to build such a community-based participatory development process will eventually create an unwelcome concentration of village interests among the local elites, making it conducive to the threat of elite capture at the cost of the poor and marginalised members of the community.

References

Aiglsperger, J. (2014). *'Yiatrosofia yia ton Anthropo': Indigenous knowledge and utilisation of MAC plants in Pirgos and Pretoria, Rural Crete: A community perspective on the plural medical system in Greece*. Ph.D. Dissertation. Leiden Ethnosystems and Development Programme. LEAD studies No. 8. Leiden University. xxvi + 235 pp. ill.

Anggriani, N. (2016). *Indonesia's village law: A step toward inclusive governance*. The Asia Foundation. http://asiafoundation.org/2016/02/17/indonesias-village-law-a-step-toward-inclusive-governance/.

Antlov, H., & Eko, S. (2012). *Village and sub-district functions in decentralized Indonesia*. Paper for the DSF's closing workshop (pp. 12–13).

Bellah, R. N. (1959). Durkheim and history. *American Sociological Review, 24*(4), 447–461.

Booth, A. (1988). *Sejarah Ekonomi Indonesia*. Lembaga Jakarta: Penelitian, Pendidikan dan Penerangan Ekonomi dan Sosial (LP3ES).

de Tocqueville, A. (1856). *L'Ancien régime et la révolution*. Paris: Lévy.

Dodani, S., & LaPorte, R. (2005). Brain drain from developing countries: How can brain drain be converted into wisdom gain? *Journal of the Royal Society of Medicine, 98*(11), 487–491.

Grootaert, C. (1999). Social capital, household welfare and poverty in Indonesia. Local level institutions. Working paper no. 6. Washington, DC: World Bank.

Indonesian Law No. 5/1979. (1979). *Undang-Undang Republik Indonesia No. 5 tahun 1979 tentang Pemerintahan Daera*. Indonesian Law No. 5/1979 on Village Goverment. Jakarta: SEKNEG.

Indonesian Law No. 22/1999. (1999). *Undang-Undang Republik Indonesia No. 5 tahun 1999 tentang Pemerintahan Desa*. Indonesian Law No. 22/1999 on Regional Administration. Jakarta: SEKNEG.

Indonesian Law No. 32/2004. (2004). *Undang-Undang Republik Indonesia No. 32 tahun 2004 tentang Pemerintahan Daera*. Indonesian Law No. 32/2004 on Regional Administration. Jakarta: SEKNEG.

Indonesian Law No. 33/2004. (2004). *Undang-Undang Republik Indonesia No. 33 tahun 2004 tentang Pemerintahan Daera*. Indonesian Law No. 33/2004 on Regional Administration. Jakarta: SEKNEG.

Indonesian Law No. 6/2014. (2014). *Undang-Undang Republik Indonesia No. 6 tahun 2014 tentang Desa*. Indonesian Law No. 6/2014 on the Village. Jakarta: SEKNEG.

Indonesia-Investments. (2016). *GINI ratio Indonesia declines: Economic inequality narrows*. Jakarta: Statistics Indonesia (BPS). http://www.worldbank.org/en./results/2011/04/05/indonesia-second-urban-poverty-project.

Ismawan, B. (2003). *Keuangan Mikro dalam Penanggulangan Kemiskinan dan Pemberdayaan Ekonomi Rakyat. Kemiskinan dan Keuangan Mikro*. Jakarta: Gema PKM.

Ismawan, B. (2013). Doing well by doing good: Turning people's enterprises around—A conversation with Bambang Ismawan of Indonesia's Bina Swadaya. *Asian Politics and Policy, 5*(3), 461–468.

Ismawan, B. (2016). Innovation in developing village-owned enterprise (Bumdes) as social enterprise. Jakarta: Indonesia-Japan Local Administration Seminar. August 25, 2016.

Jackson, K. J., & Pye, L. W. (1980). *Political power and communications in Indonesia*. Berkely LA: University of California Press.

Kompas. (2012). Hampir 54 persen Penduduk Indonesia Tinggal di Kota. *Kompas Daily Online*. August 23, 2012. http://nasional.kompas.com/read/2012/08/23/21232065/%20Hampir.54.Persen.Penduduk.Indonesia.Tinggal.di.Kota.

Kompas. (2016). Tahun Depan Tiap Desa Dapat Anggaran di atas Rp.1 Miliar. *Kompas Daily*, May 20, 2016.

Ludden, D. (2011). Development regimes in South Asia: History of the governance Conundrum. In K. Visweswaran (Ed.), *Perspectives on modern South Asia: A reader in culture, history and representation* (pp. 224–237). Wiley-Blackwell: Malden.

Maurer, B., & Mainwaring, S. D. (2012). Business: Plural programs and future financial worlds. *Journal of Business Anthropology, 1*(2): Autumn.

Rauniyar, G., & Kanbur, R. (2009). *Inclusive growth and inclusive development: A review and synthesis of the Asian development bank literature*. Occasional paper no. 8. Manila: Independent Evaluation Department, Asian Development Bank.

Resosudarmo, B., & Yamazaki, S. (2011). *Training and visit (T&V) extension vs. farmer field school: The Indonesian experience*. Working paper no. 2011/01. Canberra: Australian National University.

Rofiq, A., Salim, A., Untung, B., Laksono, I., Bulan, W.R., Arifah, U., et al. (2016). *Praktik Baik Desa dalam Implementasi Undang-Undang Desa*. Jakarta: Pusat Telaah dan Informasi Regional (PATTIRO) (pp. 1–60).

Sarosa, W. (2006). *Indonesia—urbanization and sustainability in Asia: Case studies of good practice*. Manila: Asian Development Bank.

Sartori, G. (1968). Democracy. In *International encyclopaedia of the social sciences* (Vol. 4). London: The Macmillan Publishers & The Free Press.

Slikkerveer, L. J. (1990). *Plural medical systems in the horn of Africa: The legacy of 'Sheikh' hypocrates*. London: Kegan Paul International.

Slikkerveer, L. J. (2006). The challenge of non-experimental validation of MAC plants: Towards a multivariate model of transcultural utilisation of medicinal, aromatic and cosmetic plants. In Bogers, R. J., Craker, L. E., & Lange, D. (Eds.), *Medicinal and aromatic plants: Agricultural, commercial, ecological, legal, pharmacological and social aspects*. Wageningen UR frontis series (Vol. 17).

Slikkerveer, L. J. (Ed.). (2012). *Handbook for lecturers and tutors of the new master course on integrated microfinance management (IMM) for poverty reduction and sustainable development in Indonesia*. Bandung: LEAD-UL/UNPAD/ MAICH/GEMA PKM.

Slikkerveer, L. J., & Slikkerveer, M. (1995). Tanaman obat keluarga (TOGA). In D. M. Warren, L. J. Slikkerveer, & D. Brokensha (Eds.), *The cultural dimension of development: Indigenous knowledge systems*. London: Intermediate Technology Publications Ltd.

Turner, M., Podger, O., Sumardjono, M., & Tirthayasa, W. (2003). *Decentralisation in Indonesia: Redesigning the state*. Canberra: Asia Pacific Press.

United Nations Development Programme (UNDP). (2016). *Inclusive development.* New York: Oxford University Press. http://www.undp.org/content/undp/en/home/ourwork/poverty reduction/focus_areas/focus_inclusive_development.html.
Wertheim, F. (1968). Dual economy. In *International encyclopaedia of the social sciences* (Vol. 4). London: The Macmillan Publishers & The Free Press.
Yaron, J. (1994). What makes rural finance institutions successful? *World Bank Research Observer, 9*(1), 49–70.

Benito Lopulalan is a Programme Advisor of the Sinergy Foundation Indonesia. He holds a B.A. Degree in Agriculture from Universitas Brawijaya and is also Correspondent of the Globe Media Group in Jakarta. In 2014 he joined the LEAD Programme of Leiden University in The Netherlands as Research Assistant for three months in 2015, and since then continues to work in the field of integrated microfinance management for poverty reduction and development in Indonesia.

Chapter 13
'*Kopontren* and *Baitul Maal Wat Tamwil*: Islamic Cooperative Institutions in Indonesia

Kurniawan Saefullah and Nury Effendi

> *For social development, pesantren had struggled for the orphans and elderly people; performed mass circumcision; provided scholarship for the unfortunate students; gave hands to the least advantaged; gave health services; provided rehabilitation of drug addicts; cooperated with the society for societal needs ….*
>
> (Mohamad Mustari 2013)

13.1 Introduction

Indonesia is the 4th largest populated country in the world with a population of 258,705,000 in 2016. It has a Muslim population of about 87.2%, or 225,000,000 people, substantiating the largest Muslim population in the world. Although Indonesia also has several indigenous religions, these are small in number, some of which are becoming almost extinct, i.e. *Kejawen* (East and Central Java), *Sunda Wiwitan* (Banten), *Buhun* (West Java), *Aluk Todolo* (Toraja, Sulawesi), *Parmalim* (North Sumatra), and *Kaharingan* (Kalimantan). Following the influence of Buddhism, Taoism and Hinduism, and later followed by Christianity and Roman Catholicism during colonialisation by the Portuguese, the British, and the Dutch, it was the expansion of Gujarat and Arab traders to Indonesia after the 7th century which made Islam a strong religion in the country (Table 13.1).

The influence of Islam and the Arabic culture on Indonesia varies from the language to the cultural traditions which are in use today throughout the country. The term *Adil* ('Justice'), for instance, originally came from the Arabic language. Similarly, the term *Hajat* or *Hajatan* ('Ceremony' or 'Festivity') in the Sundanese language has also been influenced by the Arabic language and religion. Schooling

K. Saefullah (✉) · N. Effendi
FEB, Universitas Padjadjaran, Bandung, Indonesia
e-mail: kurniawan.saefullah@unpad.ac.id

L. J. Slikkerveer et al. (eds.), *Integrated Community-Managed Development*, Cooperative Management, https://doi.org/10.1007/978-3-030-05423-6_13

Table 13.1 Distribution of religion in Indonesia (estimated in 2010)

Rank order	Religion	Percentage of the religion population
1	Islam	87.2
2	Christianity	6.9
3	Roman catholicism	2.9
4	Hinduism	1.7
5	Buddhism	0.7
6	Confucianism and others (including indigenous religions)	0.06
Total		100.0

Source Adapted from Indonesia-Investments (2010)

and education have also been influenced by Islam. The *Pondok Pesantren* ('Islamic Boarding School') is an educational institution which has been established and in operation for many centuries throughout the country, from Aceh to the western part of the Nusa Tenggara Province (cf. Wessing 1978, 1979; Dhofier 1982; Inkopontren 1997; Indonesia-Investments 2010).

The dual financial system in Indonesia allows both the conventional and mainstream ones, as well as the Islamic financial system. The Islamic financial institutions in the country can be divided into two categories: the banking sectors and the non-banking sectors. The banking sectors can be further sub-divided into Islamic Banks and Islamic Rural Banks. As for the non-banking sectors, these include *Takaful* ('Islamic Insurance Institutions'), and *Koperasi Pondok Pesantren* or *Kopontren* and *Baitul Maal wat Tamwil* ('Islamic Cooperative Institutions') (cf. Seibel and Agung 2005; Effendi 2009).

The *Koperasi Pondok Pesantren* ('Kopontren') and *Baitul Maal wat Tamwil* ('BMT') are two Islamic cooperative institutions which integrate a financial institution based on Islamic cosmology, involving the daily practices of Muslim Indonesians. While the *Kopontren* was a unique institution, only established in Indonesia, the BMT was an adapted form of the *Baitul Maal* institution which evolved during the Islamic expansion of the 6th century, with an additional role of *Tamwil* ('Investment'). Both institutions have contributed to the development of Indonesia, particularly in the Muslim communities. This section further describes both institutions.

According to Islamic law, the operation of Islamic cooperative institutions should follow the ideal of the social order of brotherhood and solidarity. It should also follow mutually beneficial partnerships between depositors-investors, and financial institutions with their staff and owners, including borrower-investors, who all share the risks and benefits in various ways. Speculative transactions and other morally hazardous activities, as well as most types of consumer-lending, are prohibited by Islamic values. The Islamic cooperative institutions in Indonesia, in particular *Kopontren* and BMT, are characterised by the obligation to help the poor. As such, both these instituteons are established in order to support poverty alleviation programmes.

13.2 *Pesantren* and *Koperasi Pondok Pesantren*

Islam conjures up various images, not only in the view of its believers, but also of those who are studying this religion. In Indonesia, Islam was known for its flexibility on the blending of its teachings with local culture. The dynamics of Islam and its flexibility to merge with the indigenous cultures have facilitated its dissemination throughout the country. Since its diffusion all over the country, Islamic values have been easily adapted with the indigenous values and vice versa. The Indonesian Muslims have not only implemented the religious practices of Islam, but have also integrated them with their social, economic and political ethics within the local context. In the field of education, the Islamic institutions have played a major role in educating the Indonesians for many centuries.

The *Pesantren* ('Boarding School') is an indigenous educational institution in Indonesia where Islamic values are used as the basis for its process of teaching and learning. In Malaysia, this institution is known as *Pondok* ('Cottage'), and in Pakistan, it is called *Madrasa* ('Islamic School'). The *Pesantren* has a unique system as it uses in general a boarding system, where the students are staying in the institution night and day to learn Islamic values not only from their classical and modern references, but also by practicing them in daily life. It is a kind of training center for students who would like to learn and practice Islam. Dhofier (1982) rightly defined *Pesantren* as: "*a religious boarding school in which students reside and study under the direction of a teacher known as a Kyai.*" The role of the *Kyai* in the *Pesantren* is similar to that of a *Shaikh* in the Ethiophian *Madrasa* ('Islamic Schools'). He acts as an informal leader, particularly for the Muslim communities, and plays his role as Preacher, Educator, sometimes as Healer, as well as a Public Figure in the communities. The difference between the *Shaikh* in Ethiophia and the *Kyai* in Indonesia is that the *Kyai* mostly performs his tasks in the *Pesantren*, while the *Shaikh* does not. While the teacher in the *Pesantren* is called *Kyai*, the student is called *Santri* ('Students'). The word '*santri*' is derived from the Sanskrit word *Shastri*, which means student or learner. The term *Pesantren* is constructed by adding the word *Santri* with the prefix '*pe*' and the suffix '*an*', resulting in the term *'Pesantrian'* or *'Pesantren'*. Although the historical origin of the word *Pesantren* is not clear, it has existed for more than 300 years (cf. Dhofier 1982; Slikkerveer 1990; Wagiman 1997; Madjid 1997).

Originally, the *Pesantren* was established in the rural areas, where the institutions integrated their activities with the daily life of the local people. One of the most famous methods of teaching at the *Pesantren* was the *Sorogan*, where the *Kyai* and the *Santri* were sitting together on the floor or on a carpet, studying Islamic knowledge and traditions. However, following the process of modernisation throughout the country, more *Pesantren* are now also operating in the urban areas. The method has also been transformed into the classical approach, where the *Kyai* and the *Santri* are sitting at desks and tables with a whiteboard and LCD projector available in the classroom, similar to other schools. However, the challenge has not only been in the adaptation to these material facilities, but also to the national education system, including the modernisation of the system. In the curriculum, both Islamic subjects and general subjects are also taught in the *Pesantren*. By consequence, the *Santri* have to follow not only 'secular subjects', but also subjects related to nationalism and citizenship. As Murray (1992) mentions: "*The first concern is education for general citizenship, made up of two main parts:* (1) *communication skills and basic knowledge of the social and natural environments that every citizen needs, and* (2) *patriotism, meaning a commitment to ideals that bind the nation's assorted peoples together as a unit, as reflected in the motto Bhinneka Tunggal Ika* (*Unity in Diversity*)." Therefore, the *Santri* are educated not only to become good Muslims, but also good Indonesians (cf. Murray 1992; Madjid 1997; Pohl 2006).

The *Koperasi Pondok Pesantren* or *Kopontren*, is an economic institution inside the *Pesantren* which executes the activities based on the cooperative principles and Islamic values. The management of *Kopontren* is adapted from the cooperative institution, while the institutional values are adapted from the Islamic values. The main activity of the *Kopontren* is to sustain the activities of the *Pesantren* economically. The economic activities of the *Pesantren* are in general as follows: (a) the construction and maintenance of the material facilities of the *Pesantren*; (b) the education and training process; (c) the welfare of the *Kyai* and the *Santri*; (d) the livelihood of the *Santri*; and (e) the expansion programmes of the *Pesantren*, i.e. the joint programmes with the community, the government, etc.

The funds of the *Pesantren* are coming from four different sources: *Iuran* ('school fees'), *ZIWS* (*Zakat, Infaq, Shadaqah* and *Waqaf*, all being Islamic charities), *Kopontren* and government or private contributions. As mentioned above, the activities of the *Kopontren* are integrated with other economic, social and educational activities in the *Pesantren*. While the activities of the *Pesantren* include mainly the education and training of the *Santri*, the *Kopontren* conduct community-based business activities in the *Pesantren* such as collecting money from internal and external sources with a view to supporting the main activities of the *Pesantren* (Fig. 13.1).

As regards the internal sources, the money is collected from the tuition fees of the *Santri* from the owner, i.e. the *Kyai* and the family, as well as from other internal resources such as the canteen, agricultural activities, etc. External sources include money collected from alumni contributions and sponsors, as well as from other partners of the *Kopontren*, including the government, if available. Food, uniforms, books, and daily products are also handled by the *Kopontren*. Since the main objective of the *Kopontren* is to sustain the activities of the *Pesantren*, it is not meant for commercial gain. In some cases, such as the *Pesantren Al-Ihya* in Bogor, and the *Pesantren Al-Ittifaq* in Ciwidey, South Bandung, the *Kopontren* are integrating the education and training activities with agribusiness activities. In both institutions, the *Santri* are also trained to become skilled farmers. Thus, the *Kopontren* conduct business activities which are based on cooperative principles and Islamic values, of which the profit is used to maintain and strengthen the *Pesantren*.

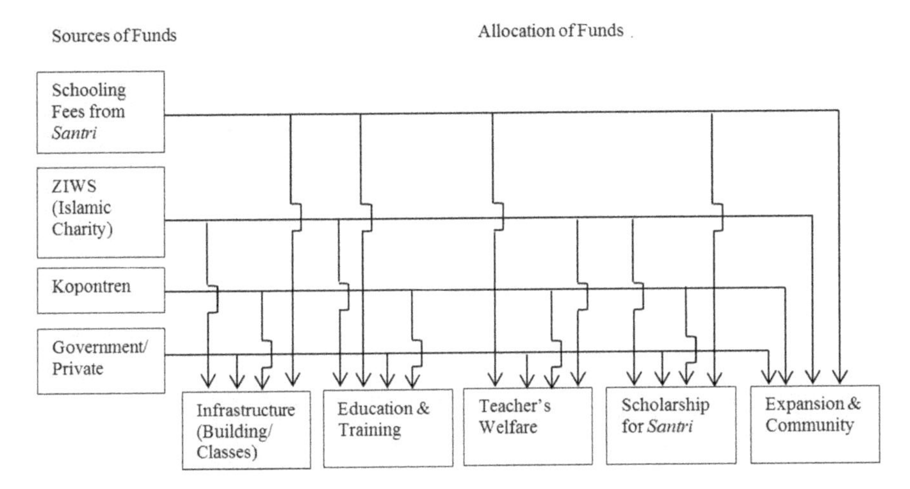

Fig. 13.1 Representation of the economic cycle of the *Pesantren*. *Source* Adapted from Yakub (1992)

The *Kopontren* has an institutional structure, following the principles of cooperative institutions. It has an Advisory Board and a Management Board, as well as Members. The Members usually include *Santri*, teachers, *Kyai*, as well as representatives of the local communities around the *Pesantren*. The Advisory and Managerial Boards are assigned by the *Rapat Tahunan Anggota* ('Annual Meeting') of the *Kopontren*, where the programmes are evaluated and profits gained are either shared among the Members of the *Kopontren* or allocated to the development of the *Pesantren*. The *Kopontren* can also establish some business units to support their commercial activities. The government could also support the *Kopontren* indirectly if the legal status of the *Kopontren* is a cooperative institution, governed by the Ministry of Cooperatives, and Small and Medium Enterprises (cf. Seibel 2007) (Fig. 13.2).

Direct support from the government can be received by the *Kopontren* through a subsidy or any other government support as the *Pesantren* itself is regulated and monitored by the Ministry of Religious Affairs. The government can also support the *Kopontren* in opening up a market for the *Kopontren* which enables it to gain access to economic resources. The local community, in which the *Pesantren* is located, is acting both as member and consumer, and it is also cooperating with the *Kopontren* in various ways. For instance, if the *Pesantren* is located in an agricultural area, the local community could cooperate with the *Kopontren* in order to support the agricultural activities. In the urban areas, the *Kopontren* provides different goods and services, such as groceries, trading, training, etc.

The financial activities of the *Kopontren* include the establishment of a savings and borrowing or lending institution, such as for instance the *Baitul Maal wat Tamwil* (BMT). In Indonesia, the BMT is not only operating under the *Pesantren*, but also independently in various sectors in the rural and urban areas. Both the

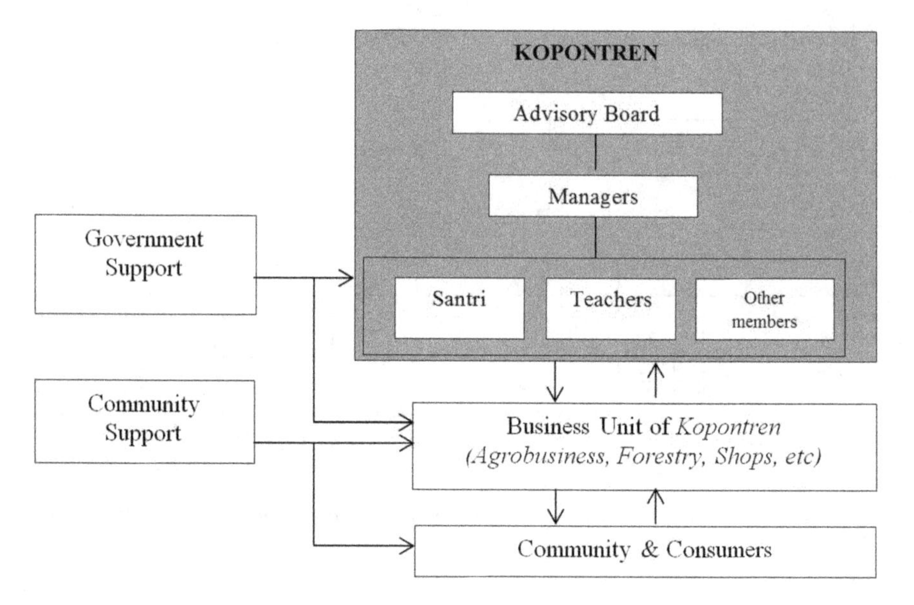

Fig. 13.2 Representation of the business activities of the *Kopontren*. *Source* Adapted from Yakub (1992)

Kopontren and the BMT are conducting their economic activities on the basis of both cooperative principles and Islamic values. The Islamic values which govern the activities of the *Kopontren* are *Shiddiq* ('Honest or Transparent'), *Ta'awwun* ('Cooperation'), *Adil* ('Justice') and *Ridha* ('Agreement-Based Activity').

There are no exact numbers available of the *Kopontren* in Indonesia, but a number can be estimated on the basis of the number of *Pesantren*. According to the statistical data from the Ministry of Religious Affairs, the number of *Pesantren* in Indonesia has reached 27,230 with a total of 3.7 million *Santri*. Assuming that there is one *Kopontren* in every *Pesantren*, the number of *Kopontren* may also amount to 27,230 institutions (cf. Inkopontren 1997; Statistics of MRA 2012).

13.3 *Baitul Maal Wat Tamwil* (BMT)

The intermediary roles of the Savings-Surplus-Units (SSU) and the Savings-Deficit-Units (SDU), which are implemented by most financial institutions, have always played an important role in the economic development of a country. By consequence, adequate capital is required to scale up the productive activities which are mostly conducted by firms or entrepreneurs. This consideration does not include the poor people whose condition and welfare would also need to be scaled up.

As financial institutions always require their clients to provide collateral in exchange for the capital lent by the institution, the poor people are kept away from the financial system. The Islamic financial institutions, however, emphasise that they not only support the economic activities of the people, but most importantly, they also support attempts to reduce poverty. Since the time of the Prophet, the institution of *Baitul Maal* has existed and played its role as a financial, religious and social institution.

The institution of *Baitul Maal* represents the 'treasurer' and provides services to those who are in financial need. The sources of its funds are both the people as well as the society, including charities, various fines and mandatory funds. In modern times, it generates many good innovations. In contrast, *Baitul Maal* in the early days of Islamic civilisation was not only regarded as a religious and social institution, but it was also a financial department, including the treasury of the government, the taxation department, and public works. It also had other functions, including its main objective of reducing poverty and disparities among the people in the community. *Baitul Maal* in Indonesia is known as *Baitul Maal wat-Tamwil* (BMT), and has a different role to play in society (cf. Hamzah et al. 2013).

BMT is the Islamic Cooperative Institution in Indonesia which provides mainly savings and lending/borrowing services. Although the government gives two legal options to the BMT, as either a small corporation or a cooperative institution, most are operating on the basis of the cooperative principles guided by Islamic values. Furthermore, as the BMT is operating in the community, its activities integrate with various local institutions. If the BMT operates in an agriculture area, the integration takes place with the farmers' associations.

If the BMT is operating in urban areas, the integration may take place with street traders, traditional markets, micro-businesses and religious institutions such as mosques which are distributing charities to the poor people in the community.

Early development of Baitul Maal wat Tamwil

The first BMT was established in Bandung in the early 1980s, under the name of *Bank Teknosa* or the *Teknosa Cooperative.* Although the name included the term 'Bank', its operation was based on cooperative and Islamic principles. The institution was established by some students who spent their extra-curricular activities in the Salman Mosque, one of the largest mosques in Bandung. The students were initially upset by the high incidence of poverty around the mosque: some poor people had no jobs, while others had become beggars. The bad conditions motivated the students to establish *BMT Teknosa*. By accumulating funds from charities, *BMT Teknosa* was able to distribute money to support the poor people in various ways: some received funding to have access to basic needs such as food and shelter; others received some capital to undertake some economic activities. In order to ensure that the poor people would use their funds in a responsible way, *BMT Teknosa* combined financial support with training and education, using Islamic principles and values. Apart from savings and borrowing money, *BMT Teknosa* started a small shop providing various products for students, such as books, apparel, bags, etc. and also provided scholarships for students from poor families.

Although *BMT Teknosa* was a breakthrough in terms of the Islamic way of helping the poor, the institution had to discontinue its activities as early as 1990, as it had run out of funding support. The number of borrowers had exceeded the number of savers or contributors, eventually limiting the existence of *BMT Teknosa* to less than 10 years.

Despite the fact that the first BMT in Indonesia was not sustained, the initiative motivated other people to follow in the footsteps of *Bank Teknosa*. In the beginning of 1990, the number of BMT increased rapidly, following the new policy which allowed the establishment of Islamic institutions as long as they were not dealing with political issues. The growth of BMT followed after the establishment of the *Indonesian Muslim Scientific Association* (ICMI) which also proposed the establishment of the *Bank Muamalat*, the first Islamic Bank in Indonesia. According to Seibel and Agung (2005), almost 3000 BMT were established, and operated from 1990 to 2005, although there were critical examinations of their sustainability. The need for an alternative financial system which follows the Islamic principles in economic transactions was the main motive behind the rapid growth of BMT in Indonesia (cf. Andriani 2005; Seibel and Agung 2005; Hamzah et al. 2013).

The principles of operation of the Baitul Maal wat Tamwil

According to the Indonesian regulations, BMT can operate as a private institution or as a cooperative institution. However, most BMT have chosen the cooperative principles guided by the Islamic values. Similar to the *Kopontren*, the BMT consists of three institutional structures: an Advisory Board, a Managerial Team, and its Members. The Advisory Board and Managerial Team are selected at the annual meeting of the BMT. The main services provided include savings and borrowing/investing activities. The term *Baitul Maal* refers to the process of collection of money from any source based on Islamic charities and economic transactions, while the term '*Tamwil*' means 'to make the money productive or use it for investment'. While *Baitul Maal* refers to any activity of collection or distribution of non-profit funds, *Baitut-Tamwil* refers to any activity which is focused on collecting and lending funds for profit activities. Thus, *Baitul Maal wat Tamwil* can be defined as an Islamic cooperative institution where money is collected by using Islamic principles, and redistributed to productive economic activities, which are allowed by the Islamic principles (cf. Yadi and Djazuli 2002) (Fig. 13.3).

As regards the *Baitul Maal* activities, the BMT uses the Islamic principle of charity, where the collected money is used to support the *Mustahiq* ('Needy Receivers'). There is, however, no obligation for receivers to pay back, unless for charity. Concerning the *Baitut-Tamwil*, the BMT uses *Nisbah* ('Profit-Loss Sharing'), *Baí* and *Ijarah* ('Trading') and *Wadiah* and *Mudharabah* ('Saving' and 'Investment'). Any business which is permissible by Islam could be supported by BMT. Unlike other micro-banking activities, the uniqueness of BMT lies in the rate of *Nisbah* ('Profit-Loss Sharing') which is negotiable, meaning that the BMT and the receivers could negotiate the percentage they would agree on for sharing the profit or loss. The principle of *Nisbah* ('Profit-Loss Sharing') renders the BMT or the Islamic Cooperative Institution more humane and lawful for both parties.

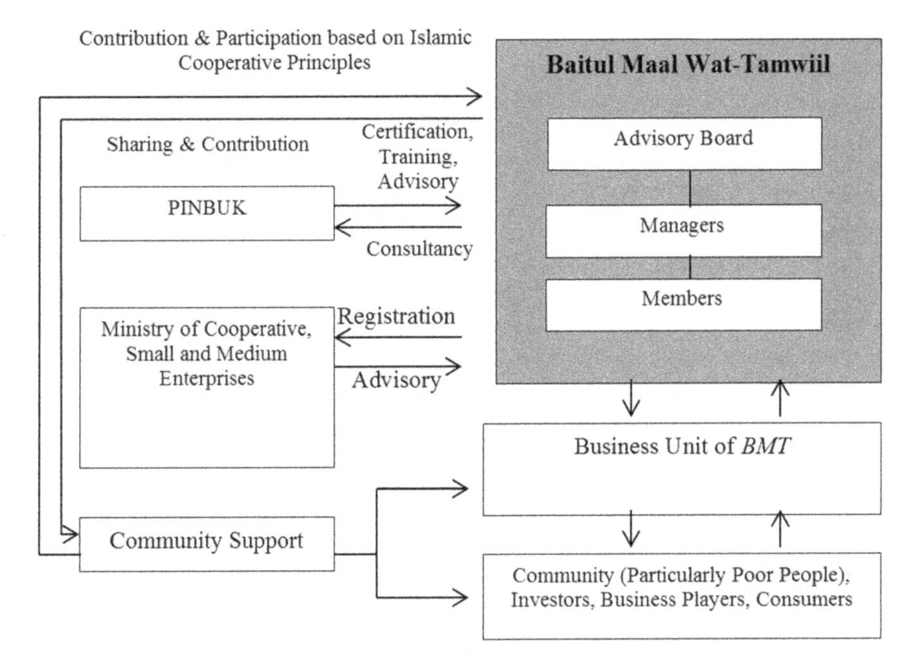

Fig. 13.3 Representation of the integrated activities of BMT with the government and the community. *Source* Adapted from PINBUK (1999)

As such, these institutions also support the concept of sustainable development, with regard to the criteria of sustainable agricultural development in such a way that they are indeed humane and just (cf. Reijntjes et al. 1992).

An important institution which supported the BMT in the early 1990s was *Pusat Inkubasi Bisnis Usaha Kecil* (PINBUK) ('Center for Micro-Enterprise Incubation') (cf. PINBUK 1999). The institution operated under the *Yayasan Inkubasi Bisnis Usaha Kecil* (YINBUK) ('Foundation for Micro-Enterprise Incubation') which was established in 1995 by the former President Mr. B. J. Habibie, while he was Chairman of *Ikatan Cendekiawan Muslim Indonesia* (ICMI) ('Association of Indonesian Muslim Scholars'), together with Mr. Hasan Bashri, the former Chairman of the *Majelis Ulama Indonesia* (MUI) ('Council of Indonesian Ulama'), and Mr. Zainul Bahar Noor, the first Director of *Bank Muamalat Indonesia*. The *Pusat Inkubasi Bisnis Usaha Kecil* (PINBUK) ('Center for Micro-Enterprise Incubation'), which was mostly operated by young people, provided services related to the activities of the BMT. It prepared people in the community for the establishment of the BMT, provided education and training for those who implemented the BMT, and also provided certification to the BMT prior to its registration at the Ministry of Cooperatives, Small and Medium Enterprises. Over the past few years, there has been no exact indication available of the number of BMT in Indonesia. The reported data by PINBUK as documented by Seibel and Agung (2005) are showing that the number of BMT have increased dramatically from 1990 to 2000, amounting to about 3000 in 2001. According to Hamzah et al. (2013), there are three factors which determine the rapid growth of the BMT: (1) the increased demand from small enterprises to obtain funding support which they could not obtain from other institutions; (2) the higher expectations from Muslims to conduct business activities which follow Islamic values; and (3) the success stories from other BMT which are motivating other people to establish similar institutions.

Although a rapid growth occurred in the number of the BMT from the 1990s up to the early 2000s, only one-fifth to one-third of those BMT were able to operate in a sustainable way. In most cases, where the BMT discontinued, the main reasons were mismanagement on the part of the managers and the inability of clients to comply with their agreements (cf. Seibel and Agung 2005).

References

Andriani. (2005). Baitul Maal Wat-Tamwil: Konsep dan Mekanisme di Indonesia. *Empirisma, 14* (2), 248–258.

Dhofier, Z. (1982). *Tradisi Pesantren: Studi tentang Pandangan Hidup Kyai*. Jakarta: LP3ES.

Effendi, N. (2009). *Financial sector analysis in Indonesia*. Paper presented in the International Workshop in Integrated Microfinance Management for Poverty Reduction and Development in Indonesia. May 18–28, 2009. Universitas Padjadjaran—Leiden University. Bandung: UNPAD/FEB.

Hamzah, H., Rusby, Z., & Hamzah, Z. (2013). Analysis problem of Baitul Maal Wat Tamwil (BMT) operation in Pekanbaru Indonesia using analytical network process (ANP) approach. *International Journal of Academic Research in Business and Social Sciences, 3*(8).

Indonesia-Investments. (2010). *Agama di Indonesia*. http://www.indonesia-investments.com/id/budaya/agama/item69.

Inkopontren. (1997). *Trend Koperasi Pondok Pesantren: Data dan Gambar Induk Koperasi Pondok Pesantren, Pusat Koperasi Pondok Pesantren, Pimer Koperasi Pondok Pesantren* (pp. 1–113). Jakarta: Inkopontren.

Madjid, N. (1997). Pesantren, Kontinuitas dan Perubahan. In A. Azra (Ed.), *Bilik-bilik pesantren: Sebuah potret perjalanan*. Penerbit Paramadina: Jakarta.

Murray, T. R. (1992). Indonesia. In T. R. Murray (Ed.), *Education's role in national development plans: Ten country cases*. New York: Praeger.

Mustari, M. (2013). The roles of the institution of pesantren in the development of rural society: A study in Kabupaten Tasikmalaya, West Java, Indonesia. *International Journal of Nusantara Islam, 1*(1), 13–35.

PINBUK. (1999). *Kajian Evaluasi Pengembangan Koperasi Pondok Pesantren dan BMT*. Working paper, Jointly Published by PINBUK and Proyek Pengembangan dan Pemantatapan Koperasi Perkotaan. Jakarta: Dirjen Koperasi Perkotaan, Dep. Kop dan Pengusaha Kecil.

Pohl, F. (2006). Islamic education and civil society: Reflections on the Pesantren tradition in contemporary Indonesia. *Comparative Education Review, 50*(3).

Reijntjes, C., Haverkort, B., & Waters-Bayer, A. (1992). *Farming for the future: An introduction to low-external-input and sustainable agriculture*. London: Macmillan Press.

Seibel, H. D. (2007). *Islamic microfinance in Indonesia: The challenge of institutional diversity, regulation and supervision*. Paper for the Symposium of Financing the Poor: Towards an Islamic Microfinance, Islamic Finance Project. Boston MA: Harvard Law School. April 14, 2007.

Seibel, H. D., & Agung, W. D. (2005). Islamic microfinance in Indonesia. In *Sector project financial system development* (pp. 1–80). Eschborn: GTZ.
Slikkerveer, L. J. (1990). *Plural medical systems in the Horn of Africa: The legacy of "Sheikh" Hippocrates*. London: Kegan Paul International.
Statistics of MRA. (2012). *Analisis dan Interpretasi data pada Pondok Pesantren, Madrasah Diniyah (Madin), Taman Pendidikan Qurán Tahun Pelajaran* 2011–2012, *Analisis Statistik Pendidikan Islam* (pp. 1–106). Jakarta: Kementrian Agama Republik Indonesia/Ministry of Religious Affairs. http://pendis.kemenag.go.id/file/dokumen/pontrenanalisis.pdf.
Wagiman, S. (1997). *The moderization of the Pesantren's educational system to meet the needs of Indonesian communities* (Ph.D. thesis). McGill University, Montreal.
Wessing, R. (1978). *Cosmology and social behaviour in a West Java settlement*. Paper for the International Studies. South East Asia Series 47. Athens, OH: Ohio University Center for International Studies.
Wessing, R. (1979). Life in the cosmic village: Cognitive models in Sundanese life. In E. M. Bruner & J. O. Becker (Eds.), *Art, ritual and society in Indonesia* (pp. 96–126). Athens, OH: Ohio University Center for International Studies.
Yadi, J., & Djazuli, A. (2002). *Lembaga-Lembaga Perekonomian Umat*. Jakarta: PT Raja Grafindo Persada.
Yakub, M. (1992). Kopontren (Koperasi Pondok Pesantren): Keberadaan dan Pengembangannya. In *Pengelola Majalah Infokop*. Balitbangkop: Departemen Koperasi.

Kurniawan Saefullah is Lecturer of the Faculty of Economics and Business, Universitas Padjadjaran, Bandung, Indonesia since 1998. He graduated from Universitas Padjadjaran and the International Islamic University, Malaysia. He has published several books and articles with his colleagues, including Introduction to Management and Introduction to Business. In 2009, he joined the International Project on Integrated Microfinance Management (IMM) for Poverty Reduction and Empowerment in Indonesia. As a Ph.D. Researcher at the Leiden Ethnosystems and Development Programme (LEAD), Faculty of Science of Leiden University, The Netherlands, he is finalising his Ph.D. Dissertation on 'Gintingan in Subang: An Indigenous Institution for Sustainable Community-Based Development in the Sunda Region of West Java, Indonesia' under the supervision of Prof. Dr. L. J. Slikkerveer. The focus of his research is on the important role of the local cosmology of Tri Tangtu in the process of sustainable community development, while he also developed a model of integration of community-support institutions on the basis of the Indigenous Knowledge Systems (IKS) in combination with the IMM and ICMD approaches, such as financial, medical, educational, communication and social services through the indigenous institutions of the local communities in the Sunda Region of West Java.

Nury Effendi is the Dean of the Faculty of Economics and Business of Universitas Padjadjaran in Indonesia. He obtained his Ph.D. from the University of Oklahoma (USA) with a specialization in Financial Economics. Besides teaching and research, he is a Member of the Advisory Board of Bank Jabar and the Indonesian Association of Economics Graduates (ISEI). He is also Senior Lecturer in the Course on *Integrated Microfinance Management* (IMM), Bandung.

Part IV
Indonesia: Indigenous Institutions for Integrated Community-Managed Development

Chapter 14
Gotong Royong: An Indigenous Institution of Communality and Mutual Assistance in Indonesia

L. Jan Slikkerveer

> *Indonesia has a challenge to develop the character of the youth through the awareness of local wisdom....one of which is 'Gotong Royong', the cultural heritage from the Indonesian ancestors. 'Gotong Royong' can be applied in facing the era of globalisation, where the youth have an important role to play as agents of change and the next generation of the nation.*
>
> Ayu Solicha Nur Kusumaningrum et al. (2015)

14.1 The Dual Spirit of Community Service and Mutual Assistance

In many cultures around the world, the phenomenon of *communality* has been observed and documented, where usually a social gathering is held to accomplish a common task often by the joint provision of manual labour by residents to the community as a whole. Using the term *communality* in the socio-cultural context of people belonging or relating to a community and not in the statistical sense of estimates of variance in correlation analyses, such social activities also include the joint organisation of local competitions of horse riding, cattle raising and gardening. In general, these events are based on a spirit or feeling of cooperation and belong to a group or community stemming from common interests and goals, and represented in a shared response to the need of the entire community to accomplish a common task: the repair of social centres, churches, mosques etc., usually at the village level. In contrast to the market mechanism of economic activities for the accrual of individual benefits, the provision of a collective input to common goals is of great importance in the countries where cultural heritage still plays an important role in community life.

L. J. Slikkerveer (✉)
LEAD, Leiden University, Leiden, The Netherlands
e-mail: l.j.slikkerveer@gmail.com

L. J. Slikkerveer et al. (eds.), *Integrated Community-Managed Development*, Cooperative Management, https://doi.org/10.1007/978-3-030-05423-6_14

An example of such a local institution of communality in North America is known as *gadugi*, a Cherokee reference to cooperative labour where men and women work together voluntarily to help the group of elderly, largely with their housing and health care. In Ireland, a similar concept of *meitheal* is used to denote cooperative labour among neighbours to execute in turn collective farming work of weeding and harvesting in the field.

In Northern Europe, where families in sparsely populated areas often tend to live in remote farms, the institution of *talkoot* is well known, referring to voluntary social work and gatherings, not only in Finland, but also in Sweden and Estonia. In Norway, this institution is known as *dugnad*, used to indicate a typical characteristic of Norway, where the execution of joint tasks is usually concluded with a common meal and drinks.

In the Mediterranean Region, the concept of *imece* denotes the village-based collaboration in Turkey, such as joining forces to build a common road, a bridge or a community center, or just coming together to participate in the organisation of any sort of ceremony.

Similarly, the institution of communal work is widely spread in Latin America, where *mink'a* is practiced by the indigenous communities in Ecuador, Bolivia, Chile and Peru, especially among the Quechua- and Aymara-speaking peoples. In the Horn of Africa, especially in Sudan, the Arabic concept of *naffīr* is used to bring villagers together to perform communal tasks of herding, building of homesteads and harvesting. The institution of *naffīr* has also recently been used in the military context of the local militia. In Sub-Saharan Africa, similar concepts exist in Kenya with *Harambee* ('All Pull Together') and in Tanzania with *Ujamaah* ('brotherhood'), both referring to the same meaning of the spirit of communality of working together and cooperation for the common good.

The root of the term *Gotong Royong* refers to the Javanese verb *Ngotong*, obviously cognate with the Sundanese term *Ngagotong*, meaning 'several people carrying out some task', while *Royong* means 'together'. Thus, *Gotong Royong* can basically be defined as an institutionalised activity by people carrying out a communal task together.

Based on his fieldwork in Central Java, the Indonesian anthropologist Koentjaraningrat (1961) categorises *Gotong Royong* into two types, namely *spontaneous help* and *mutual assistance*. Spontaneous help occurs generally in collective activities in agriculture, house building, celebrations, public works and in the event of disaster or death. Mutual assistance, however, is usually based on the principle of individual reciprocity, whether it is on the initiative of the citizens, or imposed as an expression of mutual cooperation. From a socio-cultural perspective, the value of *Gotong Royong* as an indigenous institution is spiritual, manifested in the form of individual behaviour or an action to do something together for the sake of mutual interest for the community. In the context of community development, the observation of Koentjaraningrat (1961) is important, that the institution of *Gotong Royong* renders human life in Indonesia more empowered and prosperous: "*With mutual assistance, various problems of common life can be solved easily and cheaply, as well as community development activities.*"

The wider context of its role in the local culture of Indonesia is also underscored by Nasroen (1967), who asserts that it forms: "*one of the core tenets of the Indonesian philosophy*." In his view, the concept of *Gotong Royong* is even a major philosophical principle on which the uniqueness of the Indonesian philosophy is grounded.

Geertz (1983) in his contribution on *Local Knowledge: Fact and Law in Comparative Perspective,* documents various forms of indigenous institutions in Indonesia, some of which also tend to embody such communal cooperation in work, politics, and personal relations: "*vaguely gathered under culturally charged and fairly well indefinable value-images—Rukun ('mutual adjustment'), 'Gotong Royong' ('joint bearing of burdens'), 'Tolong-Menolong' ('reciprocal assistance'), it governs social interaction with a force as sovereign as it is subdued.*"

In this context, Bowen (1986) focuses on the various forms of labour involved in collective work, where he makes a three-way distinction among bases for collective work: labour mobilised as a direct exchange; generalised reciprocal assistance; and labour mobilised on the basis of political status. While the first base refers to labour exchange in agricultural activities, the second base of generalised reciprocity underscores the obligation of community members to provide assistance to the building of a house, the marriage of a child, or the death of a relative. The third type of mutual assistance is called *Gotong Royong*, consisting of labour which is mobilised on the basis of political status or subordination. Although Bowen (1986) recognises the general meaning of *Gotong Royong* as 'assistance', he argues that it resembles corvée, when it is commandeered by local officials for the construction of district roads, bridges etc. Such an obligatory interpretation of the concept of *Gotong Royong* had been expanded into a form of a forced labour system by the Dutch in the nineteenth century, and later by the Japanese during World War II (cf. Jay 1969; Breman 1980).

The non-material aspects of *Gotong Royong* in Java are further underscored by Hahn (1999), who points to the fact that the traditional Javanese culture does not emphasise material wealth, but that there is: "*respect for those who contribute to the general village welfare over personal gain. And the spirit of gotong royong or volunteerism, is promoted as a cultural value*."

The above-mentioned studies on *Gotong Royong* indicate an extension of the meaning of communality in Indonesia, as it not only refers to the joint participation of individuals to provide a collective contribution to their community, but also to individual assistance to fellow members in need in the community in the form of either material, financial, physical, mental and spiritual help. The latter dimension of mutual assistance to neighbours refers to a strong social obligation, often accompanied by unspoken reciprocity (cf. Kamus Besar Bahasa Indonesia 2016; UCEO 2016) (Illustration 14.1).

Similar indigenous institutions are functioning along the same principles elsewhere in the country, such as in the Sundanese region in the form of *Andilan* and *Talitihan*, while in other regions, such local institutions are known as *Metulung* in Bali, *De'nyande* in Madura, *Grebek* in Central Java and *Mbecek* in East Java.

In the wider context of Southeast Asia, including Singapore, Malaysia, and Brunei, the use of the concept of *Gotong Royong* for communal work is also

Illustration 14.1 Example of the implementation of *gotong royong* in moving a house of a neighbouring family in the village. *Source* Tanoesudibjo (2016)

well-known. In Singapore the residents' joint activities in the 1970s to clear the drains at Kembangan have also been identified as an example of G*otong Royong*. In Kampong Radin Mas, where the Malay villagers were resettled in 1973, the entire life in the village was based on the common spirit of *Gotong Royong* where the villagers were rallying to work together as part of the traditional culture of the community (cf. Taylor and Aragon 1991; Coate and Ravallion 1993; Rochmadi 2012; Tahir 2013; Baland et al. 2016).

Although most authors refer to a voluntary service to the community as a whole, or to neighbours, the social control always seems to be operational in terms of the cultural background of the community. The overriding consideration, however, is that *Gotong Royong* is a traditional, institutionalised expression of communality in the spirit of selfless cooperation for community goals and mutual assistance for individuals, characteristic for the Indonesian 'village culture' (cf. Barnouw 1946; Warren et al. 1995; Forshee 2006; Situngkir and Prasetyo 2012; Saefullah 2019).

14.2 The Dual Dimensions of *Gotong Royong*: Communalism Versus Individualism

The above-mentioned assessment of *Gotong Royong* in Indonesia indicates the two interrelated dimensions at the community level: voluntary service of all members to the community and mutual assistance to neighbours in need. In his study on the nature of reciprocity and collective labour in Indonesian villages, Bowen (1986) makes a distinction between the terms *Koperasi*, referring to cooperatives as the basis of the economy, and *Musyawarah,* meaning consensus as the basis for legislative decision-making and *Gotong Royong*, focused on mutual assistance. In his view, each of these terms has to do with the obligations of the individual towards the community, the propriety of power, and the relation of state authority to traditional social and political structures. In his research in Java, Koentjaraningrat (1966) describes the related terms *Grojogan*, *Krubutan*, *Gentosan*, *Samabatan*, and *Gujuban* as alternative words for the same form of labour exchange.

A distinction between the two above-mentioned dimensions has also been the subject in the wider debate on the presumed dualism of *collectivism* and *individualism*, and in particular to their relevance for community development. While individualism is generally described as the economic, political, or socio-cultural autonomy of the individual within the community, as opposed to collectivism, the different conceptualisations of individualism ranging from egocentric to more integrative social characteristics render such opposition to collectivism as rather relative. Although collectivism as a philosophy focuses on the group, i.e. the communal, societal, or national interests in different types of economic and political systems, it is also built on the input of the individual members of the group, such as for example in cooperatives, where a collective organisation is composed of individuals who are organised to work together.

In the debate, the role of culture has recently gained much attention in research in several societies, showing the differential influence which collectivist and individualist cultures have on the willingness of people to cooperate with others in certain group activities. Basically, collectivists tend to be more likely to accommodate an individualistic culture and to change their behaviours based on their situations better than individualists (cf. Parks and Vu 1994; Wagner III 1995; Brewer and Venaik 2011).

In his cross-cultural comparison, Hofstede (2001) finds that in a collectivist culture where people value their 'ingroup' as a whole, the individual identifies him- or herself as belonging to a group, valuing the goals of their group as more important than those of individuals. Thus, he/she is more connected to his/her group, and tends to care less about personal goals as an individual, but more about combined goals as a whole group. Hofstede (2001) concludes that in a collectivist society: "*people value their* 'ingroup' *as a whole, taking into account how their actions give a positive or negative impression to 'outgroups' while staying tightly knit with their ingroup*."

In his approach to the concepts of individualism and collectivism from a cross-cultural communication perspective, Rutledge (2011) argues that there are generally two related types of cultures. In individualist cultures, individual uniqueness and self-determination are valued, and an individual is admired for being a 'self-made man', 'making up his/her own mind' or 'working well independently'. In collectivist cultures, however, people identify with groups which protect them in exchange for loyalty and compliance. Interestingly, Rutledge (2011) also points to the paradox that individualist cultures tend to believe that there are universal values which should be shared by all people, while collectivist cultures tend to accept that different groups follow different values. In this context, Rutledge (2011) concludes that: "*Many of the Asian cultures are collectivist, while Anglo cultures tend to be individualist.*"

Similar research by Oyserman et al. (2002) shows that individualism shows no significant difference among different groups, while collectivism indicates a significant distinction. For instance, Asian Americans and Latino/Hispanic Americans show a significant difference in their communal work, supporting the notion that the difference in individualism and communalism may vary across ethno-cultural groups.

The relation between the prevalence of either collectivism or individualism in a society and its process of socio-economic development is characterised by causality which runs in two directions. As Ball (2001) asserts, the collectivist or individualist character of a society will influence the course of economic development in a different way, while in turn, economic growth will alter the orientation of the society toward individualism or collectivism. Basically, the two-way causality seems to concern the interaction between economic and cultural factors in the society, where collective norms are forming the social capital upon which nations are being built: these are the networks of relationships among individuals who live and work in a particular society, enabling that society to function effectively.

Among the current theories about the connection between individualism and collectivism in relation to socio-economic development, several authors support the general notion that development is facilitated by collectivism, and impeded by individualism. Triandis (1990) asserts that in collectivist cultures, in which social behaviour is determined by goals shared with some collective: "*it is considered socially desirable to place collective goals ahead of personal goals*". The study of Putnam (1993) on the role of civic traditions in development in Italy shows that the economic and political performance of the regions tends to vary greatly depending on their social and political history. Putnam (1993: 182) also found that those regions with strong civic traditions in terms of: "*active engagement in community affairs, by egalitarian patterns of politics, by trust and law abidingness*" did perform significantly better than regions lacking those traditions. The latter regions in which political and social participation were organised vertically showed that: "*mutual suspicion and corruption were regarded as normal, involvement in civic associations was scanty* [and] *lawlessness was expected.*"

Other empirical studies have documented a similar positive influence of collectivism and social capital on economic growth in various countries, such as by Banfield (1958) and Fukuyama (1995). In their cross-country study, Knack and Keefer (1997) tests these conclusions by using data from the World Values Survey, encompassing thousands of interviews from respondents in both developed and developing nations. By measuring social capital by the degree of 'trust' among the respondents in dealing with other people, they find that their measure of trust is positively related to growth in per capita income: "*a ten-percentage point increase in* [trust] *is associated with an increase in growth of four-fifths of a percentage.*"

In the above-mentioned review of the various conceptualisations of *Gotong Royong,* its dimension of collectivism is not only dominant throughout the country, but the related accumulation of social capital is well-grounded in the indigenous norms, values and institutions which have developed over many generations.

As regards the dualist situation in Indonesia, the dimension of mutual assistance to community members in need is expressed by the high level of solidarity among neighbours in the community, being an important characteristic of the indigenous village tradition. Often, the whole community will jointly assist a family to move a house, build a new home or plant rice and harvest the yield. As mentioned before, such mutual assistance is still part of a social obligation, often accompanied by unspoken reciprocity. The overriding considerations, however, remain based on the collective side of voluntary village cooperation, as part of the indigenous culture.

14.3 *Gotong Royong*: A Distinctive Pan-Indonesian Institution

In his definition of *Gotong Royong* as an indigenous institution, Slikkerveer (2014) underscores not only the informal institutional characterisation of the socio-anthropological conceptualisation of a fundamental complex of specific cultural norms, values and behaviours of cooperation, mutual assistance and selfless service to the community, but also the underlying aspects of indigenous knowledge, beliefs and practices which are grounded in the cosmological principles of the harmonious relations between the three spheres of the local worldview—human, natural and spiritual—which have been transferred over many generations. The latter aspect clearly transpires through the above-mentioned spirit of *Gotong Royong,* which not only values the maintenance of the harmonious relationships among community members, but also strengthens the social prestige and cohesion among them.

As far as the *emic* view on such indigenous institutions is concerned, local people's view which is also related to their culture is an important factor to be considered in the analysis. As described in Chap. 7, Slikkerveer (1990, 1999) has developed an analytical model of transcultural behaviour with a view to explaining how different cultures in various settings and the related causal factors affect the local people's differential patterns of behaviour, based on their knowledge, beliefs and practices. The studies by Agung (2005), Djen Ammar (2010) and Ambaretnani (2012) provide ample evidence that particular pyscho-social factors tend to dominate the contribution to communal behaviour among ethno-cultural groups in Bali and in the Sundanese region of Indonesia.

These findings are supported by the work of Schwartz et al. (2010) who indicate that communalities which are measured by preferences of collectivism versus individualism are strongly associated with psycho-social variables. The above-mentioned studies show that indigenous knowledge and beliefs are strongly associated with the cultural values, norms and behaviour which are based on the cosmological views of the population concerned.

While Agung (2005) distinctly documents in his outstanding study the role of the Balinese cosmology of *Tri Hita Karana* in the remarkable conservation behaviour of the Balinese of the bio-cultural diversity on the island, Saefullah (2019) meticulously illustrates the influence of the Sundanese cosmology of *Tri Tangtu* on the daily life and activities of the Sundanese people in West Java. In the next contribution to this Chapter, Saefullah further elaborates on the indigenous institution of *Gintingan* and its underlying Sundanese cosmology.

These and other indigenous cosmologies underscore that well-being and happiness in life can only be achieved by behaviour and activities which maintain a harmonious balance between the human, natural and spiritual worlds. The position of the Balinese and Sundanese people in the debate on individualism *versus* collectivism would lean towards a collectivist orientation, as it supports the harmonious life among their fellow humans as part of their cultural traditions in the community. Similar evident relations between communalism and traditional cultures are documented by Jason et al. (2016), who found in their study of Latino Recovery Homes, that among Latino people, whose communities are characterised by collectivism as proxy to communalism, the preference for traditional types of houses is showing a positive interaction in comparison with modern houses (cf. Slikkerveer 1990, 1999; Agung 2005; Djen Ammar 2010; Schwartz et al. 2014; Jason et al. 2016).

As the Javanese are forming the largest ethno-cultural group in Indonesia and their language and culture have influenced many other sub-cultures in the country over the past centuries, the origins of the widely used concept of *Gotong Royong* in Java can be traced back as Javanese. The process of merging and converging among cultures has been described by Ortiz (1947) for Latin America as *transculturation*, where the acculturation between two cultures A and B evolve into a third culture C. In Sub-Saharan Africa, Slikkerveer (1990) introduced the process *of transcultural heath care utilisation*, denoting the concurrent use of traditional, transitional and modern medical systems—known as the plural medical system—by the various ethno-cultural groups living in the Horn of Africa.

Since the development of the Javanese institution of *Gotong Royong* also shows historical parallels with a similar evolvement of indigenous counterpart institutions among other ethno-cultural groups in other regions of the country, resulting in varying degrees of the process of transculturation, these indigenous institutions represent typically pan-Indonesian phenomena of a rather diversified form of the traditional Indonesian village culture.

The conceptualisation of pan-Indonesian characteristics dates back to the work of the Dutch anthropologist Van Wouden, who in 1935 documented the Indonesian culture as encompassing the existence of common principles of equilibrium, dualism and spirituality in such areas of language, law, kinship, communication, cosmology, music and songs throughout the vast Archipelago which he identified as confined within the 'Field of Ethnological Study' (FES) of Indonesia.

The identification of Indonesia as a FES has further been substantiated by various structural anthropologists belonging to the 'Leiden Tradition of Structural Anthropology', e.g. Van Wouden (1968); De Josselin de Jong (1980) and Schefold (1988).

As mentioned in Chap. 7, the concept of the FES has also served a more functionalist purpose of one of the three components of the 'Leiden Ethnosystems Approach', introduced by Slikkerveer (1990), who developed a specific IKS-oriented research methodology to document, describe, understand and explain indigenous knowledge systems in various sectors of the community. In combination with the two other components—the 'Participant's View' (PV) and the 'Historical Dimension' (HD)—the 'Leiden Ethnosystems Approach' provides an in-depth assessment of the indigenous culture from the participants' *emic* point of view, in which the concept of the FES as a regional comparative research method provides a more realistic and less normative representation of the actual situation in the field than many previous Eurocentric depictions of non-Western cultures.

In Java, the underlying traditional Javanese cosmology of the conception of three worlds—the human world, the underworld and the upper world—dates back to the Hindu influence, where the term *Triloka* in the text of *Wédhatama*—a classical Javanese moralistic work—is referring to the explicit interconnection of the three worlds.[1] Yumarma (1996) points to the related importance of maintaining *Keselarasan* ('harmonious unity') in human relationships and avoiding conflict in the communities, in which reconciliation is sought after in ceremonies of restoration of *Slamatan* and *Wayang*.

The pan-Indonesian characteristics of *Gotong Royong,* manifest in the country-wide use of the term, the dualism in its implementation and its cosmological basis of communality and mutual assistance, and as such documented for the many sub-cultures of the country, render a further assessment of some of these major indigenous institutions and their foundations outside the Javanese culture as being rather justified. The parallel traditions of the Javanese institution of *Gotong Royong* include, *i.a.*, *Metulung* in Bali and *Andilan* in the Sunda Region of West Java, equally providing a promising point of embarkation for the implementation of the integrated community-managed development approach.

[1]As Yumarma (1996) documents, the *Serat Wédhatama* is a classical Javanese text which represents a didactic moralistic work which is slightly influenced by Islam. Although it is formally ascribed to Mangkunegara IV, there are indications that there seem to be more authors involved. The script describes the existence of three kinds of worlds using the term *triloka*, which may probably originate from the Hindu terms denoting a division into three worlds. The Javanese conception encompasses the human world, the underworld and the upper world which are interconnected through direct relationships in which humans have to maintain a harmonious balance in order to safeguard harmony in human life.

14.4 The Political Construction of *Gotong Royong* for National Unity

Following the dominance of the Javanese culture in matters of central government and administration, accelerated after independence to consolidate the unity of the enormous Archipelago with hundreds of sub-cultures and indigenous languages, the socio-cultural interpretation of *Gotong Royong* in Java has been elevated to the national level as a rather dominant dictum to functionalise the balanced socio-cultural relations in traditional villages in rural areas throughout the country in the provision of of new impetus for communal labour for the common good of the entire nation.

During the foundation of the Republic of Indonesia, *Gotong Royong* soon became the slogan for national unity, bridging contrasting positions and differences in opinions about the way forward. For Soekarno, the first President of the Republic, *Gotong Royong* formed the emblem of the Indonesian nation as a dynamic force of uniting opposites (cf. Soekarno 1965). In his famous speech of 1945, known as 'The Birth of the Pancasila', President Soekarno underscored his conception of *Gotong Royong* as a national uniting principle which would bring together Christians and Muslims, rich and poor, native Indonesians and naturalised citizens in a mutually tolerant struggle of the new nation against the enemy: "*The Indonesian State that we erect must be a Gotong Royong State!*" (cf. Panitia Lima 1977:128).

With the establishment of the New Order by President Soeharto in 1965, the political significance of *Gotong Royong* shifted from the previous horizontal unifying principle among striving groups to unify the new republic to a vertical guideline for intervention by the state into village life. Bowen (1986:552–553) characterises this shift by arguing that: "*In particular, the term has accompanied and supported efforts at the top-down implementation of development programmes.*"

Good examples of such 'top-down' policies are provided by the way in which the government introduced the *Bimbingan Massal* (BIMAS) ('mass guidance') in the 1960s. The programme of rice intensification, where foreign companies were invited to intervene in the local rice-farming practices with costly, high-yielding varieties and their heavy support systems requiring the massive input of pesticices and fertilisers was facilitated by renaming the programme as 'BIMAS Gotong Royong' to functionalise the meaning of *Gotong Royong* involving cooperation and mutual assistance for the purpose of deployment of labour at the community level.

After a few years, however, the 'BIMAS Gotong Royong' Programme was discontinued after opposition and resistance by the farmers because of high external input costs and low yields (cf. Hansen 1973).

Another expression of state intervention in rural communities with an appeal to the indigenous institution of *Gotong Royong* is found in the successive national development programmes known as 'Inpres Desa', the acronym of the 'Presidential

Order for the Villages'. The programme included an annual village grant for the improvement of public roads, bridges etc., in return for which the communities are expected to provide free labour, and building materials.

The government-induced development programmes have been accompanied by a number of educational booklets, published during the 1980s by the Ministry of Education and Culture, in which the general values of communal services and reciprocity are described for different provinces on the basis of historical relations in the pre-colonial kingdoms, and their continuation during the periods of colonialism and independence. Special attention is paid to the aspect of *Kerja Bakti* ('voluntary service'), with reference to obligatory service provided to national development programmes, including 'Inpres Desa' (cf. Dokumentasi Kebudayaan Daerah 1982).

The focus on the implied meaning of the indigenous institution of *Gotong Royong* concerning the ascribed provision of free labour to government development programmes, however, has been criticised by scientists, including Koentjaraningrat (1974), Mubyarto (1982) and Bowen (1986), who point to the historical continuation of the relationships of dependency between the national state and the local communities, which during the 1960s caused some villagers to re-think their obligations to provide free *Gotong Royong* labour to national agricultural policies.

As described in Chap. 9, similar scepticism has recently been observed among the local population about the implementation of the Indonesian Law No. 6/2014 (2014), where the lack of participation in the related development policies has been identified as one of the main causes. In practice, such non-participatory attitudes among the local people towards outside interventions can be traced back to a lack of visible benefits, suspicion about the provision of 'cheap labour', corruption, elite capture and ignorance of the poor members of the local communities.

If the indigenous principles of *Gotong Royong* can be redefined and revitalised into renewed functions of the dual spirit of communality and mutual assistance at the community level in terms of a 'bottom-up' approach instead of the prolonged 'top-down' policies of the recent past, this institution can provide a solid basis for improving the lives of the villagers and reducing the gap between the rich and the poor.

Such reorientation towards the meaning of the indigenous institution of *Gotong Royong* in its original setting of the indigenous knowledge systems and the underlying indigenous cosmologies at the community level would allow for the 'bottom-up' introduction of development policies, programmes and projects which could anticipate increased participation of the participants, and as such render their contribution meaningful to attain sustainable development.

The direct relevance of such reorientation for the present approach of *Integrated Community-Managed Development* (ICMD) is embodied in its 'bottom-up' orientation, in which the indigenous institutions and the underlying cosmologies are functionalised by the integration of local and global knowledge systems in order to achieve poverty reduction and sustainable community development in Indonesia and beyond.

References

Agung, A. A. G. (2005). *Bali endangered paradise? Tri Hita Karana and the conservation of the Island's biocultural diversity*. Ph.D. Dissertation. Leiden Ethnosystems and Development Programme. LEAD Studies No. 1. Leiden University. xxv + ill., p. 463.
Ambaretnani, P. (2012). *Paraji and Bidan in Rancaekek: Integrated medicine for advanced partnerships among traditional birth attendants and community midwives in the Sunda Area of West Java, Indonesia*. Ph.D. Dissertation. Leiden Ethnosystems and Development Programme. LEAD Studies No. 7. Leiden University. xx + ill., p. 265.
Baland, J. M., Bonjean, L., Guirkinger, C., & Ziparo, R. (2016). The economic consequences of mutual help in extended families. *Journal of Development Economics, 123*(2016), 38–56.
Ball, R. (2001). Individualism, Collectivism, and Economic Development. *Annals of the American Academy of Political and Social Science, 573*, 57–84. Culture and Development: International Perspectives (January 2001).
Banfield, E. C. (1958). *The moral basis of a backward society*. New York: The Free Press.
Barnouw, A. J. (1946). Cross currents of culture in Indonesia. *Far Eastern Quarterly 5*(2), 143–151 (February 1946).
Bowen, J. (1986). On the political construction of tradition: *Gotong Royong* in Indonesia. *Journal of Asian Studies, 45*(3), 545–561.
Breman, J. (1980). *The village on Java and the early-colonial state*. Rotterdam: CASPI.
Brewer, P., & Venaik, S. (2011). Individualism-collectivism in Hofstede and GLOBE. *Palgrave Macmillan Journals, 42,* 436–445.
Coate, S., & Ravallion, M. (1993). Reciprocity without commitment: characterization and performance of informal insurance arrangements. *Journal of Development Economics, 40*(1), 1–24.
De Josselin de Jong, P. E. (1980). The Netherlands: Structuralism before Levi-Strauss. In S. Diamond (Ed.), *Anthropology: Ancestors and heirs* (pp. 243–257). The Hague: Mouton.
Djen Amar, S. C. (2010). *Gunem Catur in the Sunda Region of West Java: Indigenous communication on the MAC plant knowledge and practice within the Arisan in Lembang*. Ph.D. Dissertation. Leiden Ethnosystems and Development Programme. LEAD Studies No. 6. Leiden University. xx + ill., p. 218.
Dokumentasi Kebudayaan Saera. (1982). *Proyek Inventarisasi dan Dokumentasi Kebudayaan Daerah - Sistim Gotong Royong dalam Masyarakat Pedesaan Daerah Istimewa Yogyakarta*. Jakarta: Departemen Pendidikan dan Kebudayaan.
Forshee, J. (2006). *Culture and customs of Indonesia*. Westport CT: Greenwood Press.
Fukuyama, F. (1995). *Trust: The social virtues and the creation of prosperity*. New York: Free Press.
Geertz, C. (1983). Local knowledge: fact and law in comparative perspective. In: Geertz, C (Ed.), *Local knowledge: Further essays in interpretive anthropology* (pp. 167–234). New York: Basic Books.
Hahn, R. A. (1999). *Anthropology in public health: Bridging Differences in culture and society*. Oxford: Oxford University Press.
Hansen, G. E. (1973). *The Politics and Administration of Rural Development in Indonesia*. Research Monograph No. 9. Berkeley: University of California, Center for South and Southeast Asia Studies.
Hofstede, G. (2001). *Culture's consequences: Comparing values, behaviors, institutions and organizations across nations* (2nd ed.). Thousand Oaks, CA: Sage Publications Inc.
Jason, L. A., Luna, R. D., Alvarez, J., & Stevens, E. (2016). Collectivism and individualism in Latino recovery Homes. *Journal of Ethnicity in Substance Abuse*, pp. 1–14 (26 April 2016).
Jay, R. (1969). *Javanese villagers*. Cambridge, Mass.: M.I.T. Press.
Kamus Besar Bahasa Indonesia (KBBI). (2016). Gotong Royong, *The Indonesian Dictionary*. URL: http://kbbi.web.id/gotong%20royong.

Knack, S., & Keefer, P. (1997). Does social capital have an economic payoff? A cross-country investigation. *Quarterly Journal of Economics, 112*, 1251–1288.

Koentjaraningrat. (1961). *Some sociological-anthropological observations on Gotong Royong practices in two villages in Central Java*. Ithaca, New York: Cornell University Modern Indonesia Project.

Koentjaraningrat. (1966). Bride-price and adoption in the Kinship relations of the Bgu of West Irian. *Etrhnology, 5*(3), 233–244.

Koentjaraningrat. (1974). *Kebudayaan, Mewtalitet dan Pembangunan* [*Culture, mentality and development*]. Jakarta: Gramedia.

Kusumaningrum, A. S. N., Evi, Z., A'yun, M. Q., & Fadhilah, L. N. (2015). "Gotong Royong Sebagai Jati Diri Indonesia". In *Proceedings of the National Seminar "Selamatkan Generasi Bangsa dengan Membentuk Karakter Berbasis Kearifan Lokal"*. Surakarta: Rumah Hebat Indonesia.

Mubyarto. (1982). A fixation of agricultural policy in Indonesia. In Hainsworth, G. B. (Ed.), *Village-level modernization in Southeast Asia* (pp. 27–42). Vancouver: University of British Columbia Press.

Nasroen, M. (1967). *Falsafah Indonesia*. Jakarta: Bulan Bintang.

Ortiz, F. (1947). *Cuban counterpoint: Tobacco and sugar*. New York: Alfred A. Knopf Inc.

Oyserman, D., Coon, H. M., & Kemmelmeier, M. (2002). Rethinking individualism and collectivism: Evaluation of theoretical assumptions and meta-analyses. *Psychological Bulletin, 128*, 3–72.

Lima, P. (1977). *Uraian Pancasila [The Composition of the Pancasilal*. Jakarta: Mutiara.

Parks, C. D., & Vu, A. D. (1994). Social dilemma behavior of individuals from highly individualist and collectivist cultures. *Journal of Conflict Resolution, 38*, 708–718.

Putnam, R. (1993). What makes democracy work?. *National Civic Review, 82*(2).

Rochmadi, N. (2012). *Gotong Royong sebagai common identity dalam Kehidupan Bertetangga Negara-Negara ASEAN: Jurnal Forum Sosial*. Malang: Universitas Negeri Malang.

Rutledge, B. (2011). *Cultural differences—Individualism versus collectivism*. URL: http://thearticulateceo.typepad.com/my-blog/2011/09/cultural-differences-individualism-versus-collectivism.html.

Saefullah, K. (2019). *Gintingan in Subang: An Indigenous Institution for Sustainable Community-Based Development in the Sunda Region of West Java, Indonesia*. Ph.D. Dissertation. Leiden Ethnosystems and Development Programme. LEAD Studies No. 11. Leiden University.

Schefold, R. (1988). *Indonesia in focus: Ancient traditions-modern times*. Santa Monica: Marketing Services.

Schwartz, S. J., Unger, J. B., Des Rosiers, S. E., Lorenzo-Blanco, E. I., Zamboanga, B. L., Huang, S., et al. (2014). Domains of acculturation and their effects on substance use and sexual behavior in recent hispanic immigrant adolescents. *Prevention Science, 15*, 385–396.

Schwartz, S. J., Weisskirch, R. S., Hurley, E. A., Zamboanga, B. L., Park, I. J. K., Kim, S. Y., et al. (2010). Communalism, familism and filial piety: Are they birds of a collectivist feather? *Cultural Diversity and Ethnic Minority Psychology, 16*(4), 548–560.

Situngkir, H., & Prasetyo, Y. E. (2012). *On social and economic spheres: An observation of the "Gantangan" Indonesian tradition*. Personal RePEc Archive Paper No. 18097, Munich: MPRA.

Slikkerveer, L. J. (1990). *Plural medical systems in the horn of Africa: The legacy of "Sheikh" Hippocrates*. London: Kegan Paul International.

Slikkerveer, L. J. (1999). Ethnoscience, 'TEK', and its application to conservation. In D. A. Posey (Ed.), *Cultural and spiritual values of biodiversity: A complementary contribution to the global biodiversity assessment* (pp. 167–260). Nairobi/London: UNDP/Intermediate Technology Press.

Slikkerveer, L. J. (2014). Local Institutions in Indonesia. In *Lecture Notes*. Master Course on Integrated Microfinance Management, Bandung. Faculty of Economics and Business, Universities Padjadjaran.

Soekarno. (1965). *Dibawah Bendera Revolusi* [*Under the Flag of Revolution*], Vol. 2. Jakarta: Panitya Penerbit.

Tahir, I. (2013). *A village remembered: Kampong Radin Mas, 1800s-1973*. Singapore: OPUS Editorial Private Limited.

Tanoesudibjo, H. (2016). Membangkitkan Nilai Gotong Royong, Daerah dan Desa, *Politik Hukum Keamanan,* September 2016. URL: https://www.harytanoesoedibjo.com/id/membangkitkan-nilai-gotong-royong/.

Taylor, P., & Aragon, L. V. (1991). *Beyond the Java sea: Art of Indonesia's outer islands*. New York: H.N. Abrahams.

Triandis, H. (1990). Cross-cultural studies of individualism and collectivism. In *Nebraska Symposium on Motivation, 37*, 41–133.

UCEO. (2016). G*otong Royong and benefits of Gotong Royong for life*. Surabaya: Ciputra University. http://ciputrauceo.net/blog/2016/2/15/gotong-royong-dan-manfaat-gotong-royong-bagi-kehidupan.

van Wouden, F. A. E. (1968). *Types of social structure in Eastern Indonesia*. Leiden: Koninklijk Instituut voor Taal-, Land- en Volkenkunde.

Wagner III, J. A. (1995). Studies of individualism-collectivism: Effects on cooperation in groups. *The Academy of Management Journal, 38*, 152–172.

Warren, D. M., Slikkerveer, L. J., & Brokensha, D. (1995). *The cultural dimension of development: Indigenous knowledge systems, IT studies on indigenous knowledge and development*. London: Intermediate Technology Press.

Yumarma, A. (1996). *Unity in diversity: A philosophical and ethical study of the Javanese.* Concept of Keselarasan. Inculturation-Intercultural and Interreligious Studies No. XIX. Roma: Editrice Pontificia Universita.

L. Jan Slikkerveer is Professor of Applied Ethnoscience and Director of the Leiden Ethnosystems and Development Programme, Faculty of Science, Leiden University, Leiden, The Netherlands. He received his Ph.D. on his fieldwork in the Horn of Africa from Leiden University in 1983 and an Honorary Degree from the Faculty of Medicine of Universitas Padjadjaran in Bandung, Indonesia in 2005. He further extended advanced training and research in the newly-developing field of applied ethnoscience in several sub-disciplines, including ethno-economics, ethno-medicine, ethno-biology/botany, ethno-pharmacy, ethno-communication, ethno-agriculture and ethno-mathematics in combination with a focus on three target regions of South-East Asia, East-Africa and the Mediterranean Region. He is the supervisor of 25 Ph.D. students at Leiden University and has published more than 100 books and 300 articles on the subject. While he has also received a number of substantial subsidies from the European Union in Brussels for international development programmes in South-East Asia, East-Africa and the Mediterranean Region, he also conceptualised both the IMM & ICMD approaches in Indonesia, and is Senior Lecturer in the International Master Course on Integrated Microfinance Management (IMM) at the Faculty of Economics and Business of Universitas Padjadjaran in Bandung, Indonesia.

Chapter 15
Gintingan: An Indigenous Socio-cultural Institution in Subang, West Java

Kurniawan Saefullah

> *The problem of indigenous institutions for development agencies is their embeddedness in other aspects of life, but ultimately this may make them a more powerful resource for development....*
>
> Elizabeth E. Watson (2003)

15.1 Introduction

Subang is one of the districts in the Province of West Java of Indonesia, where people's livelihood at the community level has been influenced by the Sundanese culture. The Sundanese cosmology of *Tri Tangtu* influences people's livelihood from socio-cultural activities to economic transactions. One of the Sundanese traditions which represents the integration of economic activities and socio-cultural events is an institution named *Gintingan*. It can be defined as a socio-cultural institution, which is based on communality in terms of joint participation of individuals to provide a collective contribution to their community and of individual assistance to fellow villagers in need, known as *Gotong Royong* in the Javanese culture. *Gintingan* is practiced as a local initiative by people in the community when a particular household, which has a *Hajat* ('important need'), receives contributions from the community members through the provision of a *Gantangan*, a vessel of rice with a content of about 10 l (cf. Illustration 15.1).

In contrast to the modern microfinancing system, *Gintingan* is a typical representation of an indigenous community-managed institution, based on the local people's cosmovision of *Tri Tangtu* which influences their livelihood practices. In this way, this institution maintains the harmonious balance among the villagers during socio-cultural events known as *Hajatan,* including weddings, circumcisions,

K. Saefullah (✉)
FEB, Universitas Padjadjaran, Bandung, Indonesia
e-mail: kurniawan.saefullah@unpad.ac.id

L. J. Slikkerveer et al. (eds.), *Integrated Community-Managed Development*,
Cooperative Management, https://doi.org/10.1007/978-3-030-05423-6_15

Illustration 15.1 A *Gintingan* vessel from West Java, which contains 1 l of rice. *Source* Temuan Galeri (2012)

rituals, etc. The tradition itself has generally been implemented by the people living in the northern agricultural areas of the island of Java, including the district of Subang (cf. Saefullah 2019).

15.2 Concept and Practice of *Gintingan*

The term '*Gintingan*' originates from *Gantang* or *Ginting*, which refers to a particular wooden vessel to contain a special amount of rice. *Gantang* itself is known not only in the agricultural areas of Indonesia, but also in some other parts of South-East Asia, albeit that there are different scales of measure indicated. While in Indonesia, *Gantang* contains in general 10 l of rice according to Irawan (1999), in The Philippines, the equivalent of one *Gantang,* known as *Ganta,* contains about 3 l of rice. The United Nations (1966) estimate the contents at 8.38–8.57 l of rice, while in Sabah and Sarawak, Malaysia, such a measure indicates 2.42 kg. of rice. In Brunei, similar vessels contain 3.63–4.55 l of rice.

In addition to the above-mentioned general contents of the *gantang* in Indonesia, the scale measures are varying per region from 4.91 l in Palembang; 9.59 l in Batavia during the colonial period of time; 5.01–7.49 l in Makassar; and 13 l of rice in Kalimantan. Figure 15.1 shows a wooden *Gintigan* ('vessel') of 1 l from Java.

The concept of *Gintingan* has been described by several authors, such as Irawan (1999), who defines *Gintingan* as: '*Pola hajatan dengan penggunaan sistem arisan'* ('the practice of a ceremony using the rotation system of mutual help'). His definition, however, is mainly influenced by his observation that Gintingan is usually practiced regularly and that the contributions are based on the input—or 'buy-in'—into a common pot to be used by a member, mostly for support of a particular socio-cultural event. While the socio-cultural motivation dominates the implementation of *Gintingan*, there is also an economic aspect involved of mutual assistance and reciprocity at the community level, albeit not in an obligatory sense. Indeed, participation in the *Gintingan* is voluntary, and not restricted to any membership or particular time frame of regular meetings. The social purpose of this institution of strengthening the spirit of communality and mutual assistance is reinforced by the cultural events of the above-mentioned ceremonies. As documented by Saefullah (2019), the *Sisingaan* ('traditional dance') of Subang is

Illustration 15.2 *Sisingaan*: the traditional Sundanese dance in Subang. *Source* Saefullah (2019)

performed by families during the ceremonies in praise of the celebration. Such dances have also been reported elsewhere in the Subang Region by Irawan (1999) and Wijaya (2009) (Illustration 15.2).

Unlike the above-mentioned authors, Prasetyo (2012) defines *Gintingan* as: '*Sistem ekonomi tradisional berupa simpanan kredit beras melalui hajatan*' ('a traditional economic system, using the exchange of rice, and implemented through ceremonies'). His interpretation focuses on the economic aspects of reciprocity assuming that in practice, the institution creates the concepts of 'debtor' and 'creditor'. The community members who contribute to the family in need act as a 'creditor', while the receiving family act as a 'debtor'. The receiving family makes a record of any contribution from their relatives and neighbours in a special book, known as *Buku Beras* ('rice book') or *Buku Gintingan* ('*Gintingan* book'). The book is similar to a cash-flow book, as it records the in- and outflow of the contributions. The contributions include mainly rice, money, or other valuable materials. Although Prasetyo (2012) defines the institution of *Gintingan* from an economic perspective, the motives, the process, the events, as well as the evaluation of this institution in the Subang Region, are not based on economic motives, particularly on the profit-making point of view. The recent qualitative observations and in-depth interviews, complemented with quantitative household surveys in four villages in the district of Subang, clearly show the socio-cultural motivation of the local people in their intergenerational practice of the traditional institution. The research findings underscore the overall objective of the institution to preserve the social cohesion and communality among the members of the community (cf. Saefullah 2019). It represents the view of the local people on their relationship with their human, natural and spiritual world, based on their traditional cosmology (cf. Agung 2005; Saefullah 2019). Wijaya (2009) relates the term *Gintingan* to *Gentenan*, which means 'reciprocity'. In the Sundanese language, the term *Silih* refers to 'reciprocity', and is also expressed in the philosophical foundation of the social interaction of *Silih Asah* ('reciprocal learning'), *Silih Asih* ('reciprocal love') and *Silih Asuh* ('reciprocal care'), as further elaborated below.

The social aspect of this traditional institution is also observed in the way in which the obligation which follows the transaction is functioning. Although there is an economic aspect underlying the transaction through the contribution of a certain amount of rice among the people, and this contribution is also recorded in the above-mentioned *Buku Beras*, there is, however, no finite period of time of repayment set in terms of a reciprocal contribution system, or a formal sanction in the event of a late recompensation or failed reciprocal return of a contribution (cf. Harris 1997; Djunatan 2011).

The practice *of Gintingan,* as documented in the Subang Region can be explained as follows: If one household in the community has an important *Hajat* ('need') concerning the organisation of a social event, it will inform the community leader about their need for contributions from their fellow villagers. Since the contributions will be collected for the event, the community leader then informs all community members about the upcoming *Hajatan* ('ceremony'), and the voluntary obligation to fulfill the related needs of the household concerned, such as a wedding ceremony. Thereupon, the household sets the date of the ceremony in consultation with the community leader and the community members. Usually, the leader establishes an organising committee, which as a community-managed institution arranges the plans and preparations, as well as the implementation of the ceremony.

The organising committee then divides the tasks among the members of the community, such as collecting and administrating the contributions, organising the rituals, and preparing the cultural events surrounding the rituals and ceremonies which the household would like to conduct. The informal organising committee will then send out invitations to all households in the community. Thereafter, the people in the community will then make their contribution to the needy household in the form of rice, money or other valuable materials, with a specific measurement.

Considering that the *Gintingan* institution mainly uses rice as the form of contribution, a *Gantang* ('vessel') is used to measure the volume of rice. Although in the Subang region, the *gantang* contains 10 kg of rice, people could use either this scale or contribute more. The total of such contributions could easily amount to rather large quantities of rice or money. If, for instance, 200 households in the community adhere to the *Gintingan* to contribute to a needy household, and each household contributes one *Gantang* of rice, the needy household will receive at least about 2000 kg of rice. If a conversion is made of the total amount of collected rice with a price of 1 kg at 10,000 Rupiah, the total amount of collected rice would be worth about 20 million Rupiah, equivalent to about 1500 USD (2017) (Illustration 15.3).

The collected *Gantang* of rice from the community is used by the needy households for the organisation of the rituals and ceremonies of the event, including the support of the ceremony which follows after the event. As mentioned above, the needy household makes a record of each single contribution by the community members in the *Buku Beras* ('rice book') or *Buku Gintingan* ('*Gintingan* book'). The book is used by the household members to document how much rice or other *Gintingan* contributions they received, in the case where they should contribute to a similar need for another household in the community in the future. Interestingly, while there is no finite period of time to return a contribution, the reciprocal recompensation can only be done in a similar way if the other household has a similar need or problem.

According to Mauss (2002), this type of traditional institution is a form of reciprocal exchange, implemented by local people in a community as a positive return for what they have already received. Based on recent research in the Subang Region, *Gintingan* is usually implemented in agricultural societies. Similar traditional institutions are also operational in other parts of the Subang Region, albeit with a different name: *Andilan*. In Indramayu, the institution is known as *Josan*, while in East and Central Java, it is called *Rewangan* in Banyumas, and *Bojokan* in Boyolali (cf. Harris 1997; Mauss 2002; Saefullah 2019) (Illustration 15.4).

There are some interesting features of this indigenous community-managed institution, which are particularly shown in its practices and obligations. As regards the practice of *Gintingan,* it is based on the principle of mutual assistance through a reciprocal transaction, which is directly related to the pan-Indonesian institution of

Illustration 15.3 Family members in Subang collecting the contributions of rice in a *Gintingan*. *Source* Saefullah (2019)

Gotong Royong and its parallel institutions of *Andilan* in the Sunda Region of West Java and *Metulung* in Bali. As described in the previous Chap. 14, these indigenous institutions are operating on their cosmological basis in terms of communality and mutual assistance among the members of the local communities.

However some authors, including Irawan (1999), compare the practice of *Gintingan* with *Arisan,* being a rotating savings and credit association (ROSCA). The difference is that *Gintingan* is not held regularly at particular dates and that recompensation of the received contributions is not confined to a limited period of time. As regards the obligation of reciprocity, the household which receives the contribution from its community members should only 'repay' in the case when another household with a similar need is proposing to activate a *Gintingan* for their planned ceremony. As an example of a traditional institution which is also based on the principle of reciprocity, it complies with the characteristics of the related category of: (1) absence of the need for immediate return; (2) absence of a systematic calculation of the value of the service and products exchanged; and (3) an overt denial that a balance is being calculated, or that the balance must come out even (cf. Geertz 1956; Harris 1997; Van den Brink and Chavas 1997).

The practice of *Gintingan* has been documented in various forms. In Sukamelang in the Subang district, for instance, the community can contribute more than one *Gantang*, or in other forms such as money or goods which are needed by the household concerned. In the case where the needy household receives contributions

Illustration 15.4 Example of an invitation of *Gintingan* in Subang (2016). *Source* Saefullah (2019)

in money, then they also have to repay in the form of money. The same applies to other forms of such contributions. The custom of collecting money is also not merely used for rituals or ceremonies, and in some cases observations have been made that a household allocates some money to buy land for farming or to pay the cost of schooling, etc. Also, in the case of Cimanglid in the Subang district, *Gintingan* is not practiced for the purpose of a wedding, but for building a house.

There are different opinions on the practice of *Gintingan*. The proponents of the practice of *Gintingan* argue that the tradition is not only providing certain economic support for needy households from the community, but also demonstrating the benefits from the principle of mutual assistance of *Gotong Royong*, where personal communication and social cohesion among the members of the community are maintained and reinforced. The opponents, however, point to the burden which is somehow put on some people who are not able to return a similar contribution in the future. They advocate the idea that the contributions should not be provided on the basis of expected returns, but be based on a social obligation. This perspective has found support from the recent research by Saefullah (2019) in the Subang district, where in the village of Bunihayu a poor woman held a wedding ceremony for her daughter. As she had no money to conduct the wedding ceremony, the members of her community supported her through the practice of *Gintingan* in various ways, ranging from material contributions to the actual execution of the event. In her case, the people in the community did not expect her to repay their contributions; the case further substantiates that *Gintingan* is indeed a traditional socio-cultural institution without any pursuit of profit or economic gain which is characteristic for economic institutions (cf. Saefullah 2019; Slikkerveer: this Volume).

The origin of *Gintingan* itself is also a subject of discussion. Irawan (1999), for instance, states that it was established during the 1970s, while Wijaya (2009) supports the notion that it has been practiced since 1978. Prasetyo (2012), however, does not mention any argument on the beginnings of *Gintingan* in West Java, particularly in Subang. Anthropologist Mulyanto (*pers. comm.* 2011) from Universitas Padjadjaran in Bandung supposes that *Gintingan* could have a long history, as he relates its origin and similar practices in different areas of Indonesia, as well as in other neighbouring countries. In his view, the various similarities of the institution in Indonesia and in other countries, such as the comparable use of the terms of *Gantang* or *Ganta*, the principle of mutual assistance, and the type of the communities where the institutions are practiced, might even represent a common Austronesian heritage. *Gintingan* as an indigenous socio-cultural institution, however, is mostly practiced in Indonesia, and in particular as an important part of the Sundanese culture in West Java, where the principles of the mutual assistance and communality of the pan-Indonesian institution of *Gotong Royong* have also influenced the traditional institution of *Gintingan*.

As culture is related to peoples' cosmovision, the Sundanese philosophy of life also influences the practice of *Gintingan*. According to Irawan (1999), the Sundanese philosophy of life which influences G*intingan* includes the following:

- *akur jeung dulur sakasur* ('...should live in harmony with our blood');
- *akur jeung dulur sasumur* ('...with our neighbour'); and
- *akur jeung dulur salembur* (' with our community').

The concept of harmony used within the context of harmony with our blood, harmony with our neighbour, and harmony with our community reflects the foundation of the Sundanese cosmovision of *Tri Tangtu*, which includes the triadic structure of human existence. Since *Tri* means *Three* and *Tangtu* means *Realms*, *Tri Tangtu* refers to three realms, also known as *Buana* in the Sundanese language. The term *Buana* is similar to the Balinese term *Bhuwana*, which also refers to realms. In the Sundanese context, the three realms of the Sundanese cosmology of *Tri Tangtu* include the following:

- Upper realms: The sacred-spiritual realms of God (*Buana Nyungcung/Mayapadha*);
- Middle realms: The human realms (*Buana Panca Tengah/Madyapadha*)
- Lower realms: The realms of the earth and the environment (*Buana Rarang/Arcapadha*).

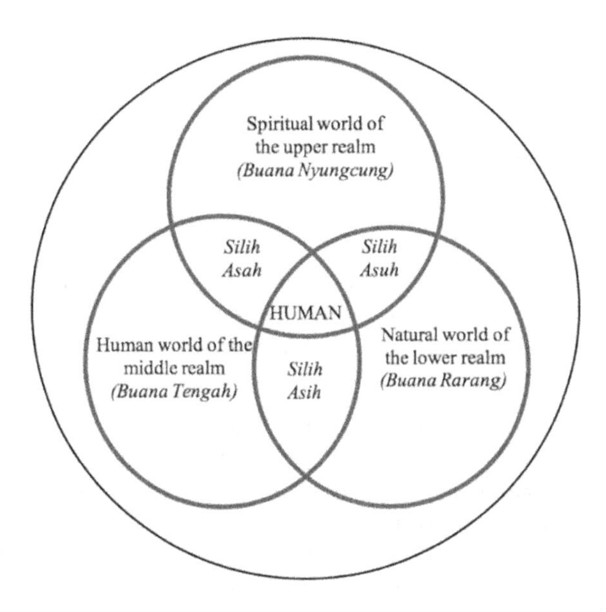

Fig. 15.1 Schematic representation of the Sundanese cosmovision of *Tri Tangtu*, encompassing the three realms of the spiritual, human and natural worlds *Source* Saefullah (2019)

The concept of *Tri Tangtu* is used by the indigenous Sundanese population within the context of their traditional beliefs and livelihood. Accordingly, the Sundanese believe that in order to reach a 'harmonious life', people should maintain the balance between the three realms as the harmonious balance between the gods, humans and the earth determine happiness in life. Thus, humans (*Middle Realm*) should maintain a harmonious relation with the spiritual realms (*Upper Realm*), which is expressed in the concept of *Silih Asah* ('reciprocal learning') in the Sundanese cosmology. Similarly, humans (*Middle Realm*) should also maintain a harmonious relationship with the earth and the environment (*Lower Realms*), which is expressed in the concept of *Silih Asih* ('Reciprocal Love') in the Sundanese cosmology. Eventually, the state of *Silih Asih* can be achieved if humans maintain a harmonious relationship with both the spiritual realm and the realm of the earth and the environment. The achievement of such an overall state of harmony is reflected in the concept of *Silih Asuh* ('Reciprocal Care') of the Sundanese cosmology (cf. Wessing 1978; Sumardjo 2010; Djunatan 2011). Figure 15.1 shows a schematic representation of the Sundanese cosmovision of *Tri Tangtu*, which shows a clear similarity with the traditional Balinese cosmovision of *Tri Hita Karana,* as described in Chap. 3, and the African cosmologies, as described in Chap. 4.

The term '*Silih*' in the Sundanese language refers to a mutual or reciprocal relationship. Sundanese cosmology prescribes that a harmonious balance can only be achieved by the practice of reciprocity. For instance, nature will treat humans well, if people treat nature in a respectful and responsible way. *Gintingan* provides an excellent example of how the three harmonies of *Silih Asah, Silih Asih,* and *Silih Asuh* have been put into practice by the people of the Sundanese community in

Subang. In this context, *Gintingan* indeed fulfills the underlying principles as follows:

- the community leader represents the first element in the *Buana Nyungcung* ('Spiritual World') of *Tri Tangtu* as the community leader who puts the spiritual belief into practice by taking care of the people in the community;
- the household, or the individual member of the community in need represents the second element in the *Buana Tengah* ('Human World') of *Tri Tangtu,* where they are interacting with each other on the basis of reciprocity; and
- the rice, money or any other materials which are contributed within the practice of *Gintingan* represent the third element of *Buana Rarang* ('Natural World') of *Tri Tangtu*.

The traditional principle of the voluntary provision of a 'contribution' to community members or the community as a whole has been an important part of the Indonesian culture, and is not only practiced in the Sundanese region, but also in other regions of the country. In line with the conception of *Gotong Royong* as a pan-Indonesian institution, described in the previous Chap. 14, the traditions of *Jagong* in Central Java, *Nyande* in Madura, *Mbecek* in East Java, and *Talitihan* and *Andilan* in West Java are practicing the same principles of voluntary contribution and reciprocity (cf. Slikkerveer: this Volume).

It is interesting to observe that the principle of the harmonious balance among the three realms of the cosmology has also recently been taken into account in the current development paradigm based on the concept of *endogenous development* through the re-conceptualisation of '*human well-being*' as the objective of sustainable development.[1] As it is now intensively discussed among development scholars, the 'materialist-economic' view of development has been criticised as its approach and implementation are lacking the inclusion of the important cultural dimension.

For a long time, the development of the people and the community has only been analysed on the erroneous basis of the outsiders' perspective, and not from the insiders' view. Warren, Slikkerveer and Brokensha (1995) were among the first scholars who criticised the dominance of the outsiders' perspective in socio-economic development. While most earlier development experts and planners were usually neglecting the local peoples' views in their interventions from outside the communities, such as policies, programmes and projects, these applied-oriented scientists proposed by the end of the 1980s a shift from a normative to a more realistic paradigm based on the emic view of the local participants—*insider's perspective*—in contrast to the previous etic view—*outsider's perspective*. Their strategy, implemented in the 'bottom-up' approach, was soon followed by others

[1]According to White (2009), there are three dimensions embodied in the concept of 'well-being':

1. the material dimension, which concerns practical welfare and standard of living;
2. the social dimension, which concerns social relations and access to public goods;
3. the human dimension, which concerns capabilities, attitudes to life and personal competence.

including Chambers et al. (1989), Richards (1985), Posey (1999), Woodley et al. (2006) and Loeffelman (2010).

The practice of *Gintingan* has not only improved the life and well-being of local people in terms of financial material circumstances, but also in terms of their social and human interactions. In a social respect, people could maintain their culture of 'providing care' in rural areas, while such cultural behaviour is rarely found in the urban areas. Although some modern institutions have been established in order to address the socio-economic problems at the community level, such as financial institutions and government agencies, they are usually lacking the human dimension of compassion, empathy and care for fellow humans. In contrast, the traditional institution of *Gintingan* continues to maintain the elements of personal communication and social interaction among the local people in the community, where banking transactions, for instance, seek to replace the personal dimension to become impersonal and commercially-oriented, often resulting in a lack of participation and non-compliance with outside interventions as mentioned above.

By consequence, the local people in the communities tend to become alienated from their fellow humans, neighbours and other members of their communities, living together, but like strangers; in the words of a Bangladeshi villager who rightly defined the philosophy of 'ideal well-being': '*bhat, kapor on shonman niye shukhey thakbo*' ('we live in happiness with rice, clothes and respect').

Thus, *Gintingan* can be considered as a traditional institution which can actively make an important contribution to the newly-proposed community-managed development efforts in order to achieve a state of well-being for all participants in the community—including the poor and marginalised families and individuals—and as such provide a major impetus to the achievement of sustainable community development. Indeed, *Gintingan* provides a most promising example of a traditional socio-cultural institution in the Sudan region with great potential to achieve human well-being and health in the rural areas through a renewed strategy which is based on its significant principles of mutual assistance and communality. Human well-being as the ultimate objective of sustainable development cannot only be measured by the material dimension of rice and clothes, but also by the spiritual dimension of peace and happiness, and by the socio-cultural dimension of mutual respect and assistance.

In this way, the traditional socio-cultural institution of *Gintingan* has become crucial as an integrated community-managed development institution where people in the community are not only able to fulfill their material needs, including the cost of housing, food and education, as well as the performance of ceremonies and rituals, but also their spiritual needs of fullfillment and a happy life, helped by other members of their community and mutual respect by maintaining their culture of 'taking care' of all members—rich and poor—in the community (cf. Warren, Slikkerveer and Brokensha 1995; Woodley et al. 2006; White 2009; Loeffelman 2010; Saefullah 2019).

References

Agung, A. A. G. (2005). *Bali endangered paradise? tri hita karana and the conservation of the island's biocultural diversity*. Ph.D. Dissertation. Leiden ethnosystems and development programme. LEAD studies No. 1. Leiden University. xxv + 464 pp. ill.

Chambers, R., Pacey, A., & Thrupp, L. A. (1989). *Farmer first: Farmer innovation and agricultural research*. London: Intermediate Technology Publications Ltd.

Djunatan, S. (2011). *The principle of affirmation: An ontological and epistemological ground of interculturality*. Ph.D. Dissertation, Rotterdam: Erasmus University.

Geertz, C. (1956). *The rotating credit association: A 'middle rung' in development*. Cambridge, MA: Massachusetts Institute of Technology, International Studies Center.

Harris, M. (1997). *Culture, People, Nature: An Introduction to General Anthropology*, Pearson Higher Education. URL: https://www.pearsonhighered.com.

Irawan, E. (1999). *Sistem Gintingan dalam Hajatan terhadap Kelangsungan Ekonomi dan Perkembangan Kesenian Tradisional di Daerah Kabupaten Subang*. Laporan Penelitian, Bandung: STISI.

Loeffelman, C. (2010). *Gender and development through western eyes: An analysis of microfinance as the west's solution to third world women, poverty, and neoliberalism*. DissertationsWashington, D.C.: George Washington University.

Mauss, M. (2002). *The gift: The form and reason for exchange in archaic societies*. London: Routledge Classics.

Mulyanto, D. (2011). *Gintingan and social structure of sundanese people, an interview by the author with Dede Mulyanto*. Field Note, Nov 2011.

Posey, D. A. (Ed.). (1999). *Cultural and spiritual values of biodiversity: A complementary contribution to the global biodiversity assessment*. Nairobi/London: UNEP/Intermediate Technology Press.

Prasetyo, J. E. (2012). *Komersialisasi sosial di pedesaan: Studi terhadap modal sosial gantangan di tiga desa miskin di kabupaten subang*. MA Thesis, Bogor: Program Studi Sosiologi Pedesaan, Institut Pertanian.

Richards, P. (1985). *Indigenous agricultural revolution: Ecology and food production in West Africa*. London: Hutchinson.

Saefullah, K. (2019). *Gintingan in Subang: An Indigenous Institution for Sustainable Community-Based Development in the Sunda Region of West Java, Indonesia*. Ph.D. Dissertation. Leiden Ethnosystems and Development Programme. LEAD Studies No. 11. Leiden University.

Sumardjo, J. (2010). *Estetika paradoks*. Bandung: Sunan Ambu Press.

Temuan Galeri. (2012). *Picture of a wooden vessel of gantang*. URL: http://temuangalery.blogspot.nl/2014/10/literan-beras-1-gantang-450-rb.html.

United Nations. (1966). *Ganta measure from different sources*. URL: https://sizes.com/units/ganta.htm.

Van den Brink, R., Chavas, J. P. (1997). *The microeconomics of an indigenous african institution: The rotating savings and credit association. Economic development and cultural change*. Vol. 45, No. 4 (July 1997).

Warren, D. M., Slikkerveer, L. J., & Brokensha, D. (1995). *The cultural dimension of development: Indigenous knowledge systems*. IT Studies on Indigenous Knowledge and Development, London: Intermediate Technology Press.

Watson, E. E. (2003). Examining the potential of indigenous institutions for development: A perspective from borana Ethiopia. *Development and Change, 34*(2), 287–309. Blackwell Publishing.

Wessing, R. (1978). *Cosmology and social behavior in a west javanese settlement*. Papers in International Studies, *Southeast Asia Series* No. 47. Athens, OH: Ohio University Center for International Studies, Southeast Asia Programme.
White, S.C. (2009). Analysing well-being: A framework for development practice. *Development in Practice* 20(2) (Mar 2009).
Wijaya, K. (2009). Kondangan sistem Narik Gintingan dalam Perspektif Sosiologi Hukum Islam: Studi Kasus di Desa Citrajaya Kecamatan Binong Kabupaten Subang (unpublished thesis).
Woodley, E., Crowley, E., Pryck, J. D. de, Carmen, A. (2006). Cultural Indicators of Indigenous Peoples' Food and Agro-Ecological Systems, SARD Initiatives, *Paper on the 2nd Global Consultation on the Right to Food and Food Security for Indigenous Peoples*, 7–9 Sep 2006, Nicaragua.

Kurniawan Saefullah is Lecturer of the Faculty of Economics and Business, Universitas Padjadjaran, Bandung, Indonesia since 1998. He graduated from Universitas Padjadjaran and the International Islamic University, Malaysia. He has published several books and articles with his colleagues, including Introduction to Management and Introduction to Business. In 2009, he joined the International Project on Integrated Microfinance Management (IMM) for Poverty Reduction and Empowerment in Indonesia. As a Ph.D. Researcher at the Leiden Ethnosystems and Development Programme (LEAD), Faculty of Science of Leiden University, The Netherlands, he is finalising his Ph.D. Dissertation on 'Gintingan in Subang: An Indigenous Institution for Sustainable Community-Based Development in the Sunda Region of West Java, Indonesia' under the supervision of Prof. Dr. L. J. Slikkerveer. The focus of his research is on the important role of the local cosmology of Tri Tangtu in the process of sustainable community development, while he also developed a model of integration of community-support institutions on the basis of the Indigenous Knowledge Systems (IKS) in combination with the IMM and ICMD approaches, such as financial, medical, educational, communication and social services through the indigenous institutions of the local communities in the Sunda Region of West Java.

Chapter 16
Grebeg Air Gajah Wong: An Integrated Community River Management Project in Java

Martha Tilaar

> *Holistic views of water and the environment and of human health and society need to inform and influence the way that resources and services are managed in indigenous areas and beyond.*
>
> —Jimenez et al. (2014)

16.1 Integrated Approaches to River Water Resources Conservation

The importance of a river cannot be overstated. It represents the course of water which originates in the mountains and flows downwards until it reaches the sea. Rivers are among the most obvious and significant features of the landscape of Planet Earth. They transport water by gravity, from the headwaters to the ocean, where river water meets sea water. On its perpetual journey, water from the river crosses hills, grounds and plains, encompassing both rural and urban areas. Along its entire route, a river forms a basic resource not only for human life but also for the flora and fauna. Unfortunately, the condition of some rivers has become rather critical in some areas, obstructing its original destination. In some urban areas in developing countries, rivers have deteriorated by the pollution from factory waste, public garbage and dirty sewer systems, posing a serious threat to human and animal health.

In Indonesia, there are many cases which are challenged by the ecological problem of the rivers. Overall, about three-fourths of the rivers in Indonesia have become polluted as a result of several factors. By using *Kriteria Mutu Air* (KMA), the quality of water is regulated by the Indonesian Law No. 82/2001, established by the Directorate-General of Environmental Pollution and Damage of the Ministry of

M. Tilaar (✉)
Martha Tilaar Group, Jakarta, Indonesia
e-mail: martha@marthana.co.id

L. J. Slikkerveer et al. (eds.), *Integrated Community-Managed Development*,
Cooperative Management, https://doi.org/10.1007/978-3-030-05423-6_16

Environment in 2015 (cf. Sapariah 2015). Soon thereafter, Wendyartaka (2016) reports that about 66% of the rivers in 33 provinces in Indonesia are highly polluted, 24% have a medium pollution level, and 6% are less polluted. Only 2% of the rivers in the country are considered clean.

The damage to the rivers has been caused by the recent increase of *fecal coli, total coliform*, Biological Oxygen Demand (BOD), Chemical Oxygen Demand (COD), and Hydrogen Sulfide (H2S) in the river waters. In turn, the contaminated elements affect the increase of many diseases such as intestinal diseases, skin diseases, as well as lung diseases. The negative impacts on human life do not include environmental disasters such as floods, avalanches, and other ecological hazards.

Notwithstanding, the main causes of river damage are domestic and industrial waste. Surprisingly, domestic waste is the top contributor to river damage as there are about 26% of households or circa 16 million households in Indonesia, which do not have proper toilet and cleaning facilities. Most of these households use rivers as their 'giant trash bin'. In several big cities, such as Jakarta, Yogyakarta, and Bandung, BOD is even higher than the generally accepted limit. The situation indicates that the rivers are not only damaging, but also dangerous to all species: humans, flora and fauna. The deteriorating condition of the rivers is also threatening human life, including daily activities. For instance, in November 2016, there was a flood in Bandung City in West Java, caused by heavy rains, but aggravated by many factors, including domestic waste in the river which blocked the river's flow, and many houses and vehicles were damaged. In Garut, another district in the Southern part of West Java, a similar flood destroyed hundreds of houses and farms, and many people were reported missing or dead.

Such critical conditions can, of course, no longer be tolerated. The problems of rivers have been designated as a serious environmental hazard and an ecological threat, not only at the national level of Indonesia, but also worldwide. Supported by the *Pacific Institute* (PI) in Oakland, USA, and the *United Nations Environmental Programme* (UNEP) in Nairobi, Kenya, Ross (2010) published an important document about *'Cleaning the Waters'* as an attempt to socialise the importance of having clean water and how to make every country pursue solutions to solve the damage to water sources, including the problem of polluted rivers. The document identifies some fundamental impacts of poor water quality on the human and the natural environment: its impact on human health, the environment, water quantity, vulnerable communities, and on livelihoods in general. Coad (2005) reports that about 2.5 billion people around the globe are lacking improved sanitation facilities and almost 1 billion people still use unsafe water sources, including polluted rivers. In addition, people who have access to drinking water tend to receive unsafe and inadequate services. Therefore, it is not surprising that the *Post-2015 Agenda of the Sustainable Development Goals* (SDGs) of the United Nations (2015) ranks 'Clean Water and Sanitation' as number 6 in the important SDGs to be achieved by the year 2030, as follows:

6.1 achieve universal and equitable access to safe and affordable drinking water for all;
6.2 achieve access to adequate and equitable sanitation and hygiene for all and substantially increase water-use efficiency across all sectors.

According to Ross (2010) there are at least three types of efforts which can be executed: education and awareness building, monitoring/data collection on the water quality and tightening governance and regulations which are related to preserving water quality. A comprehensive approach as a solution to the problem of river pollution needs to be implemented. One of the promising efforts is the integration of a 'bottom-up' approach with a 'top-down' programme. In this context, a *Public-Private Partnership* (PPP) can be implemented as a tool which is available to the government to solve the problem, considering that such a solution requires the involvement of many parties and huge financial and non-financial resources.[1] Such an approach could even be more effective if it were based on the integration of local and global knowledge, practices and beliefs regarding adequate community-based management of water from river resources (cf. SPES 1994; Coad 2005; Ross 2010).

16.2 MTG: Implementing Four Strategic Corporate Social Responsibilities

Rivers are important for the equilibrium and maintenance of the balance between the environment and biodiversity. In some big cities around the world, particularly in the developing world, the condition of their rivers has deteriorated extremely, due to the irresponsible behaviour of people towards the function of rivers.

In view of the above-mentioned deteriorating conditions of rivers worldwide, there is an urgent need to consider the rivers as a most precious good, to be guarded and protected from all forms of pollution or excessive exploitation. In some cultures, such as the *U'wa Indians* of Columbia, the river is referred to as: "*a living organism: a blood that nourishes the earth*"' (cf. UNESCO 2014).

The Martha Tilaar Group (MTG) is one of the largest companies in Indonesia, and it aims at promoting the great indigenous heritage of Indonesia worldwide. It practices a *green business philosophy* in which all business aspects of the company take the balance between humans and the environment seriously into account. MTG applies a model of Corporate Social Responsibility (CSR) (cf. Martina Berto 2012,

[1]A Public–Private Partnership (PPP) refers to a cooperative arrangement between two or more public and private sectors, usually for the long term. Although governments have used such a mix of public and private endeavours throughout history, a clear trend towards governments across the globe making greater use of various PPP arrangements has developed over the past decades. As a commercial legal relationship, PPP has been defined by the Government of India in 2011 as: "*an arrangement between a government/ statutory entity/ government-owned entity on one side and a private sector entity on the other, for the provision of public assets and/or public services…*" (cf. Ministry of Finance, India 2005).

2013). In general, CSR refers to a form of corporate self-regulation which is integrated into a business model, whereby a company monitors and ensures its active compliance with the spirit of the law, ethical standards and national or international norms (cf. Rasche et al. 2017).

The MTG model, however, goes a step further to promote actions which seek to contribute to the social good beyond the interests of the company and the basic requirements of the law in a positive manner. In this way, it is not only an increase in long-term profits and shareholder confidence which is sought through good public relations and high ethical standards by taking social responsibility for all corporate actions. MTG has developed a strategy to implement four pillars in all of its business operations to make a positive impact on both the environment and the stakeholders, including employees, consumers, investors, local communities, and indigenous people, as shown in Fig. 16.1.

Based on the principles of women's empowerment, caring for green business, culture and education, the four pillars of CSR of the MTG include the following:

1. *Beauty Green* is a programme of caring for nature, involving green production, environmental protection, natural resources conservation, and developing the strategy of a green company for a green nation.
2. *Beauty Education* is the programme of supporting national education, both formally and informally through the implementation of special projects. It provides facilities and opportunities to village students to continue their study to acquire special skills which can be used to improve their profession such as spa therapist, massage therapist, make-up expert and hair dresser. The programme also provides scholarships as an opportunity for students from poor families to pursue a higher degree in education. In addition, the programme also establishes teaching communities in many areas and subjects, based on the local peoples' needs which are directly relevant to their daily life.
3. *Beauty Culture* is a programme for supporting the conservation of the culture of Indonesia and exploring the potential of local communities to contribute to sustainable community development. The preservation of the local culture is also important to safeguard the local knowledge and wisdom for humans to live in a healthy environment.
4. *Empowering Women (Beauty in Women)* has also played an important role in the community of Indonesia as many local cultures have come to believe that the role of women is compelling evidence as the strong support for family life. By implementing this programme, the company believes that by extending the potential of women, it is possible to further improve family welfare and to bring about important changes in the concepts of the market and business of the company.

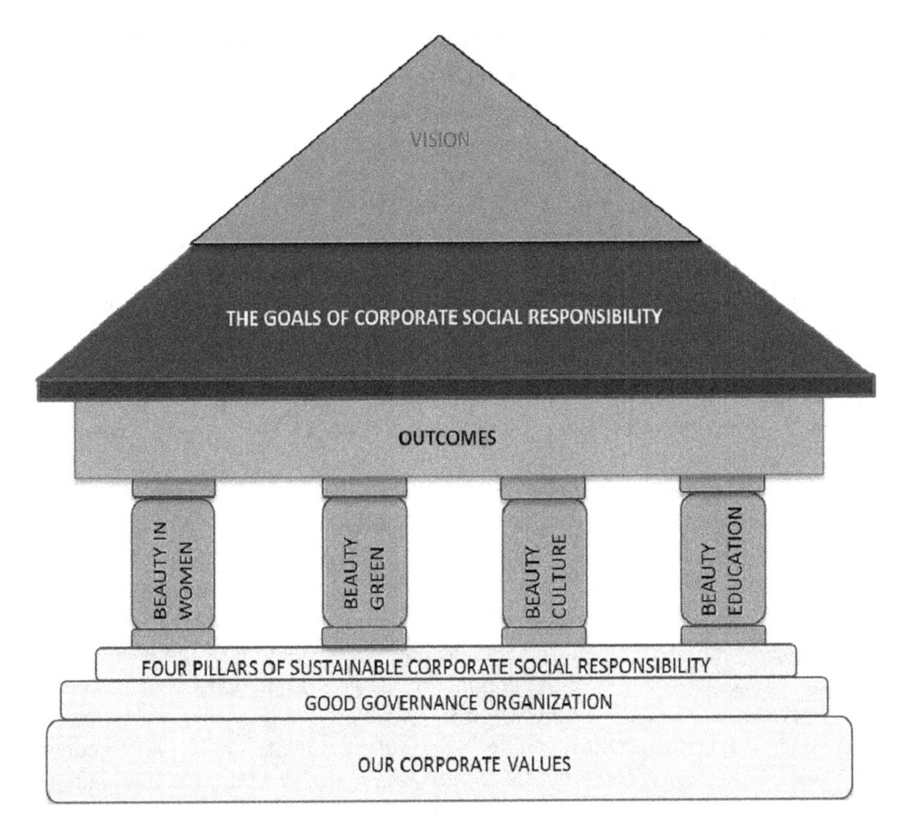

Fig. 16.1 Representation of MTG model with the four pillars of CSR Strategy of Martina Berto, Jakarta. *Source* Martina Berto (2012). *Note* **Vision**: to be one of the world's leading companies in cosmetics and spa industries with natural nuances and eastern values through product innovation and modern technology to optimise added value to stakeholders. **Mission**: (a) to increase the welfare of rural and forest communities, (b) to collaborate with multi-stakeholders in green programmes, (c) to empower women for better well-being, (d) to be a partner of the government and the United Nations bodies. **Outcomes**: (a) more companies and institutions are willing to provide capacity building to rural and forest communities, (b) the achievement of Millennium Development Goals in Indonesia, (c) promotion of sustainable partnership or collaboration among multi-stakeholders. **Corporate values**: discipline, integrity, innovation, persistence, perseverance, credible products, conservation of nature, culture preservation, customer focus

16.3 *Grebeg Air Gajah Wong*: An Integrated River Conservation Project

The word Grebeg refers to the traditional Javanese expression of welcoming the special days of the year, such as Mulud ('birth of the Prophet'), Syawal ('tenth month of the Islamic calendar') and Idul Adha or Suro ('New Year of Java'). Grebeg Suro is a special annual festive event celebrated to annually welcome the Year of East Java. *Grebeg Air Gajah Wong* refers specifically to the programme of

Integrated Sustainable Water Resource Management in Central Java, implemented through the traditional institution of the community and promoted by MTG on the basis of its first principle, being the pillar of Beauty Green as a programme of caring for nature. It involves green production, environmental protection, natural resources conservation, and developing the strategy of a green company for a green nation.

In 2012, in response to the call for input from companies in Indonesia to the society, MTG initiated the programme with a view to revitalising the function of the river through the local institution of *Grebeg* with the voluntary joint input of local people in collaboration with the local government, NGOs as a PPP towards integrated community development, and as such contributing to the achievement of one of the *Sustainable Development Goals* (SDGs) of the United Nations (2015) on 'Clean Water and Sanitation'. The initiative of the *Grebeg Air Gajah Wong* programme is to revitalise the river in the Sleman District and Depok Sub-District of Yogyakarta through the *Grebeg* institution of communal activities in order to clean the river and make its water usable and accessible again for people in the community.[2]

The programme started in 2012, where MTG, together with the District Government of Sleman, took the initiative to implement an integrated water sanitation project for the river in Sleman. It involved the participation not only of the private sector and government officials, as an implementation of the PPP scheme, but also the NGOs, people in the community as well as students and teachers. It combined the concept of a 'cleaning the river' programme with education and training as well as the establishment of a special community group to maintain the river water's sanitation, namely *Komunitas Peduli Gajah Wong* ('Community for *Gajah Wong*'). The programme started with an initial discussion between representatives from MTG, the Sleman people and the local government of the Sleman District of Yogyakarta on the importance of revitalising the River *Gajah Wong*. During the meetings, the wider importance of not only rivers, but also water sanitation, conservation and water resources management were discussed. Although there are three rivers in the area: the Rivers *Gajah Wong*, *Winongo* and *Code*, the case which was discussed focused on the pollution and ecological problems of the River *Gajah Wong* in the Sleman District of Yogyakarta.

The main consideration was that the River *Gajah Wong* has more functions for the benefit of the people, ranging from irrigation to domestic uses. The discussions were interactive throughout the process from identifying the causes to the possible solutions. One of the outcomes from the discussions was that the management of the cleanliness and maintenance of the river cannot be left only to the government,

---

[2]The United Nations (2017) estimates that approximately 70% of all water abstracted from rivers, lakes and aquifers is used for irrigation, while at the same time more than 80% of wastewater resulting from human activities is discharged into rivers or the sea without any pollution removal. Since clean water and sanitation are key conditions for achieving the United Nations Sustainable Development Goals (SDGs) by 2030, the civil society is called upon for not only investment in water research and development, but also promotion for the inclusion of women, youth and indigenous communities in water resources management.

Illustration 16.1 The practice of *Gotong Royong* in the Grebeg Air Gajah Wong Project. *Source* Kelompok Peduli Gajah Wong (KPGW) (2012)

without the full participation of the members of the local community, from private entities, government agencies, NGOs as well as educational institutions, in order to increase the awareness of all stakeholders through education and training.

The participants who were coming from different groups decided to implement a collaborative project, which they called '*Grebeg Air Gajah Wong*', which refers to the restoration of the function of the River *Gajah Wong*. The programme consists of workshops for training and education on the general importance of preserving the environment and the dangers of household waste and pollution, particularly for the River *Gajah Wong*. The importance of healthy houses and aromatic plants was also introduced during the workshops. As a follow-up from the workshops, all participants went down to the river and jointly cleaned it by practicing the institution of *Gotong Royong* to revitalise the function of the river (Illustration 16.1).

The first results are impressive as the river became cleaner than it used to be. The integrated approach also showed to be successful during the implementation of the programme. Moreover, the awareness of the local people regarding the importance of the sanitation of the water in the River *Gajah Wong,* which links up with their local knowledge and ptractice of sustainable water management, has also increased. The *Grebeg Air Gajah Wong* programme is an example of how integrated community development can be implemented through collaboration between the people in the community, the local government, and non-government institutions, by taking the local peoples' indigenous knowledge and institutions into account. Similar projects on water resource quality improvement have been initiated elsewhere around the world, supported by the *United Nations Environment Programme* (UNEP) and the *Pacific Institute* (PI) in Oakland, CA, USA, which are both motivating various groups and institutions concerned with protecting the limited freshwater resources and preserving their fundamental role in maintaining the health of humans and the ecosystem on Planet Earth. While Palaniappan et al. (2010) seek to focus on Water Quality Solutions, it is hoped that joint efforts in the area will pay particular attention to the integration of local peoples' knowledge and traditional institutions in order to enhance local peoples' full participation in community-based river management and conservation in the future.

Slikkerveer (2007) also underscores the importance of an integration between local institutions and government agencies which encompasses various sectors and disciplines, despite the fact that such integration should principally be initiated by the communities themselves in order to safeguard the implementation of the 'bottom-up' approach from indigenous knowledge and institutions as the ultimate model to increase local peoples' participation in comparable programmes and projects on sustainable water management systems in Indonesia and elsewhere (cf. Slikkerveer 2007; Ross 2010).

References

Coad, A. (2005). *Private sector involvement in solid waste management: Avoiding problems and building on successes*. Washington, DC: World Bank.

Jimenez, A., Cortobius, M., & Kjellen, M. (2014). Water, sanitation, and hygiene and indigenous peoples: A review of the literature. *Water International, 39*(3), 277–293.

Kelompok Peduli Gajah Wong (KPGW). (2012). *Grebeg Air Gajah Wong*: An Integrated River Conservation Project.

Martina Berto. (2012). *Annual report*, Jakarta: PT Martina Berto, Tbk.

Martina Berto. (2013). *Communication on progress: Corporate sustainability report*. PT Martina Berto, Tbk. Jakarta: PT Martina Berto, Tbk. URL: CoP-Sustainable-Report-MBTO.pdf.

Ministry of Finance (India). (2005). *Scheme for Support to Public Private Partnership in Infrastructure*. New Delhi: Department of Economic Affairs, Ministry of Finance, Government of India. www.pppinindia.gov.in/documents/20181/21751/VGF_GuideLines_2013.pdf.

Palaniappan, M., Gleick, P., Allen, L., Cohen, M., Christian-Smith, J., & Smith, C. (2010). *Clearing the waters: A focus on water quality solutions*. Nairobi: UNEP.

Rasche, A., Morsing, M., & Moon, J. (2017). *Corporate social responsibility: Strategy, communication, governance*. Camebridge, UK: Camebridge University Press.

Ross, N. (Ed.). (2010). *Clearing the waters: A focus on water quality solutions*. Nairobi: PI/UNEP/UNON, Publishing Services Section.

Sapariah, S. (2015). Inilah Para Petinggi Baru Kementerian LHK, Apa Pesan Menteri Siti?, Mongabay-Indonesia. (29 May 2015).

Slikkerveer, L. J. (2007). *Integrated Microfinance Management and Health Communitcation in Indonesia. Cleveringa Lecture,* 7 December 2007, Jakarta: Trisakti School of Management & The Sekar Manggis Foundation.

SPES. (1994). *Economy and Ecology in Sustainable Development*. Gramedia Pustaka Utama and SPES Foundation.

UNESCO (2014). *Drops of Water 1: What is a River*. Paris: UNESCO. URL: http://www.unesco.org/fileadmin/MULTIMEDIA/FIELD/Venice/pdf/special_events/bozza_scheda_DOW01_1.0.pdf.

United Nations (UN). (2015). *The 2030 agenda for sustainable development and the sustainable development goals (SDGs)*. New York: United Nations.
Wendyartaka, A. (2016). Air Sungai di Indonesia Tercemar Berat, Kompas (15 December 2016), URL: http://print.kompas.com/baca/2016/04/29/Air-Sungai-di-Indonesia-Tercemar-Berat.

Martha Tilaar is the Director and Chair of the Martha Tilaar Group and Honorary Chair of the Martha Tilaar Foundation in The Netherlands. She received an Honorary Degree from the World University in Tucson, Arizona, USA, and is supporting many projects to educate, develop and support local womens' groups thoughout Indonesia. She is also a Guest Senior Lecturer in the Course on *Integrated Microfinance Management* (IMM), Bandung.

Chapter 17
Warga Peduli AIDS: The IMM Approach to HIV/AIDS-Related Poverty Alleviation in Bandung, West Java

Prihatini Ambaretnani and Adiatma Y. M. Siregar

> *Identifying community members' perceptions of barriers to HIV prevention efforts presented by local social institutions and perceived barriers to the utilisation of HIV services provided by social institutions will help us to develop interventions that incorporate approaches to deal with these challenges.*
>
> —Akers et al. 2010

17.1 HIV Epidemic in Indonesia and Poverty

According to the United Nations Programme on HIV/AIDS (UNAIDS 2014), Indonesia has become one of the countries with the fastest growing human immunodeficiency virus (HIV) epidemic rates in Asia. Although the incidence of the epidemic of HIV in Indonesia is relatively low, the rate shows a yearly increase (cf. Reddy et al. 2012). As of 2009, the cumulative death toll due to HIV has amounted to nearly 17,000 people of whom 42% were reported to be related to Intravenous Drug Use (IDU), while hetero- and homosexual transmission accounts respectively for 48 and 4% (cf. Ministry of Health of Indonesia 2009). Recently, however, the trend has changed. The epidemic is now spreading faster through sexual intercourse and is shifting to the general population. Within the same period of time, the percentage of HIV transmission through IDU decreased from 53% in 2001–2005 to 34% in 2011, and heterosexual transmission increased from 37 to 71% (cf. National AIDS Commission [NAC] 2012). One of the results of this new trend of the epidemic in Indonesia is that the position of women has become crucial.

Another concern in Indonesia, albeit often neglected, regards the impact of HIV on poverty. There is a positive correlation between poverty and poor health and the relationship between HIV and under-development is said to be bi-directional, so

P. Ambaretnani (✉) · A. Y. M. Siregar
FISIP, Universitas Padjadjaran, Bandung, Indonesia
e-mail: nprihatini@gmail.com

L. J. Slikkerveer et al. (eds.), *Integrated Community-Managed Development*, Cooperative Management, https://doi.org/10.1007/978-3-030-05423-6_17

that poverty leads to higher HIV transmission and HIV further increases poverty (Pronyk et al. 2005; Stratford et al. 2008). Eventually, this relationship creates a vicious circle of poverty and health which may hamper the overall economic development. As such, income and savings have become crucial as households of People Living with HIV (PLHIV) struggle to build and protect their economic resources, while the socio-economic impact of HIV becomes just as important as the health impact of the disease (cf. Donahue 2000).

17.2 Going Beyond Microfinance to Combat HIV Epidemic

Microenterprise represents a potential intervention model for HIV prevention, because the existence of poverty may lead to higher risk of contracting HIV infection. Individuals at risk for HIV often have histories of trauma, drug abuse, incarceration, unemployment, poor education and homelessness. These kinds of histories may, to some extent, be alleviated by economic empowerment programmes (cf. Stratford et al. 2008). Indeed the idea of integrating HIV prevention and economic interventions holds great promise. For example, certain microfinance initiatives may produce skills which empower women to negotiate safer sex in terms of assertiveness, recognition of gender norms, challenging gender norms, and practicing new enactments of agency and independence (cf. Dworkin and Blankenship 2009).

Supporting microfinance institutions (MFIs) to move beyond the conventional financial practice, therefore, may produce important implications for areas where the HIV epidemic hinders both health and development (cf. Pronyk et al. 2005). MFIs could create opportunities for high-risk groups to manage their own business, generate their own income and strengthen their financial footing. Such an approach can be used to prepare them for and to cope with crises, and may increase their financial access to health care. MFIs may also provide or facilitate non-financial HIV services, such as training, advice, or even health care. It is important, however, for microfinance institutions to retain their main focus as financial institutions, regardless of what kind of HIV activities they undertake. Neglect of this focus may draw MFIs away from their long-term mandate and compromise the achievement of long-term financial goals (cf. Parker 2000). As such, MFIs may need to collaborate with other HIV organisations. For instance, addressing the social impact of HIV is the objective of a HIV organisation, not a MFI. However, mitigating the social impact of HIV may help MFI clients to remain useful clients. As such, a form of collaboration and integration between both organisations is needed (cf. Donahue 2000). One example of the role of microfinance in mitigating the impact of HIV is the experience of *CARE International* in Cote d'Ivoire, providing village savings and a loan programme. When appropriate medical treatment is available, the PLHIV are capable of participating in, and benefiting from certain microfinance activities. This approach has increased the economic self-sufficiency of HIV-positive clients (cf. Holmes et al. 2010).

It is obvious that in the challenge of the HIV epidemic, the application of conventional microfinance practices will not result in successful poverty alleviation for PLHIV. A more realistic analysis of the needs and the socio-economic and medical backgrounds of PLHIV is required in order to set up an appropriate microfinance scheme. Also, there is a need for a scheme which goes beyond the financial aspects, i.e. tailored to accommodate the special situation of this group of the population. It is important to build a kind of local financial scheme which reflects the characteristics, as well as specific cultural and economic challenges which PLHIV are facing, while at the same time the community needs empowerment. In this context, Bateman (2010) argues that: "*Through the auspices of proactive state and non-state community-based interventions, local finance can become instead a powerful tool that can really help build and underpin sustainable local economic and social development, and so bring about sustainable poverty reduction.*"

Embarking on the notion that if such a kind of specific microfinance scheme would exist, the next questions are: "*What is the example of such a scheme?*" and "*Who should receive the assistance?*" PLHIV form a diverse group of individuals, raising the question if the scheme should start by providing the service to all groups of PLHIV. The next two paragraphs describe a case study of the required microfinance scheme, and the argument concerning which group should be prioritised in the beginning of the scheme, based on their importance in the current HIV epidemic in Indonesia, as well as their capability to undergo a microfinance programme successfully.

17.3 Community Empowerment and Microfinance: A Case Study

The example of a combination of health promotion and community empowerment with microfinance is provided by the initiative of the *City AIDS Commission* (CAC) in Bandung, West Java. CAC has developed a scheme to provide money to eligible PLHIV as capital to start a new business. In this way, CAC acts as a kind of microfinance institution (MFI) to alleviate HIV-related poverty which is experienced by the recipients of the funds and their family. At the same time, CAC may also provide health supervision to their clients. The interviews of a pilot survey conducted in 2015 indicate in general that the programme claims to be relatively successful and that the lives of the clients have improved.

This type of MFI has applied a kind of integrated approach, in the sense that it does not act only as a provider of loans, but also as an institute which provides measures to prevent HIV. Such a model, however, is not widely available in Indonesia.

17.3.1 The Local CAC Scheme

The grant which is provided by CAC to selected recipients has resulted in a number of income-generating activities which tend to improve their welfare status. The income-generating activities undertaken by the recipients show a wide variety, ranging from businesses for cell phone credit and kiosks, to the sale of food and clothing. Among all activities, food businesses are said to be the most promising since the required capital is relatively small, while the profit may be more than 100%. Most of the recipients who were interviewed reported hardly any obstacles in running their business, but their difficulties are mostly related to their health problems. The sudden need for medicines and/or treatment may consume the capital which these clients received from CAC. Such uses of funds, however, are only allowed after the supervisor gives permission. According to the respondents, the grant from CAC was extremely helpful for the recipients in improving their welfare and finances for health care. A summary of the important elements of the scheme are listed below.

a. *The initiative*

The idea of providing funds to PLHIV has independently been developed by CAC in Bandung with the aim to support eligible recipients financially and morally. Both goals are aimed at supporting the recipients through their crises caused by HIV, rendering HIV-related poverty the main target of the programme. In this way, their financial access to HIV care for support and treatment is also provided. In order to avoid misuse of the funds, recipients are closely monitored by appointed supervisors, who keep an eye on them to make sure that the fund recipients will produce something useful and expectantly sustainable for the purposes of their welfare and survival.

The grants are made available from the local CAC budget, amounting to Rp. 5,000,000,- per grant. Due to the limited budget which is currently available for this programme, the selection of recipients has been strict in order to check that the funds are being used effectively and efficiently. Following applications made by 50 participants, 10 have been selected as eligible recipients on the basis of their proposal and interview. After selection, the recipients received some training to support them in creating and running income-generating activities, including entrepreneurship and motivation. CAC also conducted an assessment of the needs of the recipients in order to provide further support.

Furthermore, specific supervisors from selected NGOs were appointed by CAC to support each recipient in the management of their business. The supervisors observed, gave advice, and, if necessary, intervened in the recipients' business. The supervisors also acted as mediators between CAC and the recipients, where the recipients would discuss their problems firstly with the supervisors before consulting CAC. Also, CAC itself supervised the recipients through its consultations with the supervisors, its support on the development of monthly financial reports and random visits. The rather simple monthly financial reports are required by CAC as proof that the recipients are using the funds properly. The recipients are obliged to save Rp 30,000,- every month in CAC for contingency funds. Interestingly, the supervision by CAC and the supervisors of the recipients will continue even after the funds are depleted and/or the recipients have passed away. In one of the interviews, such a case was encountered, where the recipient had died from the disease, and the money was managed by her family.

Currently, CAC is considering requesting the assistance of a voluntary organisation of individuals involved in a programme called *Warga Peduli AIDS* (WPA) ('People Care for HIV'). The work of this organisation may provide a significant improvement to the integration of the recipients into society in the future. In fact, WPA is a voluntary organisation of people operating in the civil society, i.e. between the government and profit-oriented organisations and companies, with the aim to assist HIV-infected people through cooperation with CAC at any level, i.e. national, provincial, and city/district.

CAC has the confidence that, by building the capacity of WPA, the HIV-infected group of the population will have a better chance of survival from the disease through improved access to health care and integration in the society with a reduced stigma of the disease in the future. The latter point is very important since the stigma on PLHIV and HIV is still rather strong in Indonesia. In this respect, WPA plays an important role in the integration of PLHIV in their society in the future. Furthermore, CAC is considering making the programme independent and operating in its own way. Figure 17.1 shows a representation of the CAC scheme.

b. *Fund recipients*

Currently, all recipients of CAC funds are housewives who mostly got infected from their HIV-positive husbands, who may have already passed away. The decision to apply for funding has been taken because they need a resource to support their families through such crises. The funds have enabled them to set up

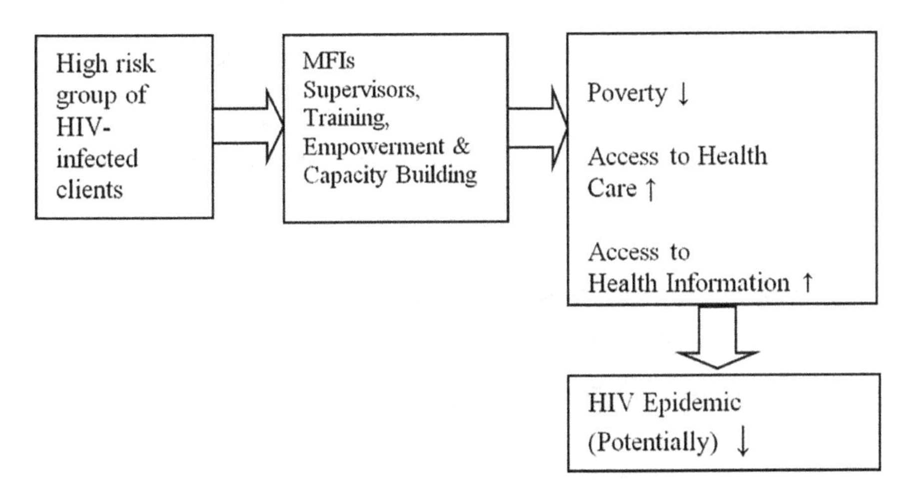

Fig. 17.1 Schematic representation of the scheme of the City AIDS Commission (CAC) in Bandung, Indonesia. *Source* Siregar (2014)

small-scale businesses ranging from food stalls and kiosks to cellphone credit stalls. The income which is generated from these businesses has enabled them to gain certain benefits for maintaining their business, sending their children to school, paying the fees for primary health care services, covering transportation costs for consultation and obtaining antiretroviral (ARV) and other necessary medicines and referral to the HIV clinic.

A reported obstacle in the scheme is the monthly financial reporting required by CAC in the form of a clear and legal transaction record. Some of the problems include the submission of legal receipts for every purchase, usually because PLHIV obtain their goods from stores which do not provide such receipts. The recipients also require more training on the special strategies to do business in order to survive.

c. *Supervisor*

The tasks of the supervisors include closely monitoring the recipients, keeping records and disseminating the funds provided by CAC. Every decision made by the recipients to use the money should be in consultation with the supervisor, who determines whether the use is feasible or not. There are two potential problems with the current supervisors. Firstly, they have not received any formal training for supervising small-scale enterprises. This lack of appropriate training may become a problem for the recipients in their efforts to maintain and develop—and potentially expand—their business. Secondly, the supervisors are the representatives of the NGOs, and are not from the same community as the recipients. As such, the empowerment may be limited exclusively to the recipients. In this context, appropriate training of the supervisors is important, and the extension of the CAC scheme to include WPA members as supervisors may have prolonged more sustainable effects on the entire CAC scheme, particularly regarding ownership, local empowerment and the reduction of the stigma.

Although the initiative has brought much benefit to the recipients, there is still room for improvement. For example, the initiative has tried to empower the community, in this case the PLHIV/fund recipients, but it has failed to involve the community in which the PLHIV are actually living to support these recipients. This point is further discussed below, where the recently introduced approach of *Integrated Microfinance Development* (IMM) could be instrumental in a potentially forward way (cf. Slikkerveer 2012).

17.4 Women as the Backbone of Community Empowerment

17.4.1 Women and the HIV Epidemic in Indonesia

From unpublished data of the *Community-based Prevention and Care for Most-at-Risk Females* (COMPAC-Female) Programme, 4 out of 13 cities/districts which have been surveyed in West Java have shown that more females were infected with

HIV compared to males. Indeed, according to the Ministry of Health of Indonesia (2008), housewives and sexworkers may become two of the key population sub-groups—apart from males having sex with males (MSM)—which will dominate the PLHIV group. Never before have women become more threatened by HIV infection in the Indonesian epidemic than today.

Facing this challenge, women should participate more actively in the HIV prevention programmes. In spite of individual endeavours to prevent infection, their activities should also be supported institutionally. Women should increase their activities to prevent the spread of HIV, starting with their own families. They should increase their knowledge of the danger of HIV transmission, and should be followed by increased community care through activities, such as providing relevant information routinely and sustainably in order to constrain the transmission of HIV.

Women also need support for the protection of their reproductive rights within their families and communities. Strengthening their position and role implies bringing to an end the socio-cultural values which have contributed to the sexual harassment of women, and protecting women's rights concerning their basic needs, such as access to information, health services and improved education. Also, women need support from community-based services to reduce the burden of women and girls who are taking care of members of the family who are infected by HIV.

The *Beijing Platform for Action* (BPFA) underscores that HIV infection consequently influences women's health in their roles as mothers who care for their families as well as providers of economic contributions to the household (cf. Beijing Platform for Action 1995).

Women and girls tend to accept the responsibilities of the burden which are caused by the HIV epidemic. All over the world, women are executing the household tasks and providing care for the sick members of the household, but at the same time they have to generate income to keep their families alive when their husbands fall ill or die because of HIV. The disease itself also increases the burden of the families, often leading to poverty and creating more suffering for women because of the loss of work and the incapability to pay for health services. In these families where a member of the household is sick and needs money for treatment, a daughter may have to stop going to school. In such cases, HIV is also a threat to the basic education of children and has an impact on young girls in primary school. Moreover, women in their reproductive age can also transmit HIV to their babies, posing a serious problem, since these women are more vulnerable to HIV infection through sexual intercourse in comparison with men. The vagina is biologically fragile and can easily get scratched during sexual activities, allowing the HIV virus to penetrate into the blood circulations and cause infection.

As regards the current HIV epidemic in Indonesia, it is clear that women will suffer from the largest burden. They currently represent the more vulnerable gender, while at the same time they are left with a larger responsibility once their spouses are no longer able to provide support for their families. The next section will discuss the position of women within the community in Indonesia, providing additional perspectives on the actual position of women in the current HIV epidemic in Indonesia.

17.4.2 The Position of Women

In Indonesia, the status of women within the household and the community is not equal to men, substantiating *gender inequality* in the society. The status of men is that they are regarded as the leaders in the households, which means that they are also leaders in the society, and as such more highly regarded than women. Women possess hardly any autonomy to control the sexual behaviour of their husbands or permanent partners, and no autonomy over their own sexual life. Furthermore, they have little knowledge and less access to acquire health information and services for the treatment of HIV, rendering them rather susceptible to HIV infection.

The local culture also influences the status of women such that they are expected to serve every single need of their partners or spouses. It is considered a sin if they do not perform their roles to serve their husbands, while as housewives, they are obliged to execute the domestic tasks. Moreover, for female *Pekerja Seks Komersial* (PSK) ('sexworkers'), they scarcely have any bargaining positions with their male clients to have save sex by using condoms.

The *Beijing Platform for Action* (1995) states that the HIV epidemic will influence women's health in performing their role as mother and caretaker of the household members, as well as contributor to the economy of the family.

It is important to view the social consequences of the HIV epidemic from the gender perspective, since men and women occupy their different roles and status in the society.

Biologically, women are more likely to become infected with HIV through unprotected heterosexual intercourse than men. In many countries, women are less likely to be able to negotiate the use of condoms and more likely to be subjected to non-consensual sex, but in Indonesia, the bargaining positions of women for healthy reproductive activities are almost impossible (cf. United Nations Programme on HIV/AIDS [UNAIDS] 2014). The conditions which women face include the following: (1) biological factors—the reproductive tools of women are more open compared to those of men. In such cases, women can easily be contaminated by Sexually Transmitted Diseases (STDs); (2) economic factors—poverty will cause women to work in jobs with a high risk of getting infected by STDs, *e.g.* prostitutes. The economic dependency of women to men causes women to have less control over themselves in terms of avoiding infection by STDs; and (3) socio-cultural factors—the unequal socio-cultural position of women and men tends to result in a situation where women are always getting blamed as the cause of the infection. In fact, many women have been infected by HIV from their sexual partners. Moreover, women are expected to adhere to their cultural status in the household and the society, while an open discussion about sex is taboo.

The *Beijing Platform for Action* (1995) stresses that equal roles and responsibilities between men and women are very important to attain welfare for themselves, their families, their society and their country. The *Beijing Platform for Actions* (1995) also shows that equal roles and responsibilities are critical to the impact of the HIV epidemic on women, especially among poor women, women and education, women and health, women and the economy and women in decision-making.

17.5 Integrated Microfinance Management: The Way Forward?

A short review of the main arguments in the discussion so far seems to pave the way for an assessment of the newly-developed approach of *Integrated Microfinance Management* (IMM).

- microloans have the potential to improve the quality of life of PLHIV, but they need to go beyond the conventional practice;
- CAC has provided a basic scheme which can be built on, while the programme prefers to operate independently;
- supervisors monitor the recipients well, but they are not trained appropriately;
- the supervisors do not come from the same community as the recipients, reducing the effect of community empowerment and sustainability;
- the WPA should be utilised to become a part of the scheme, e.g. as supervisors, and the current empowerment effect has the potential to be enhanced and sustained by involving the WPA; and
- women, as the caretakers of the family, play a key role in community empowerment.

The above-mentioned summary of the arguments clearly shows a strong need for an integrated approach to microfinance, known as *Integrated Microfinance Management* (IMM). This new approach focuses primarily on the major themes of poverty reduction and women's empowerment through the revaluation, participation and interaction of local people and their traditional institutions and organisations in diverse areas encompassing traditional herbal medicine, microcredit, bio-cultural diversity conservation, voluntary credit organisations and integrated maternal and child health (cf. Slikkerveer 1999; Agung 2005; Gheneti 2007; Leurs 2009; Djen Amar 2010; Ambaretnani 2012). The IMM approach, therefore, aims at poverty reduction and women's empowerment through the introduction of a combination of financial and non-financial services through the intermediary of traditional institutions in Indonesia, an exact approach which is now urgently required to address the current problems of HIV and poverty throughout the country (cf. Slikkerveer 2012).

By consequence, the following integrated model is proposed which seeks to combine the previous CAC scheme with the IMM approach which is represented in Fig. 17.2. The idea is to alter the supervising process, which is currently handled by the representatives from the NGOs, and consign it to the WPA, whose members come from the same community as the fund recipients. Such alterations would not only empower the PLHIV, but also the community as a whole in terms of reducing the HIV stigma and accepting the existence of PLHIV within their community. The stigma of HIV in Indonesia is one of the main obstacles in the eradication of the disease, as it causes people to shy away from getting tested to see if they are HIV-positive and being open about their HIV status. Self-stigma also causes PLHIV to believe that the public will never accept them as a part of the community, if their HIV status or their risk-taking behaviour is publicly known. These obstacles create difficulties for health workers to identify PLHIV or people running the risk of getting infected because they are, in essence, 'hiding'. The involvement of the local community through the WPA programme still has a long way to go before the society is prepared to accept PLHIV and HIV interventions, which eventually may reach the goals of eradicating the stigma.

It is also proposed that the organisation of such a scheme should come from the community itself, e.g. a *koperasi* ('cooperative association'), as it will also empower the community—including the PLHIV—to independently manage their resources. Lastly, it is proposed that women should become the fund recipients given their vital position in the current epidemic, and their proven ability to manage the microfinance programme. It goes without saying that this scheme can later be extended to the other high-risk groups, depending on the progress of the programme. The proposed model of integrated microfinance management of the HIV epidemic is expected to have potential in assisting the PLHIV to survive the disease financially, and beyond. Although it is not the goal of the scheme, the reduction of

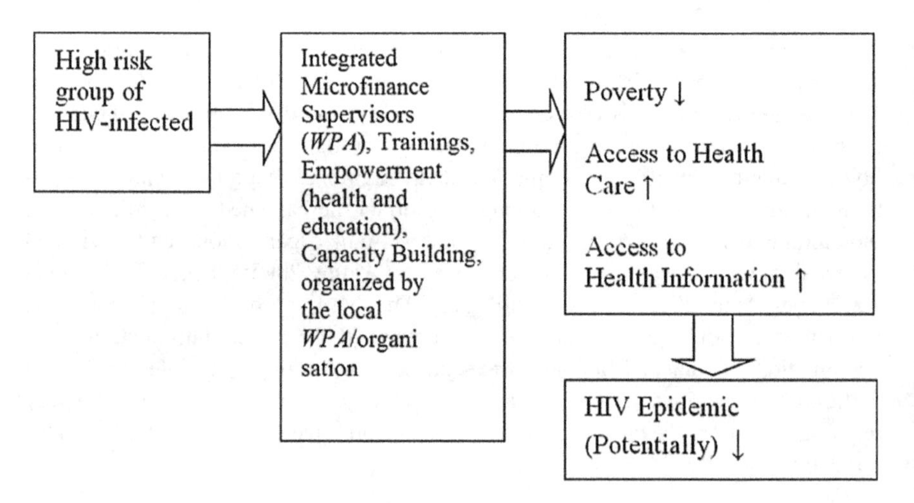

Fig. 17.2 Proposed integrated microfinance scheme. *Source* Siregar (2014)

the HIV epidemic can also be one of the positive externalities of the programme. By replicating such a scheme in other settings, a greater number of communities could be exposed to the HIV-related information, which, in turn, could further reduce the stigma and motivate people to take a HIV test and consider their illness or risky behaviour.

17.5.1 Integrated Microfinance for Women with HIV and Living with PLHIV

The definition of poverty for women in development places women as *the poorest of the poor* (cf. Moser and Mcllwaine 1997; Remenyi et al. 2007). In addition, families with women being the head of the household are the group of the poorest among the poor (cf. Rose 1992). It is assumed that increased access to resources will release women from poverty, and, as such, create the opportunity to fulfil the women's needs. As Remenyi et al. (2007) indicate, it is no wonder that women have become the target group of microcredit. Indeed, experience shows that the repayment rate of women in microcredit programmes with women as target groups is above 90%. For example, the repayment rate of women in Indonesia is 91% as compared to men, which is 80% (cf. United Nations 1998). When the management of microfinance programmes is efficient and effective, it will also become the basic milestone for women to face their unequal status in comparison to men within the society. The Integrated Microfinance Management approach is expected to result in new, sustainable pathways to women's empowerment and a better life for their families, in particular for women living within the context of HIV infection.

The role of women in the households and the communities as managers provides them with a favourable position to be empowered with the ability to increase their income through the Integrated Microfinance Management approach. When the income of the household is increasing, it will also strengthen the bargaining position of the women towards their spouses or partners, in this case, women with HIV or women living with PLHIV in low-income households. The provision of integrated microfinance services to these women will not only address their socially and economically disadvantaged conditions through increased opportunities for self-employment, but also improve their accessibility to goods, education, health care, communication and socio-cultural services. While these low-income and poor households usually have limited access to financial services, the demand for safe, reliable and convenient deposit services is overwhelming, as shown by poor people's saving capacity, desire and willingness to save for emergencies, social obligations, family and children's health, children's education and many other purposes (cf. Asian Development Bank 2000).

The *Beijing World Women Conference Declaration* of 1995 recommends that achieving gender equality and equity requires access to financial services as a means of empowering the impoverished and marginalised women (cf. Beijing

Platform for Action 1995). However, according to the *United Nations Population Fund* (UNPFA 2011), even after more than 20 years, women still constitute the majority of the poor and lack access to many of the resources utilised by men. This problem seems to be more severe among women with HIV or living with PLHIV. As regards the contrast between urban and rural areas, it seems that women in rural areas are facing socio-cultural, political, legal and economic barriers, preventing them from appropriate access to education, finance and health care. As the HIV epidemic also reaches the rural areas, it may have a more severe impact on women in these areas.

The society consists of men and women, and globally, the cultures concerned tend to assign to them different functions, where men are the head of the household as 'bread winner', while women act as the managers of the household. The different functions bring women closer to their families, while men have to work outside the house. The two positions within a society tend to result in different functions and different needs. Even illiterate women have shown to possess adequate informal financial management skills in comparison to men, largely because of their position in the society as managers of their households. Moreover, resources managed by women increase the quality of investment activities, financed by microfinance, as well as the level of expenditure of the income generated at the individual and family level.

As the *United Nations Population Fund* (UNPFA 2011) documents, women are the gateway to household security, as they generally invest in the welfare of the family. In this respect, the health and education of women—especially those who are affected by HIV infection—and the children and their families can be improved through the integration of activities of income generation, health, education and other promotional activities for women through the IMM approach. Through the implementation of the concept of IMM, the process of designing and planning integrated programmes will be more effective in targeting and achieving sustainable income for women, family health and education, especially for women living with HIV infection and/or in rural areas. In this way, the empowerment of women, in combination with the promotion of improved access to health, education and socio-cultural services in terms of training focused on preventive health interventions and enhanced opportunities for adequate education among girls, will eventually yield a greater impact.

As Kabeer (1999) mentions, empowering women is referred to as the power to determine choices and abilities to choose. One impact of microcredit on women is that their daughters could have more opportunities for schooling, compared to families in which the mothers are not receiving microcredit. Productive income for the family where women receive microcredit will increase the possibility for the family to have improved access to other social institutions including health services. According to the *United Nations Population Fund* (UNFPA 2011), the empowerment of women includes four main components:

1. the right to have the power and control over one's own life outside and inside household settings;
2. the right to access family and community opportunities and resources;
3. the right to determine and pursue individual choices; and
4. a sense of self-worthiness.

The four components are exactly in line with the IMM approach to achieve its objectives, including the empowerment of women through the functionalisation of a variety of services through traditional institutions in Indonesia. In rural areas, where generally women are in a disadvantaged position in all aspects of life in comparison with men, investment in the economic empowerment of rural women may further improve this condition in the form of supporting women in implementing community-based business ideas, entrepreneurships and the development of basic financial capacities and skills. The combination of health and education, the promotion of maternal, child and family health and nutrition, the training focusing on preventive health interventions, and the promotion of girls' education is expected to yield a greater impact on health and education than programmes which only provide microfinance. An appropriate example is provided by the *Kalibagor* health center (KIA) in the Banyumas District.[1]

Thus, empowering women with HIV or living with PLHIV should encompass the access and use of various services, including health insurance, training and education, business development and consulting services. It is clear that the improvement of women's access to financial resources and other services such as education and health care will consequently contribute to the economic empowerment of their families, as it will not only increase their income, but also improve their well-being and socio-political position within the family and the society.

17.6 Concluding Remarks

As documented by the above-mentioned case study and the related arguments, microfinance only has the potential to improve the life of PLHIV if the form of microfinance goes beyond the common practice of access to financial services and encompasses, among others, improved access to health, education and social services. Here, the new IMM approach can provide a substantial contribution. Obviously, integrating microfinance with health and educational services will

[1]The Kalibagor Health Center (KIA) is one of the region's health facilities in the district of Banyumas. According to the Kalibagor Puskesmas book, health care coverage until October 2009 was 65%. Although the assessment of the results of the Maternal and Child Health (MCH) of the Kalibagor Puskesmas indicates that they are below the target coverage of the Banyumas Health Office, the coverage increased by September 2010 to 77.80%, while coverage of the first visit of pregnant women (K1) was 79.09%, allowing the coverage of MCH to be categorised as 'good', and this coverage was similar to the K1 coverage (cf. http://journal.managementinhealth.com/index.php/rms=/article/viewFile/364/1026).

generate a better quality of life for PLHIV, at least in terms of increasing their access to health services in terms of both the financial and medical respects. The integration of microfinance with education about the nature of the disease, its prevention, its treatment, and the surrounding stigma, combined with its underlying systems of people's knowledge and practices, will increase the quality of life not only of the PLHIV, but also of their family members and the community in which the WPA can play an important role at the community level. In the Indonesian HIV epidemic, women now have the opportunity to play one of the most important roles.

Given their positive records in their performance in microfinance, it is only logical that the government and related organisations and institutions should prioritise the input of this special group of women within the context of HIV to operationalise the CAC scheme in which the IMM approach can further improve the already running CAC scheme to become more sustainable and to have a greater positive impact on the position of the PLHIV, their families and their communities.

References

Agung, A. A. G. (2005). *Bali endangered paradise? Tri Hita Karana and the conservation of the Island's biocultural diversity*. Ph.D. Dissertation, Leiden Ethnosystems and Development Programme. LEAD Studies No. 1. Leiden University. xxv + ill., pp. 463.

Akers, A. Y., et al. (2010).Views of Young, Rural African Americans of the Role of Community Social Institutions' in HIV prevention. *Journal of Health Care for the Poor and Underserved*, 2010 May; 21(2 Suppl), 1–12.

Ambaretnani, P. (2012). *Paraji and Bidan in Rancaekek: Integrated medicine for advanced partnerships among traditional birth attendants and community midwives in the Sunda Area of West Java, Indonesia*. Ph.D. Dissertation. Leiden Ethnosystems and Development Programme. LEAD Studies No. 7. Leiden University. xx + ill., pp. 265.

Asian Development Bank (ADB). (2000). *Annual report*. Manila: Asian Development Bank.

Bateman, M. (2010). *Why doesn't microfinance work? The destructive rise of local neoliberalism*. London/New York: Zed Books.

Beijing Platform for Action. (1995). *Fourth world conference on women*. New York: United Nations Entity for Gender Equality and the Empowerment of Women.

Djen Amar, S. C. (2010). *Gunem Catur in the Sunda Region of West Java: Indigenous Communication on the MAC plant knowledge and practice within the Arisan in Lembang*. Ph.D. Dissertation. Leiden Ethnosystems and Development Programme. LEAD Studies No. 6. Leiden University. xx + ill., pp. 218.

Donahue, J. (2000). *Microfinance and HIV/AIDS : It's time to talk.* Washington, DC: United States Agency for International Development (USAID).

Dworkin, S. L., & Blankenship, K. (2009). Microfinance and HIV/AIDS prevention: Assessing its promise and limitations. *AIDS Behavior, 13*, 462–469. URL: http://doi.org/10.1007/s10461-009-9532-3.

Gheneti, Y. (2007). *Microcredit management in Ghana: Development of co-operative credit unions among the Dagaaba*. Ph.D. Dissertation. Leiden Ethnosystems and Development Programme. LEAD Studies No. 2. Leiden University. xvii + ill., pp. 330.

Holmes, K., Winskell, K., Hennink, M., & Chidiac, S. (2010). Microfinance and HIV mitigation among people living with HIV in the era of anti-retroviral therapy: Emerging lessons from Cote d'Ivoire. *Global Public Health, 9,* 1–15.

Kabeer, N. (1999). *Institutions, relations and outcomes*. Delhi/London: Kali for Women Publisher/ ZBooks.
Leurs, L. N. (2009). *Medicinal, aromatic and cosmetic (MAC) plants for community health and bio-cultural diversity conservation in Bali, Indonesia*. Ph.D. Dissertation, Leiden Ethnosystems and Development Programme. LEAD Studies No. 5. Leiden University. xx + ills., pp. 343.
Ministry of Health of Indonesia. (2008). *Mathematic model of HIV epidemic in Indonesia*. URL: http://spiritia.or.id/Doc/model0814.pdf.
Ministry of Health of Indonesia. (2009). *Case statistics of HIV/AIDS in Indonesia report*. Jakarta: Directorate General of Communicable Disease Control and Environment Health.
Moser, C. & Mcllwaine, C. (1997). *Household responses to poverty and vulnerability, volume 2: Confronting crisis in angyalfold,* Budapest, Hungary. Urban Management Program Policy Paper No. 22. Washington, DC: World Bank.
National AIDS Commission (NAC). (2012). *Republic of Indonesia Country report on the follow-up to the declaration of commitment on HIV/AIDS* (UNGASS), Reporting period 2010–2011. Jakarta: National AIDS Commission (NAC).
Parker, J. (2000). *The role of microfinance in the fight against HIV/AIDS*, UNAIDS Background Paper. The Joint United Nations Program on HIV/AIDS (UNAIDS).
Pronyk, P. M., Kim, J. C., Hargreaves, J. R., Makhubele, M. B., Morison, L. A., Watts, C., et al. (2005). Microfinance and HIV prevention—Emerging lessons from rural South Africa. *Small Enterprise Development, 16*(3), 26–38.
Reddy, A., Htin, K. C. W., & Ye Yu Shwe, Y. Y. (2012). HIV and AIDS data Hub for Asia Pacific: A regional tool to support strategic information needs. *Western Pacific Surveillance and Response Journal, 3*(3), 18–21.
Remenyi, D., Money, A. & Bannister, F. (2007). *The effective measurement and management of IT costs and benefits*. Wokingham, UK: CIMA Publishing.
Rose, G. (1992). *The strategy of preventive medicine*. Oxford: Oxford University Press.
Siregar, A. Y. M. (2014). *Economic evaluation of HIV/AIDS control in Indonesia*. Ph.D. Dissertation. Radbound University of Nijmegen, The Netherlands.
Slikkerveer, L. J. (1999). Ethnoscience, 'TEK' and its application to conservation. In D. A. Posey (Ed.) *Cultural and spiritual values of biodiversity: A complementary contribution to the global biodiversity assessment*. London/Nairobi: Intermediate Technology Press/United Nations Environment Programme (UNEP).
Slikkerveer, L. J. (Ed.). (2012). *Handbook for lecturers and tutors of the new master course on integrated microfinance management for poverty reduction and sustainable development in Indonesia (IMM)*. Leiden/Bandung: LEAD/UL/UNPAD/MAICH/ GEMA PKM.
Stratford, D., Mizuno, Y., Williams, K., Courtenay-Quirk, C., & O'leary, A. (2008). Addressing poverty as risk for disease: Recommendations from CDC's consultation on microenterprise as HIV prevention. *Public Health Reports, 123*, (Febr. 2008), 9–20.
United Nations Programme on HIV/AIDS (UNAIDS). (2014). *Global aids response progress reporting 2014: Construction of core indicators for monitoring the 2011 United Nations Political declaration on HIV and AIDS*. Geneva: UNAIDS.
United Nations (UN). (1998). *The role of microcredit in the eradication of poverty: Report of the secretary-general*. New York: United Nations.
United Nations Population Fund (UNPFA). (2011). *UNFPA annual report: Delivering results in world of 7 Billion*. New York: UNFPA.

Prihatini Ambaretnani is a Senior Lecturer in Anthropology at Universitas Padjadjaran in Indonesia. She received her Ph.D. Degree in Ethnoscience from Leiden University in The Netherlands on her study of the collaboration between traditional and modern birth attendants in the Sunda Region of West Java in 2005. She is an expert in Women & Empowerment and is also Senior Lecturer in the Course on *Integrated Microfinance Management (IMM)*, Bandung.

Adiatma Y. M. Siregar is Lecturer and Programme Secretary of the Undergraduate Study Programme at the Faculty of Economics and Business of Universitas Padjadjaran, Bandung, Indonesia. He received his Ph.D. Degree from the Radboud University in Nijmegen, The Netherlands and specialised in the field of economics and health care development. He is also Senior Lecturer in the Course on *Integrated Microfinance Management* (IMM), Bandung.

Part V
The New Paradigm of Integrated Community-Managed Development

Chapter 18
Strategic Management Development in Indonesia

L. Jan Slikkerveer

> *The establishment of links between business and higher education will increase the effectiveness of both and allow managers and academics to teach and learn from one another.*
> Kolb et al. (1986)

18.1 Management Development Theory and Practice

During the design of the first Master Course on *Integrated Microfinance Management* (IMM) in Indonesia, developed by the LEAD Programme of Leiden University and its Partners in the *International Consortium of Integrated Microfinance Management* in 2009, it soon became clear that knowledge and practice of management development was essential for the appropriate education and preparation of the new cadres of Integrated Microfinance Managers (IMMs), where the training institute would bring together a complex of models and methods focused on the selection, education, training and development of future managers operating in indigenous institutions at the community level (cf. Slikkerveer 2012).

As such, the mandate for the development of the new Master Course Programme encompassed a dual purpose of, on the one hand, the managerial adaptation of the learning institution, i.e. the Faculty of Economics and Business of Universities Padjadjaran (FEB/UNPAD) in Bandung, Indonesia to create a motivating institution for the tutors and staff in order to deal with the changing situation in the socio-economic and ecological environment of the country in the near future; and on the other hand, the innovative management-oriented education and training of the new cadres of community-based managers for their future work as managers of the indigenous institutions at the community level in Indonesia. Since the new course would embark on the understanding of the principles underlying the process

L. J. Slikkerveer (✉)
LEAD, Leiden University, Leiden, The Netherlands
e-mail: l.j.slikkerveer@gmail.com

L. J. Slikkerveer et al. (eds.), *Integrated Community-Managed Development*, Cooperative Management, https://doi.org/10.1007/978-3-030-05423-6_18

of strategic management, both management theory and the development of management thinking would form core themes in the curriculum.

At the institutional level, the theory of management basically supports the clarification of the interrelationships between the development of the theory, the behaviour in institutions and the practice of management in the field. History documents that the beginning of the systematic development of management thinking at the end of the nineteenth century coincided with the development of industrial organisations, after which a number of authors launched different approaches towards the study and analysis of the structure and management of organisations. In his categorisation of the various approaches, Skipton (1983) distinguishes eleven schools of management theory, while the current analysis departs on a framework of four major approaches: *classical*, *human relations*, *systems* and *contingencies*, which can be extended with several sub-categories.

The classical approach to concepts such as organisations, structures and management focused largely on the technical requirements of the organisation, the principles of management and the rational behaviour of managers in their duties and responsibilities. Although the classical approach has been criticised over the fact that the set of principles of management would not be effective in practice, it delivered two interesting sub-categories of, respectively, 'scientific management' and 'bureaucracy'. The early principles of scientific management were formulated during the mid-twentieth century by Taylor (1947) in order to guide management in a scientific way with the purpose of developing methods for co-ordination and control of individuals in an efficient and productive organisation. Later on, however, this approach received much criticism, as it takes a rather instrumental view of human behaviour in organisations and applies a rigid control of the organisation by removing decision-making from the employees to the managers.

Although recently, more knowledge has accumulated about the socio-cultural influences within organisations and the values of motivation, satisfaction and performance, the essence of Taylor's scientific approach to management is still relevant and has further inspired the development of management thinking. While there is general agreement that most of Taylor's theories are now largely outdated, Stern (2001: 87) argues that: "*The 'scientific management' of Frederick Taylor...with its twin goals of productivity and efficiency - still influences management thinking 100 years on.*"

Bureaucracy, as the other sub-category, had been elaborated by Weber (1964) who introduced the concept of 'bureaucratic structures' to characterise an 'ideal type' of organisation, based on standard rules of expertise and administration. According to Stewart (1999), the main features of bureaucracy can be summarised to include specialisation, hierarchy of authority, system of rules and impersonality. While bureaucracies generally tend to be built on formal, well-defined and hierarchical structures, Cloke and Goldsmith (2002) show that following current changes of empowerment, egalitarian structures and the changing environment, the demand for more flexible management principles for organisational democracy is on the rise.

Especially among organisations in the public sector, the standardisation of treatment, regularity of procedures and detailed documentation are examples of the present-day principles of most bureaucracies. Although bureaucracies have become less attractive models for organisations, Green (1997: 18) argues that: "*there is still a place for bureaucracy in parts of most organisations and especially public sector organisations such as local authorities and universities*".

Ridderstrale (2001) notices that in contrast to the hierarchical structures of large organisations of the past, successful organisations now react to today's increasingly complex knowledge systems with novel solutions which no longer follow the principles of the bureaucratic model. The author refers to a more decentralised organisation, a less permanent position of people, and the conversion to learning organisations in order to benefit from advanced knowledge and a wider use of knowledge and expertise for better coordination.

The second approach towards the study and analysis of the structure and management of organisations is known as 'human relations' which evolved during the Great Depression of the 1920s. Then, several authors including Mayo (1933) started to pay more attention to human relations within organisations, such as the socio-cultural factors and related patterns of human behaviour. The approach later led to the development of the 'neo-human relations' approach, pioneered by Maslow (1943), Herzberg et al. (1959) and McGregor (1987), who focused on individual personality development and motivation, based on individual psychological needs. The approach has generated much research on communication, participation and motivation, as factors of job content and satisfaction. Meanwhile, the systems approach also received increased attention for the study of complex organisations and their inner structures as whole entities, based on the general systems theory, outlined by von Bertalanffy (1951).

Later on, Miller and Rice (1967) adapted the systems approach to compare organisations with biological organisms. As an extension, the 'contingency approach' perceives the optimal system of management to be dependent on the contingencies of the situation, emphasising the importance of situational influences on the management of organisations (cf. Pugh and Hickson 1976).

Recently, some authors criticised the various approaches and models of the management theory and practice for their failure to involve the important socio-historical factors and cultural backgrounds of the individuals in the organisations. Similarly, Schneider and Barsoux (2002) draw attention to the fact that different approaches to management tend to reflect different cultural contexts of the organisations concerned. The lack of attention to the role of culture in the development of theories of management of institutions and organisations is further illustrated by the work of Cheng et al. (2001) who argue that while there may be some universality to organisation structures, different cultures provide these structures with different meanings.

While it is clear that the study of management development has evolved towards a more scientific approach, where science and technology have become major instruments, still many queries on the socio-cultural context of different organisations and institutions—particularly of indigenous institutions—remain, which can only be answered by a useful combination of science and philosophy.

18.2 The Expediency of Strategic Management Development

A major consideration of the new IMM Master Course has been that the new managers would need access to a complex of specialised knowledge and skills which would enable them to cope with the organisational problems of management of indigenous institutions for community services. In view of the practical setting, their own experience and views acquired during their vocational or fieldwork training should also be integrated into the attainment of valuable training towards formulating management solutions.

Indeed, while the theory of management development has further been elaborated over the past decades to pertain to the adoption of a scientific approach, the need for a strategic perspective to cope with future social change and environmental uncertainty has also become increasingly manifest for institutions and organisations concerned with education and training in management. Every organisation or institution is faced with both external opportunities and threats, and internal strengths and weaknesses, which have to be analysed in order to formulate alternative strategies. However, the conventional matrix management model used for such analysis seems no longer adequate for current complex configurations (Fig. 18.1).

Initially, the matrix management model tended to focus on the management of individuals in a matrix organisation structure in which individuals with similar skills are pooled for specific assignments, as in the field of cross-functional project management. The model has spread through both the public and private sector as a visual representation of multiple authority where people work both in project and non-project settings. A fine example of a matrix management model is the *Strengths-Weaknesses-Opportunities-Threats Matrix* (SWOT) which helps managers to develop four types of strategies: 'Strengths-Opportunities Strategies' (SO), 'Weaknesses-Opportunities Strategies' (WO), 'Strengths-Threats Strategies' (ST) and 'Weaknesses-Threats Strategies' (WT)[1] (cf. Humphrey 2005; Walker 2017).

[1]The *Strengths-Weaknesses-Opportunities-Threats* (SWOT) Matrix is a framework for managers to develop four types of strategies: 'strengths-opportunities strategies' (SO), 'weaknesses-opportunities strategies' (WO), 'strengths-threats strategies' (ST) and 'weaknesses-threats strategies' (WT), which was introduced by Albert Humphrey from the Stanford Research Institute (SRI). The SWOT analysis has been adopted by organisations of all types as an aid to making decisions for identifying and analysing the internal and external factors which can have an impact on the viability of a project, a product, a place or a person (cf. Humphrey 2005; Walker 2017).

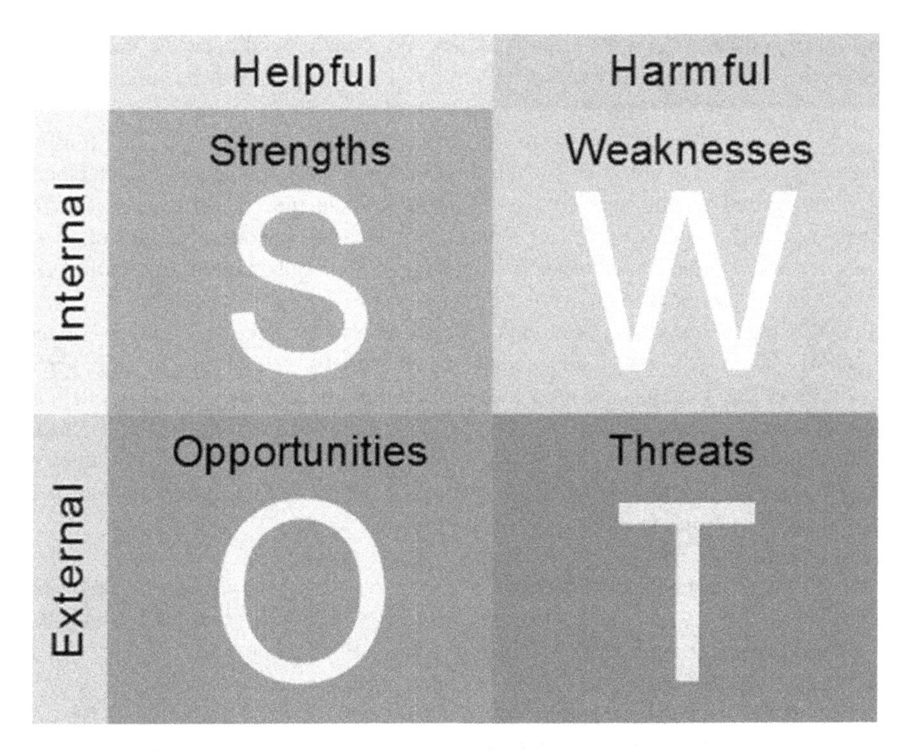

Fig. 18.1 Representation of the SWOT matrix of strengths, weaknesses, opportunities and threaths. *Source* Walker (2017)

Although the advantage of the application of the matrix organisation model becomes clear in its function to develop individuals with broader capabilities to cope with more complex and connected environments, the main disadvantages refer to its static assessment of the environment, where in reality the dynamic situation changes over time. Moreover, the SWOT analysis tends to focus on one single factor, without revealing the significant interrelationships which develop among various factors of the process.

In view of the advances of science and technology which started to accelerate with the development of complex organisations in the course of the 1980s, the growing need for both *differentiation*—specialisation—and *integration*—internal coordination—posed a challenge to the development of new management strategies.[2] In particular, the solution to problems of interdepartmental collaboration

[2]*Horizontal integration* refers to a strategy where an organisation creates or acquires production units for outputs which are alike—either complementary or competitive. One example would be when a company acquires competitors in the same industry doing the same stage of production for the creation of a monopoly. An example of horzontal integration would be McDonalds buying out Burger King. In microeconomics and management, *vertical integration* is an arrangement in which the entire supply chain of an organisation is owned by that company (http://www.businessdictionary.com/definition/horizontal-integration.html).

within organisations and institutions, where too many crucial decisions have to be taken through the regular lines of hierarchy at the top, needed the input of integrative roles. Here, the concept of integration refers to the achievement of a unity of efforts among the specialists in an organisation, with the aim to resolve inter-departmental conflicts and reach joint decisions. Among several management studies focused on the problems of coordination and integration is the work by Lawrence and Lorsch (1984) for managers on how to achieve a unified effort in complex R&D organisations, where they suggest the establishment of a 'full-scale integrating department' to solve the integration problem (cf. Fig. 18.2).

Following the recent acceleration of the advances in science and technology, the challenge of the adequate management of increasingly complex organisations and institutions has gradually shifted away from the matrix management model, as formal structures have become less functional in realising the objectives in a matrix; the attention has shifted towards the 'soft structure' of networks, communities, teams and groups which need to be set up and maintained in an innovative managerial way in order to achieve realistic goals.

An additional need in the management of change and uncertainty further requires more specialised knowledge to be involved in the solution of organisational problems, where specialised perspectives should be integrated into the learning process, particularly in education institutions in order to reach effective solutions. As Kolb et al. (1986) contend in their classic article 'Strategic Management Development' published in the *Journal of Management Development*, such an integrative learning process would require a re-examination of the role of the teachers and trainers to the extent that they should be able to manage the process of learning in a way which would not only facilitate the students to learn from their own experience, but would also integrate such knowledge of experience into their special managerial skills. According to the authors, a key function of strategic management development at the integrative level is: "*...to provide managers with access to knowledge and relationship networks which can help them become life-long learners and cope with the issues on their continually changing agendas.*"

Thus, in order to assess and develop managerial competencies in modern institutions, experimental learning theory is used where 'integrative knowledge'—also referred to as 'wisdom'—is introduced into areas specifically concerned with education and training for management development for the preparation of future managers for the complex roles which they will have to perform.

In order to cope with the complicated problems of organisations to survive in a complex and uncertain environment, managers should learn not only how to manage such environmental complexity, but also about the complex institutional forms specifically developed to cope with the environment. As the classical organisational structures and matrix management models designed to deal with environmental complexity and change are no longer effective, a strategic approach to management development has been introduced. As Kolb et al. (1986) notice, this strategic perspective anticipates the emerging requirements of management

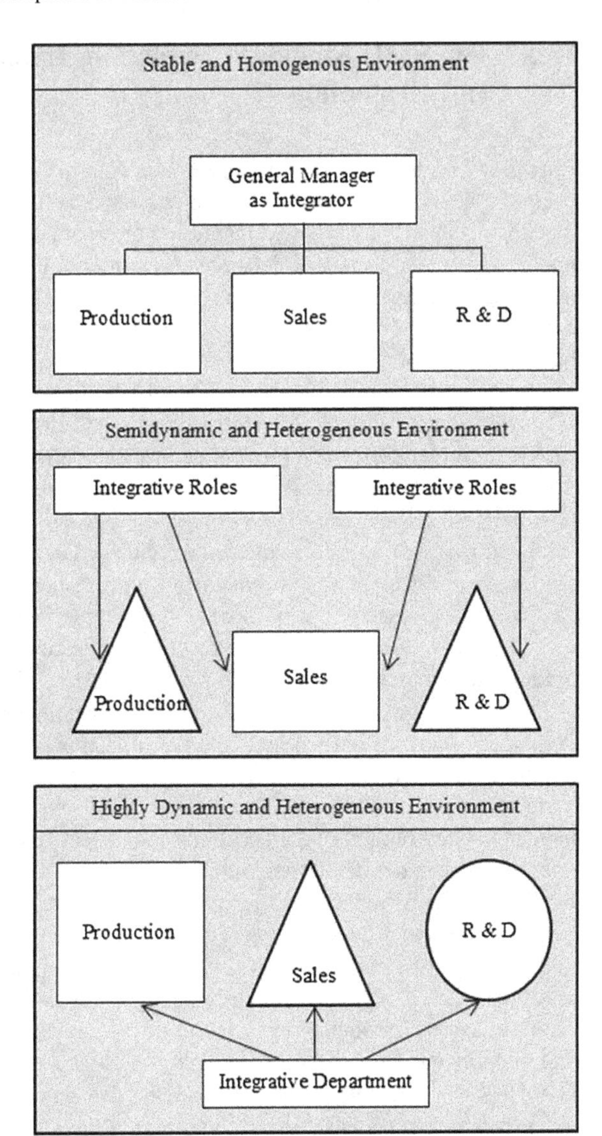

Fig. 18.2 Structural solutions to the organisational integration problem. *Source* Lawrence and Lorsch (1984)

development by analysing strategic projections for the organisation's future environment, and develops educational programmes to help managers to prepare in advance for these more complex responsibilities. The mechanism to implement such an endeavour is provided by the method of integrative learning which is more concerned with *learning how to learn* rather than with the mere acquisition of skills.

18.3 The SMD Model: Formulation, Implementation and Evaluation

Embarking on the conceptualisation of strategic management development so as to encompass a science-based step-wise approach to formulate, implement and evaluate the decision-making of an institution or organisation in order to achieve future goals, the strategic management process consists of different stages: strategy formulation, strategy implementation and strategy evaluation.[3] David (2011) has extensively studied the process of strategic management and designed a comprehensive model of Strategic Management Development (SMD) which allows for an analytical and practical representation of the three above-mentioned stages. The relationships among the major components of the strategic management process are shown in the model, which provide an answer to three important questions: *Where are we now?*, *Where do we want to go?*, and *How are we going to get there*? (cf. Fig. 18.3).

As shown in Fig. 18.3 representing the SMD model, the process of strategic management development is dynamic, where a change in one of the components can call for a change in one or more of the other components, rendering the process in a continual mode. Such changes could be triggered by the environment of the organisation, requiring an adaptation of its mission, vision or long-term strategies.

Since strategic management development facilitates institutions and organisations to be more proactive than reactive in shaping their future, it allows them to initiate certain activities which enable them to take control over their own destiny.

In the model of David (2011), *strategy formulation* as the 'inception stage' includes developing a vision and a mission, identifying an organisation's external opportunities and threats, determining internal strengths and weaknesses, establishing long-term objectives, generating alternative strategies, and choosing particular strategies to pursue. In general, *vision statements* answer questions such as: "*What does our organisation want to become?*" while *mission statements* answer questions such as: "*What is our organisation about?*"

As in any institution or organisation of great significance where managers share and support the basic vision formulated to be achieved in the long run, Drucker (1974)—known as the 'Father of Modern Management'—has formulated a set of guidelines for strategists, who in his opinion have the responsibility to develop a clear vision and mission of the organisation, underscoring both its purpose and *raison d'etre*. Most organisations have both a vision and mission statement, but the vision statement should be established first: it should be short, preferably one

[3]*Strategic management* is commonly used to refer to the formulation, implementation and evaluation of a strategy, while *strategic planning* refers only to the formulation of a strategy. The purpose of strategic management is to exploit and create new and different opportunities for *tomorrow*, while *long-range planning*, in contrast, tries to optimise for tomorrow the trends of *today* (cf. David 2011).

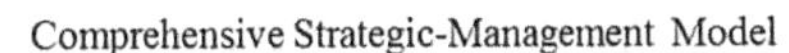

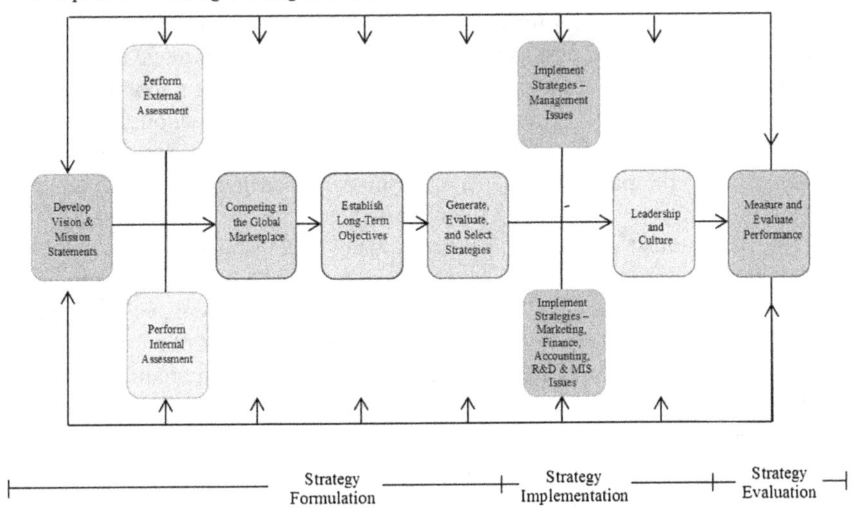

Fig. 18.3 Model of strategic management development (SMD). *Source* Adapted from David (2011)

sentence, while as many managers as possible should have an input into developing the vision statement.

Although initially, strategic management development has enabled organisations to formulate better strategies through the use of a more systematic, logical, and rational approach to take strategic choices, the study of Langley (1988) shows that the process, rather than the decision or document, is a more important contribution to strategic management: through the involvement in the process, i.e. through dialogue and participation, managers become committed to supporting the organisation, underscoring the assumption that dialogue and participation in the process will increase the managers' commitment to the mission and vision.

As indicated above, the *mission statement* is a declaration of an organisation's 'reason for being', providing an answer to the fundamental question "*What is our organisation?*" A clear mission statement as the first step in the strategic management process is essential for the effective formulation of objectives and strategies.

Strategy implementation as the 'action stage' refers to developing a strategy-supportive culture, creating an effective organisational structure, redirecting marketing efforts, preparing budgets, developing and utilising information systems, and linking employee compensation to organisational performance. Although inextricably interdependent, strategy formulation and strategy implementation are characteristically different. As David (2011: 243) explains: "*In a single word, strategy implementation means change*." Indeed, formulating the right strategies would not be enough, because managers and employees must also be motivated to implement those strategies. Such implementation activities used by most organisations include establishing annual objectives, devising policies, and allocating resources, so that the formulated strategies can be executed. Strategy implementation also includes the development of a strategy-supportive culture and the utilisation of information systems. Since this implementation stage is also called the 'action stage' of the strategic management process, it means mobilising managers and staff with a view to putting formulated strategies into action. As this often requires personal discipline, commitment and sacrifice, strategy implementation is considered to be the most difficult stage in strategic management.

Finally, *strategy evaluation* as the 'rethinking stage' is the final step in the strategic management development process for obtaining information about whether strategies work well. Strategy evaluation is essential to ensure that formulated objectives are being achieved. In most organisations, strategy evaluation is simply an appraisal of how well an organisation has performed. The three fundamental strategy-evaluation activities are: (1) reviewing external and internal factors which are the bases for current strategies; (2) measuring performance, and (3) taking corrective actions. Under the continuing conditions of change and uncertainty of internal and external events, the evaluation of strategies is needed since the success of today is no guarantee for the success of tomorrow.

Most strategists agree, therefore, that strategy evaluation is vital to the proper functioning of institutions and organisations where timely evaluations have the potential to alert the management to problems or possible problems before a situation could become critical. According to David (2011), strategy evaluation includes three basic activities: (1) examining the underlying bases of a strategy, (2) comparing expected results with actual results; and (3) taking corrective actions to ensure that performance conforms to plans.

Strategy evaluation applies to all sises and kinds of institutions and organisations, not only to initiate managerial questioning of expectations and assumptions, but also to review objectives and values, and to stimulate creativity in generating alternative approaches.

Following Langley's (1988: 40) contention that: "*Historically, the principal benefit of strategic management has been to help organisations formulate better strategies through the use of a more systematic, logical, and rational approach to strategic choice*", the important benefit of strategic management development for Integrated Microfinance Managers in Indonesia is its contribution to the objective to realise the appropriate management of indigenous institutions engaged in the provision of various community-based services throughout the country.

18.4 SMD for Indonesia's Educational and Indigenous Institutions

Recent advances in the utility of the Strategic Management Development (SMD) approach also have beneficial implications for Indonesia's institutions and organisations, at both the institutional and educational level. In line with the contention of Langley (1988: 40) that: "*Communication is a key to successful strategic*

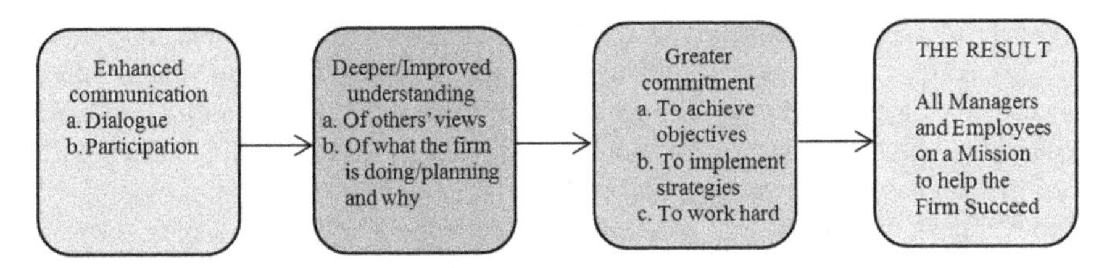

Fig. 18.4 Benefits for organisations involving managers in the process of strategic management development (SMD). *Source* Adapted from David (2011)

management", the understanding and involvement of managers in the formulation and execution of the institution's vision and mission strengthens their commitment to the institution. Figure 18.4 represents the intrinsic benefit of an institution which implements the concept of strategic management, where managers and staff are engaged in enhancing communication to increase their commitment, which in turn tends to result in a more successful functioning of the organisation.

As described in the previous section (18.3), the development of strategic management is rather specific as it aims towards leading, driving and motivating managers in- and outside the institutions and organisations with a view to formulating, implementing and evaluating relevant activities in order to achieve future goals. The implications of the SMD model for organisations and institutions considered as academic learning systems, such as colleges and universities, largely include the changes of the organisation and its environment, where the change agents comprise different categories of actors such as students, faculty and staff, network members, public and private administrators, and the media.

Linking up with the results of the recent study of the European University Association (EUA) on '*Strategic Management and Universities' Institutional Management* by Tabatoni, Davies & Barblan (2002: 5) who contend that: "... *strategic management prepares people to project themselves into the future, i.e., to face new situations in the near future, at the cost of risk and uncertainty, when dealing with changes in structures, models of action, roles, relations and positions*", the first experience with the newly-developed Master Course on Integrated Microfinance Management (IMM) at the Faculty of Economics and Business of Universitas Padjadjaran (FEB/UNPAD) in Bandung, Indonesia confirms that the complicated management of both teachers and students in the classroom as well as in their vocational training experience outside the institution is confronted with innovative change, complexity and uncertainty, requiring an innovative strategic management instrument to accumulate, study and analyse relevant data as useful information for the managers and staff concerned in order to cope with future challenges.

The collection, storage and documentation of such information has to be structured and universally accessible in order to provide all participants—managers, administrators, strategists, staff and students—with information conducive to the appropriate adaptation of strategies and policies. It is clear that in view of the innovative curriculum development, where alternative approaches and models are introduced to train the new cadres to implement the strategy of integrated micro-finance management through indigenous institutions as a contribution to poverty reduction, the challenge is to provide special education and training for information-based development and change.

Several studies show that strategic management in this respect also provides managers with the opportunity of empowerment. Indeed, empowerment strengthens individuals to value and appreciate their position and input, and increase their participation in different social activities which, in the case of strategic management, continues to be a major benefit for managers and their institutions and organisations (cf. Burkhart and Reuss 1993; Lee and Hudson 2001; Hardina et al. 2007; Adams 2008).

The basic assumption of the empowerment approach for the management of institutions and organisations is that the managers' involvement in decision-making is found to foster feelings of personal control and accomplishment. In addition, the empowerment approach also identifies several alternatives for top-down approaches to organisational and institutional management, which has contributed to the growing attention towards the management of educational and indigenous institutions in Indonesia today (cf. Allison and Kaye 2005).

In spite of the common global-local antithesis between strategies and their actual implementation, which has prompted many learning institutions to accept decentralisation in order to maintain their up-to-date position, the subject matter of the new IMM Master Course, being based on a philosophy of 'development from the bottom', has facilitated the integration of relevant data from the field not only towards the development of new management strategies of the university, but also in the innovative curriculum itself as part of the perspective of the underlying indigenous knowledge systems. As a result, from the perspective of interest from the central government in the development of alternative strategies and policies for poverty reduction in Indonesia, and on the other hand the positive attitudinal change towards indigenous communities and their institutions accelerated by recent changes in their socio-economic situation, the related educational institutions have successfully integrated these events in their implementation of strategic management at the educational and institutional level.

The first experience of the new IMM Master Course in Indonesia supports the contention of Tabatoni et al. (2002: 5) that in learning organisations and institutions, strategic management becomes the educating process of change agents acting as: "*the institutional actors whose behaviour can significantly influence change in the organisation and its milieu.*"

In line with the effective application of the strategic management development process in the educational institutions, a growing number of governmental organisations, departments and agencies have also recently adopted the SMD model successfully, resulting in improved performance and efficiency.

In the process, government agencies are experiencing that civil servants appreciate new opportunities to participate in the strategic management process, which in turn has a positive effect on the organisation's mission, objectives, strategies and policies (cf. www.strategyclub.com).

At the educational level of the *managers-to-be*, the new IMM Master Course embarks on the understanding of the principles underlying the process of strategic management, where both management theory and development of management thinking would form core themes in the curriculum. The innovative strategic management-oriented training of the new cadres of Integrated Microfinance Managers for their future work as managers of the indigenous institutions in Indonesia already shows that the focus of strategic management on the study and understanding of the special challenges and complex environments of indigenous institutions renders the tasks of implementing the chosen strategy of the integration of various financial, medical, educational, socio-cultural and communications services at the community level operational, practical and rather promising.

In addition to the teaching method of case study analysis, where students get the chance to learn how to apply theoretical concepts, evaluate practical situations and formulate service-oriented strategies, the substantial vocational or fieldwork training of students provides an excellent opportunity for them to develop practical skills and resolve implementation problems of the new policy in a changing environment on the basis of strategic management development from within the indigenous institutions (cf. Slikkerveer 2012). The adoption of such a 'grassroots' perspective by the newly-trained IMM and ICMD managers during the training process will enhance the operationalisation of the indigenous institutions in interacting with regional and national development agencies pertaining to increased local participation and added value to the integrated policy planning and implementation process throughout the country.

References

Adams, R. (2008). *Empowerment, participation and social work*. New York: Palgrave McMillan.

Allison, M., & Kaye, J. (2005). *Strategic planning for nonprofit organizations* (2nd ed.). Wiley.

Burkhart, P. J., & Reuss, S. (1993). *Successful strategic planning: A guide for nonprofit agencies and organizations*. Newbury Park: Sage Publications.

Cheng, T., Sculli, D., & Chan, F. (2001). Relationship dominance—Rethinking management theories from the perspective of methodological relationalism. *Journal of Managerial Psychology, 16*(2), 97–105.

Cloke, K., & Goldsmith, J. (2002). *The end of management and the rise of organizational democracy*. San Francisco, CA: Jossey-Bass.

David, F. R. (2011). *Strategic management: Concepts and cases*. Boston: Prentice Hall.

Drucker, P. (1974). *Management: tasks, responsibilities and practice*. New York: Harper & Row.

Green, J. (1997). *Is bureaucracy dead? Don't be so sure*. Chartered Secretary. January 18–19, 1997.

Hardina, J., Middleton, Montana, S., & Simpson, R. A. (2007). An empowering approach to managing social service organization. New York: Springer Publishing Company.

Herzberg, F. W., Mausner, B., & Snyderman, B. B. (1959). *The motivation to work* (2nd ed.). London: Chapman & Hall.

Humphrey, A. (2005). SWOT analysis for management consulting. *SRI Alumni Newsletter*, December, 2005. https://www.sri.com/sites/default/files/brochures/dec-05.pdf.

Kolb, D. A., Lublin, S., Spoth, J., & Baker, R. (1986). Strategic management development: Using experiential learning theory to assess and develop managerial competencies. *Journal of Management Development, 5*(3), 13–24.
Langley, A. (1988). The roles of formal strategic planning. *Long Range Planning, 21*(3), 40–50.
Lawrence, P. R., & Lorsch, J. W. (1984). New management job: The integrator. *Army Organizational Effectiveness Journal, 1.*
Lee, J. A. B., & Hudson, R. E. (2001). *The empowerment approach to social work practice: Building the beloved community*. New York: Columbia University Press.
Maslow, A. H. (1943). A theory of human motivation. *Psychological Review, 50*(4), 370–396.
Mayo, G. E. (1933). *The human problems of an industrialised civilization: The early sociology of management and organizations*. London: Routledge.
McGregor, D. (1987). *The human side of enterprise*. London: Penguin Books.
Miller, E. J., & Rice, A. K. (1967). *Systems of organization*. London: Tavistock Publications.
Pugh, D. S., & Hickson, D. J. (Eds.). (1976). *Organizational structure in its context: The aston programme I* (reprinted 1978). Westmead, England: Saxon House.
Ridderstrale, J. (2001). Business moves beyond bureaucracy. In Pickford, J. (Ed.) *Financial times mastering management 2.0. Financial times* (pp. 217–220). London: Prentice Hall.
Schneider, S. C., & Barsoux, J.-P. (2002). *Managing across cultures*. Upper Saddle River, New Jersey: Prentice Hall.
Skipton, M. D. (1983). Management and the organization. *Management Research News, 5*(3), 9–15.
Slikkerveer, L. J. (Ed.). (2012). *Integrated microfinance management for poverty reduction and sustainable development in Indonesia (IMM): A handbook for lecturers and tutors of the new IMM master course*. Leiden, Bandung: UL-LEAD/UNPAD-FEB.
Stern, S. (2001). *Guru guide. Management today*. October 83–84, 2001.
Stewart, R. (1999). *The reality of management*. Oxford: Butterworth-Heinemann.
Tabatoni, P., Davies, J. L., & Barblan, A. (2002). *Strategic management: A tool of leadership—concepts and paradoxes*. Brussels: European University Association.
Taylor, F. W. (1947). *Scientific management*. New York: Harper & Row.
von Bertalanffy, L. (1951). Problems of general systems theory: A new approach to the unity of science. *Human Biology, 23*(4), 302–312.
Walker, T. (2017). *How to come up with a value proposition when what you sell isn't unique.* Austin, TX: Conversionxl Agency. https://conversionxl.com/agency/.
Weber, M. (1964). *The theory of social and economic organization*. New York: Collier Macmilan.

L. Jan Slikkerveer is Professor of Applied Ethnoscience and Director of the Leiden Ethnosystems and Development Programme, Faculty of Science, Leiden University, Leiden, The Netherlands. He received his Ph.D. on his fieldwork in the Horn of Africa from Leiden University in 1983 and an Honorary Degree from the Faculty of Medicine of Universitas Padjadjaran in Bandung, Indonesia in 2005. He further extended advanced training and research in the newly-developing field of applied ethnoscience in several sub-disciplines, including ethno-economics, ethno-medicine, ethno-biology/botany, ethno-pharmacy, ethno-communication, ethno-agriculture and ethno-mathematics in combination with a focus on three target regions of South-East Asia, East-Africa and the Mediterranean Region. He is the supervisor of 25 Ph.D. students at Leiden University and has published more than 100 books and 300 articles on the subject. While he has also received a number of substantial subsidies from the European Union in Brussels for international development programmes in South-East Asia, East-Africa and the Mediterranean Region, he also conceptualised both the IMM & ICMD approaches in Indonesia, and is Senior Lecturer in the International Master Course on Integrated Microfinance Management (IMM) at the Faculty of Economics and Business of Universitas Padjadjaran in Bandung, Indonesia.

Chapter 19
Advanced Curriculum Development for Integrated Microfinance Management

Nury Effendi, Asep Mulyana and Kurniawan Saefullah

Against the backdrop of continuing poverty, where the income per capita remains around $3,000, a structural problem seems to point to the inability of the poorest of the poor and their families to participate in order to improve their socio-economic status, where education and training have shown to be among the main triggers.

L. Jan Slikkerveer (2012)

19.1 The Need for Advanced Training of Integrated Microfinance Managers

Indonesia is the fourth largest country in the world encompassing 18,491 islands, subdivided into 34 administrative provinces, and populated by 260 million inhabitants in 2016, with an average annual population growth rate of 1.3%. The recent decentralisation process in Indonesia, known as the policy for regional autonomy, which has been implemented since 2001, seems to be unprecedented in view of the rapid transformation and the expansion of the geographic administration, which is expressed in the increasing number of local governments over the past 15 years. There are currently 514 districts and municipalities, 6998 sub-districts, and 81,308 villages in Indonesia (cf. United Nations Statistics Division 2016).[1]

According to the latest *Human Development Report* of the United Nations Development Programme (UNDP 2015), the *Human Development Index* (HDI) for 2015 for Indonesia, categorised as a country with 'medium human development', is 0.684, ranking the country in the 110th position out of 188 countries in the globe.

[1]See http://nomor.net/_kodepos.php?_i=provinsi-kodepos&daerah=&jobs=&perhal=60&urut=&asc=000011111&sby=000000.

N. Effendi (✉) · A. Mulyana · K. Saefullah
FEB, Universitas Padjadjaran, Bandung, Indonesia
e-mail: nuryeffendi@hotmail.com

L. J. Slikkerveer et al. (eds.), *Integrated Community-Managed Development*, Cooperative Management, https://doi.org/10.1007/978-3-030-05423-6_19

In comparison with the previous years, with a HDI of 0.684 and the rank of 108th position out of 187 countries in 2014, preceded by a HDI of 0.629, ranking in the 121st position out of 186 countries in 2013, and a HDI of 0.617, ranking in the 124th position out of 187 countries in 2011, after an initial upsurge in the global position from 124 in 2011 to 108 in 2014, at present, there is a decline in the country's position back to the same rank order of 110 in 2015. Although the HDI shows an initial rise from 0.629 in 2013 to 0.684 in 2014, it has remained the same in 2015, i.e. 0.684 (cf. UNDP 2011; UNDP 2013; UNDP 2014).[2]

The *Asia-Pacific Human Development Report* of the United Nations Development Programme (2016) shows a similar stagnating picture for Indonesia for 2015, with a HDI of 0.684, while its position in the world ranking remains at 110.

The *Multidimensional Poverty Index*, which determines poverty beyond income-based lists, shows that in 2012, about 14.5 million Indonesians were suffering from multidimensional poverty, with a measured intensity of deprivation of 41.4%. The population living below the income poverty line as measured in 2012 by the *Purchasing Power Parity* (PPP) of $1.25 a day was 16.2%, while in 2014 it measured 11.3% of the population, showing that the contribution of deprivation to the overall poverty in 2014 for education was 24.7% (cf. UNDP 2014).

As shown in the previous Chapters, poverty and sustainable development are strongly intertwined: poverty is particularly manifest in sectors such as health care and education, where the poor and low-income families are hardly capable of escaping from the vicious circle of poverty to secure better health and education. Although the complexity of the staggering problems of poverty has virtually blocked most efforts to find a satisfactory solution, it is clear that education as a human investment for future employment is playing a major role in poverty reduction in the long term. Several economists argue that one of the major problems in Indonesia is the lack of adequately skilled and trained human resources, and the Government of Indonesia is still struggling to find new ways of improving the country's human resources, not only in the major sectors of agriculture and public services, but also in health care and education. Against the background of a ratio of employment to the population, estimated at 63.5%, i.e. the percentage of the employed at the age of 15 years and above, the total unemployment rate indicating the percentage of the labour force was 6.2% in 2016. Considering that education shows such a strong relationship with poverty and development, the improvement in education and training of human resources deserves high priority in the country's policy planning and implementation process. It is not surprising that in general, education contributes one-fourth to the overall poverty measures, underscoring that any policy related to the improvement in education could contribute to poverty reduction. With the number standing at 13 of expected years of schooling and a

[2]HDI classifications are based on HDI fixed cut-off points, which are derived from the quartiles of distributions of component indicators. The cut-off points are HDI of less than 0.550 for low human development, 0.550–0.699 for medium human development, 0.700–0.799 for high human development and 0.800 or higher for very high human development (cf. UNDP 2015).

public expenditure of 3.6% on education in Indonesia in terms of the percentage of the GDP, a greater effort seems to be needed in the coming years to increase the supply of adequately trained and skilled human resources, particularly in the area of sustainable community development in the rural areas of the country (cf. United Nations 2015b).[3]

Following the previous success of microcredit to provide support to the poor since the 1970s on the basis of subsidised loans and low interest rates, the following 'new wave' microfinance has initially also been regarded as a useful tool for the alleviation of poverty. However, the rather disappointing results of its contribution to poverty reduction after the 1990s, described in the previous Chapters, have led to the general conclusion that microfinance as a mere economic-financial instrument with the underlying commercial interests of its providers—MFIs, NGOs and commercial banks, largely supported by the World Bank—could only in some cases improve the position of the middle class, but in essence, it fails to reduce poverty among the poor and very poor, particularly in developing countries.

As documented before, an increasing number of criticists of microfinance, including Bateman (2010), Roodman (2011), Duvendack et al. (2011) and Slikkerveer (2012), have underscored the overall failure of microfinance in poverty alleviation and there is an urgent need to develop alternative approaches to achieve poverty reduction in Indonesia and the rest of the world. Since education has in general been found to contribute one-fourth to poverty, such new approaches should specifically focus on non-financial, but more human-oriented policies, in which the education institutions have to play a leading role.

In his recent research on poverty and health in Indonesia, Slikkerveer (2007) introduced a new perspective on poverty reduction, especially at the community level, in which he embarks on the indigenous systems of knowledge, practice and institutions of the population, which he proposes to integrate with global systems with a view to reaching a more participatory form of community development. Later, Slikkerveer (2012) developed an integrated model of poverty reduction in which five approaches should be combined in order to realise a substantial reduction of poverty, including the following:

1. the initiation of an integration process of community initiatives in the management and development of human-oriented policies and regulations which goes beyond the provision of (micro-)financial services to include other social and cultural services;
2. the active involvement of indigenous institutions in terms of strategising development-based activities in sectors such as health, education, and communication through a neo-endogenous development approach;
3. the execution of comparative research in different regions and communities, focused on traditional and modern institutions and agencies, in alleviating poverty in order to understand the role of various independent and intervening factors in the management and use of resources at the community level;

[3]See http://hdr.undp.org/en/content/table-1-human-development-index-and-its-components.

4. the development of an integrated community-managed development model to reach sustainable community development through the coordination and management of medical, educational and social facilities in conjunction with the cultural dimension of the community; and
5. the development of an advanced curriculum on integrated microfinance management with a view to educating and training a new cadre of *integrated microfinance managers* particularly focused on the improvement of the situation of the poor and underprivileged members of the community.

As the previous chapters of this Volume also indicate and document, the recent progress and experience in the field of such integrated approaches and the reassessment of the crucial role of indigenous knowledge, practices and institutions within the context of sustainable development have led to the extension of the IKS-based model to introduce and develop a similar advanced curriculum on integrated community management with a view to educating and training a new cadre of *integrated community managers*.

Thus, advanced education and training are becoming important goals in training new cadres of community-based managers capable of supporting policies and programmes of local poverty reduction and community empowerment. As the need for integrated microfinance institutions, operational at the community level, is rising, the demand for training in integrated microfinance management is becoming urgent. Until 2011, there was no university in Indonesia which provided a specific programme to study and support poverty alleviation within the context of sustainable development, both at the graduate and post-graduate level. So far, academic attention for poverty alleviation has merely been received from an economic perspective, while the current state in international knowledge and experience in this important subject is showing the need for a multidimensional approach to encapsulate the complex of the non-economic aspects which are operational in poverty and inequality. As mentioned before, the more realistic *Multidimensional Poverty Index* (MPI) measuring deprivation in the three dimensions of the *Human Development Index* (HDI) provokes due attention for the further study and analysis of health, education and living standards in order to design adequate poverty reduction strategies (cf. UNDP 1996; Tsui 2002; Alkire et al. 2014).

As regards the complicated situation in Indonesia, Slikkerveer and Ismawan (2012), in their analysis of the current development process in the country, point to its dual dimension: the *financial-economic dimension* of the constrained building of an inclusive financial sector to reach the rural poor and low-income groups of the population, and the *socio-educational dimension* of the impaired development of adequate education and training programmes in human resources development.

Based on the two analytical studies by Slikkerveer and Ambaretnani (2010a, b), respectively, on the *Societal Needs Assessment* and the *Educational Needs Assessment for IMM in Indonesia*, an urgent need is underscored of adequate education and training programmes in human resources to supply the development-oriented local with well-trained managers capable of extending an innovative system of integrated financial and social services to the poor (cf. Slikkerveer 2007).

The higher education system in Indonesia is built up of thousands of universities and higher degree institutions, consisting of 82 government institutions, more than 2800 private institutions, and 1 open university which is managed by the government. In total, the universities accommodate about 4.5 million students out of 25 million people who are between 18 and 24 years old. These figures show that the level of student participation in Indonesia is less than one-fifth. This number lags well behind data from other Asian countries, such as Malaysia (32.5%), Thailand (42.7%), The Philippines (28.1%), and China (20.3%). The highest participation in Asia is South Korea with a 91% participation rate.[4]

The Directorate-General of Higher Education (DGHE) of the Ministry of National Education of Indonesia has developed a long-term strategic plan called the *Higher Education Long-Term Strategy* 2003–2010 (HELTS) (DGHE 2003). The strategy designates universities as key institutions which are expected to strengthen the nation's competitiveness through improving the quality of higher education and developing the potential of students in an optimal way.

In addition, the objectives of the strategy are to empower universities to be able to strengthen the nation's competitiveness and anticipate the changes, while being autonomous and well-governed. The competitiveness of universities is further encouraged by self-reliance in developing study programmes, mobilising and utilising resources in an efficient mode. Partnerships and cooperation among universities are encouraged in order to attain a healthy and competitive form of synergy. Through these partnerships, it is expected that resources shared among universities can be functionalised to improve the quality of higher education. The *Directorate-General of Higher Education* (DGHE) (2003) has launched three 'pillars of development' in order to improve the system of higher education in the country, which can be summarised as follows:

1. equality and wider access for everyone;
2. quality, relevance and competitive development; and
3. strengthening good governance, accountability and public image.

One of the positive impacts of globalisation in higher education is that every country can share its experiences with other countries in order to improve the education quality in each country at an international level. Hence, one of the results of globalisation is that foreign institutes of higher education are allowed to cooperate with national institutes of higher education in Indonesia. As authenticated by the Indonesian Law No. 20 (2003) concerning the National Educational System, Article 65, Chapter XVIII mentions that since 2003, foreign higher education can be established in Indonesia.

The possibility of international higher educational institutes (HEIs) to establish a branch in Indonesia can be perceived as an opportunity for national institutions to collaborate with them. The Presidential Regulation No 103 Year 2007 (2007) on The Regional Convention on the Recognition of Studies, Diplomas, and Degrees in

[4]See http://hdr.undp.org/en/content/table-1-human-development-index-and-its-components.

Higher Education in Asia and the Pacific represents a continuation of the UNESCO Convention to recognise degrees from Asia and the Pacific, for which the Indonesian system of higher education has no other option than to comply (cf. UNESCO 2011).

There are various models of international collaboration between different systems of higher education. In the European Union, the collaboration is conducted under the Bologna Declaration (1999), which has reformed higher education by proposing the establishment of a 'European Higher Education Area' in which students and graduates can move freely between countries, rendering prior qualifications in one country as acceptable entry requirements for further study in another country. The principal objectives of the Bologna Declaration (1999) include the following:

1. *Adoption of a system of easily readable and comparable degrees*, meaning that the countries concerned should adopt common terminology and standards; and
2. *Adoption of a system essentially based on two main cycles: undergraduate and graduate*. Access to the second cycle shall require successful completion of first-cycle studies, lasting a minimum of three years. The degree awarded after the first cycle shall also be relevant to the European labour market as an appropriate level of qualification. The second cycle should lead to the Master and/or doctorate degree as applicable in many European countries.

Based on this scheme, a student from one university can also follow part of his/her courses at another university in another country. The other model which has been adopted by *Erasmus Mundus* refers to a consortium of various universities in Europe. Here, students are expected to be enrolled in at least three different universities from three different countries to complete their degree. This model of collaboration is implemented in order to enhance the quality of each participating university. Eventually, such a study programme will also lead to the development of the system of higher education in the participating nations.

The need for education in poverty reduction programmes and the opportunity to conduct an international collaborative mode of education and training in Indonesia have pertained to the establishment of an *International Consortium of Integrated Microfinance Management* (ICIMM) in 2009. The consortium has been initiated by the *Leiden Ethnosystems and Development Programme* (LEAD) at the Faculty of Science of Leiden University in The Netherlands with its main objective to develop an integrated strategy for poverty reduction and sustainable development in Indonesia through advanced education of specially-trained integrated community managers. In collaboration with the other Members, including the Faculty of Economics and Business (FEB) of Universitas Padjadjaran in Bandung, Indonesia, the Mediterranean Agronomic Institute of Chania (CIHEAM-MAICh) in Crete, Greece, and the Indonesian Movement for Microfinance (GEMA PKM) in Indonesia, the consortium has developed a new Master Course on *Integrated Microfinance Management* (IMM) in 2009 with a view to training a new cadre of *integrated microfinance managers* capable of managing both financial and non-financial services with a special focus on poverty reduction and sustainable development at the community level. With the introduction of the concept of *Integrated Microfinance Management*, a new impetus is provided to the strategy of vertical integration in the process of policy planning and implementation, in which the 'bottom-up' approach is based on the integration of local and global systems of knowledge, beliefs and practices.

19.2 IMM/UNPAD: A New Master Course for Integrated Microfinance Management

19.2.1 Historical Background

Universitas Padjadjaran (UNPAD) is a state university in Indonesia established in 1957, and located in Bandung, the provincial capital of West Java. Named after the historical Kingdom of Padjadjaran in the Sunda Region, and ruled by the King of Prabu Siliwangi, UNPAD has gained the most applicants and highest passing grades in the National Selection of State University Entrance (SNMPTN) since 2013. In 2014, the university was officially set as a State University of Legal Entities and accredited with 'A' by the *Badan Akreditasi Nasional Perguruan Tinggi* (BAN-PT). It ranks as the fifth university in Indonesia according to the *Webometrics University Ranking* in 2014 (with a rank of 1036 in the world). The Faculty of Economics and Business (FEB/UNPAD) ranks in 77th place in Asia according to the RePEc/IDEAS Ranking, and named UNPAD as the best institution in Indonesia in 2015.[5]

Given the long-term inter-university cooperation with the *Leiden Ethnosystems and Development Programme* (LEAD) of Leiden University since 1987, UNPAD joined the IMM Consortium (ICIMM) in 2009 in order to substantiate its role in poverty reduction and community services, being a main part of its mission as an institute of higher education in Indonesia. As initiator of the ICIMM, LEAD is an interdisciplinary programme based in the Faculty of Science of Leiden University and a Member of the Microcredit Summit Council of Educational Institutions. In 2009, the Members of ICIMM—UNPAD, MAICh and GEMA PKM—agreed with the initial proposal of the LEAD Programme to design and develop a new Master Course on *Integrated Microfinance Management* (M.IMM) which has been established at the Faculty of Economics and Business (FEB) of Universitas Padjadjaran in Bandung in 2012. Sponsored by the Ministry of Economic Affairs (EVD/INDF) in The Hague in The Netherlands, the new M.IMM Master Programme was officially opened by H.E. Dr. Sharif Hasan, Minister of

---

[5]See http://www.unpad.ac.id/2015/06/feb-unpad-masuk-jajaran-institusi-penelitian-terbaik-di-asia-versi-repecideas-ranking/.

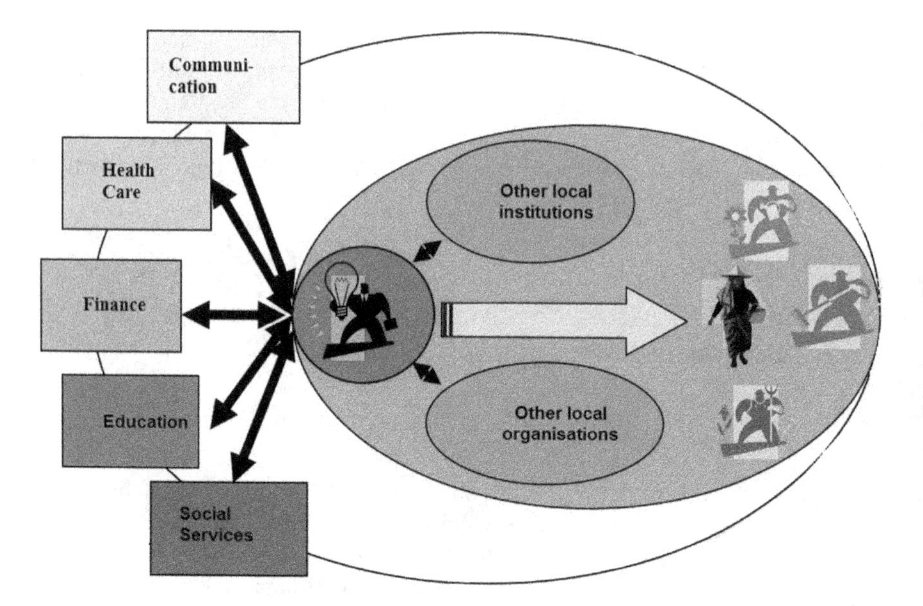

Fig. 19.1 Schematic representation of the Position of the Integrated Microfinance Manager within the Community. *Source* Slikkerveer (2012)

Cooperative, Small and Medium Enterprises of the Republic of Indonesia on the 19th of September 2011, and it started its Master programme in January 2012.

In addition to the contribution of the new M.IMM Master Course to the academic curriculum development in the country, it also responds to the growing need of Indonesia's society for a new cadre of specialised managers who will contribute to the advanced institution-based management of integrated financial, health, education, social, and communication services in order alleviate poverty and improve the health and well-being of the population of Indonesia. Figure 19.1 shows a representation of the multitasking activities and networks of the newly-trained integrated microfinance managers as they are located in the center of their local communities.

19.2.2 Mission and Objectives of the M.IMM Master Programme

Unlike other education programmes in the country, the Master Course in *Integrated Microfinance Management* has its unique approach as it is not only focusing on the provision and management of integrated financial and non-financial services and policies, but also on poverty reduction and empowerment for sustainable community development.

The IMM education and training programme aims to deliver graduates capable of managing the intervening services and policies at the community level in various sectors, and as such, to also provide a contribution to developing and advancing the new field of Integrated Microfinance Management through the following:

- training of graduates capable of identifying and solving actual community-based problems related to failing poverty reduction and empowerment; and
- training of graduates able to implement the concept of Integrated Microfinance Management to alleviate poverty and improve community prosperity within the framework of sustainable development.

In addition, the IMM Master Programme has developed the following objectives:

- to train a new cadre of integrated microfinance managers who will lead local institutions in cooperation with government agencies, NGOs, SMEs and MFIs, to provide participatory community-based services to the entire population;
- to deliver skilled professionals who will manage integrated financial, social, medical, educational and communicational services to the poor and low-income families at the community level; and
- to provide integrated microfinance managers who are capable of contributing to the realisation of the Sustainable Development Goals (SDGs) of the United Nations (2015a), set to be achieved by 2030, particularly the eradication of extreme poverty and hunger throughout Indonesia.

19.2.3 Target

The students of the M.IMM Course are expected not only to be candidates from among different development-oriented undergraduate university programmes, but also from Government Agencies, NGOs, SMEs and MFIs, in order to be trained to provide adequate financial and non-financial community-based services to the local population to overcome the burden of poverty and hunger.

19.2.4 Brief of the Master Programme

This new International Master Course with an 18-month duration embodies an international class leading up to acquiring a Master's Degree (M.IMM) from Universitas Padjadjaran, together with the International IMM Certificate on the basis of completion of the regular International Workshops and Seminars of the M.IMM Course given at the Faculty of Economics and Business of Universitas Padjadjaran in Bandung, by visiting staff from the LEAD Programme of Leiden

University in The Netherlands on relevant subjects of Integrated Microfinance Management.

The output of the Master Programme is to deliver a new cadre with a M.IMM degree who are capable Integrated Microfinance Managers that can play their role as leaders of the indigenous institutions in order to provide financial and non-financial services to all members with a focus on poverty reduction and empowerment. The graduates are also trained to master both conceptual and practical problems and situations related to the intervention of policies and services at the community level.

19.2.5 Curriculum and Structure of the M.IMM Master Programme

The advanced learning methods applied in the M.IMM Master Programme include 6 teaching modules, a 3-month supervised fieldwork/vocational training project and the writing of a Master thesis, enriched with workshops, formal presentations, discussion groups, fieldtrips and fieldwork training research. The IMM Master Programme at MM FEB/UNPAD is internationally recognised as the first Master Programme in this newly-developing field in the world, so that graduates of the course are actually pioneers in the development of integrated microfinance management in Indonesia and the rest of the globe.

The core elements of the new Master Course include the graduation of a cadre of integrated microfinance managers who are trained to manage an integrated 'package' of financial and non-financial services to the community, which can mathematically be represented in a formula, as shown in Fig. 19.2.

The Master Programme offers in total 42 credit hours of interdisciplinary subjects, combining classical lectures, group discussions, presentations, seminars and workshops, field work training research, as well as writing a thesis. The academic year of studies for the M.IMM is divided into three semesters (trimester) which include the following:

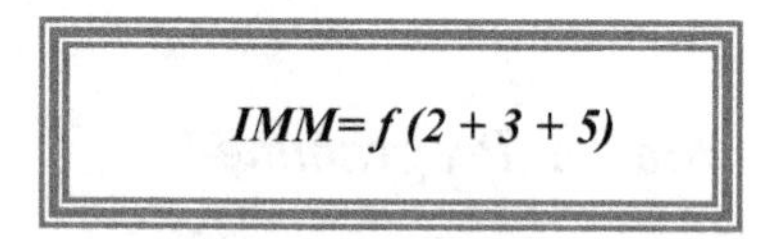

IMM = integrated microfinance management strategy
2 = the major themes: poverty reduction and empowerment
3 = the principles of output, quality and outcomes
5 = the community-based services: 1) inclusive finance, 2) health, 3) education, 4) communication, and 5) socio-cultural services

Fig. 19.2 Mathematical representation of the 'package' of financial and non-financial services provided through the integrated microfinance management strategy. *Source* Slikkerveer (2012)

- *first trimester*: the students are required to take four matriculation (Pre-IMM) courses in related subjects, such as Business Management, Mathematics and Statistics for IMM, Academic Writing, and Society & Culture;
- *second* and *third trimester*: the students are required to take one general university subject—*Philosophy of Science*—with 7 main courses, i.e. 'Introduction to Integrated Microfinance Management', 'Micro-Macro Economics of Integrated Microfinance Management', 'Regulation and Governance in Integrated Microfinance Management', 'Community-Based Management', 'Integrated Microfinance Management for Poverty Reduction and Sustainable Development', 'Sector Analysis of Integrated Microfinance Management', and 'Research Methodology for Integrated Microfinance Management'. In addition, the students are required to follow a subject in 'Economic Anthropology' and 'Entrepreneurship and Cooperatives in IMM'.
- *fourth trimester*: the students are required to prepare a project proposal and undertake their practical fieldwork or vocational training in a non-academic institution, organisation or community, supervised by the Professor of the Master Course; and finally,
- *fifth trimester*: the students are required to finalise their study by writing up their Master thesis, which is based on the results of their fieldwork or vocational training (Illustration 19.1).

The combination of the theoretical education with lectures, seminars, workshops, discussion groups and presentations with practical fieldwork/vocational training provides the students with a unique experience and a comprehensive perspective on their profession, for implementing the integrated microfinance

Illustration 19.1 The International IMM Workshop at FEB/UNPAD (2016). *Source* Mady Slikkerveer (2016)

management approach. The interdisciplinary orientation gives students a distinctive value in comparison to other study programmes, rendering the M.IMM Master Course a unique place in academic education and training. The practical fieldwork/ vocational training, required for all M.IMM students, provides them not only with the ability to gain a better understanding of community life, but it also strengthens their observations of local people and their indigenous institutions in identifying and solving practical problems, working with groups, and last but not least their ability to integrate local knowledge, practices and institutions into the process of sustainable community development (Illustration 19.2).

Since its launch in 2012, every academic year, the M.IMM Master Course conducts an *International IMM Workshop* in collaboration with the LEAD Programme of Leiden University in The Netherlands, where Lecturers and Professors jointly present workshops and seminars to the students. At the end of the International IMM Workshop, successful students receive an International Certificate from Leiden University upon their completion of the workshop's assignment, including an individual presentation and term paper.

In addition, selected students are also offered the opportunity to complete their *fifth trimester* of writing their Master thesis, which is based on the results of their fieldwork or vocational training in Leiden, providing them with an extra international academic experience. On average, the M.IMM Master Course can be finished in 18 months, including the pre-M.IMM Courses/Matriculation subjects.

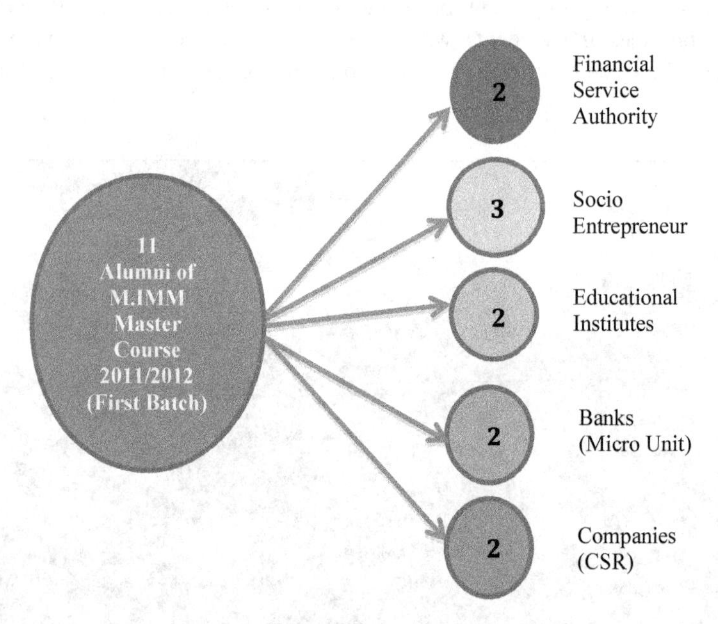

Illustration 19.2 A student of the M.IMM Master Programme involved in participant observation during his fieldwork training in a community in West Java. *Source* M.IMM (2016)

Table 19.1 Number of students enrolled in the M.IMM Master Programme

Academic year	Number of enrolled students
2011/2012	12
2012/2013	9
2013/2014	12
2014/2015	13
2015/2016	30[a]
Total	76

Source M.IMM (2016)

[a]An estimated number of 30–40 students are expected to join through the 'Scheme of Affirmation Programmes' of the Ministry of Higher Education, Research and Technology in Jakarta

19.2.6 Students and Alumni of the M.IMM Programme

Since its establishment in September 2011 and its formal operation in 2012, the M. IMM Master Programme has admitted 46 students. The annual numbers of students who have followed the Master Programme from 2011/2012 to 2015/2016 are as shown in Table 19.1.

The students who have been registered in the M.IMM Master Programme in the last three years come from various institutions, both academic undergraduate students, as well as experienced workers. Every year, the M.IMM Programme conducts a *Promotional IMM Seminar* which enables the programme to enhance its network beyond the university to include private and government agencies and institutions.

By consequence, in the academic year of 2015/2016, the Ministry of Higher Education, Research and Technology offered the M.IMM Master Programme to accommodate 2 classes under its 'Scheme of Affirmation Programmes'. With the students who joined in 2015/2016, this amounted to a total number of students of 86. For post-graduate level, the interest is rather surprising, as the positive progress in numbers not only indicates the interest of students joining the interdisciplinary approach of IMM, but it also holds a promise for the future of sustainable community development in Indonesia through advanced education and training. If the candidates are assigned to specific community-based institutions after their graduation, they will be able to exert much influence on the local people and their institutions which, in turn, will also contribute to the realisation of poverty reduction and sustainable community development in Indonesia, in the near future.

19.2.7 Different Backgrounds of the M.IMM Graduates 2011/2012

Since its establishment, the M.IMM Master Programme has granted the M.IMM Master Degree to 11 of the 12 enrolled students (first batch), who thereafter have

Fig. 19.3 M.IMM Graduates (2013). *Source* M.IMM (2016)

been employed in different kinds of institutions, both private and public. Some graduates are working in a Non-Government Organisation (NGO), while others have been employed in the private sector, mostly in Corporate Responsibility (CSR) divisions of companies, educational institutions, banking institutions, and government agencies. Some of the graduates have started to work as entrepreneurs (Fig. 19.3).

19.2.8 The Future of the M.IMM Master Programme

The implementation of the M.IMM Master Programme has shown rather positive results since its establishment in 2011. The challenge for the future is to estimate to what extent this new Master Course contributes positively to poverty reduction and sustainable community development. As mentioned above, education has shown to contribute about one-fourth to the effect of multidimensional poverty reduction measures. So, a relevant question is: "*To what extent could the M.IMM Master Programme contribute to poverty reduction in Indonesia in the near future?*"

From the brief assessment of the employment of the first group of graduates/alumni, the integrated microfinance managers are working in different sectors of society: both private and public institutions and organisations. Some are working in policy-oriented institutions, which have the potential to influence some important changes in the community or society. As the recent impact of the IMM Master Course seems to justify the input of educational and training efforts concerned for a wider context of the nation, there is a need to extend the model of IMM to the entire country through the implementation of a nationwide programme. In this context, the IMM Master Programme is challenged by an interesting agenda for the near future, which can be summarised as follows:

- The need for the establishment of a special *Centre of Excellence* in Integrated Microfinance Management should be met in order to accumulate relevant information and communication facilities, which could act as a study center and clearing house on the achievement and progress of the new strategy for poverty reduction and sustainable development in Indonesia and beyond.
- The need to set up relevant databases and accumulate cases on various local institutions within the context of poverty reduction, and development should be met in all the provinces of Indonesia.
- By consequence, the M.IMM Master Programme needs to invite more students coming from different regions of Indonesia, bringing their local data and cases, which relate to local institutions, poverty numbers and progress of development.
- The current need to establish appropriate publication media, which will be distributed nationally or even internationally, should be met with a view to promoting the importance of the concept of IMM in poverty reduction and sustainable development by learning from relevant cases in Indonesia. The media could include international academic journals, which publish the results of any research from different regions of Indonesia and the world, showing the way in which Integrated Microfinance Management is implemented and evaluated in different areas, each in their various forms and contexts.
- The growing need to set up and extend a network among different institutions throughout the country in order to render the M.IMM Master Programme capable of influencing public policies and programmes for poverty reduction and community empowerment, including the establishment of public-private partnerships.

19.3 Towards a Nationwide Poverty Reduction Strategy of Integrated Microfinance Management

In May 2011, the Government of Indonesia launched a master plan for 15 years of economic development, known as the Master Plan *Percepatan Pembangunan Ekonomi Indonesia* 2011–2025 (MP3EI 2011) ("Acceleration and Expansion of Indonesian Economic Development 2011–2025"). The prime purpose of MP3EI is to enable Indonesia to become a developed and prosperous country with a national GDP of around $4–4.5 trillion by 2025, thus becoming the 9th largest economy in the world. The short-term goal of MP3EI (2011) is to achieve a GDP per capita of $4700 for each corridor by 2014, increasing to $15,000 by 2025.[6] The objective seems rather ambitious considering that there is still a relatively high number of poor and low-income families in the country. Furthermore, the new challenge of the *Post-2015 Agenda of Sustainable Development Goals* of the United Nations (2015a) urges Indonesia to consider not only the implementation of income-related measures to meet the Sustainable Development Goals, but also the factors of sustainability which have to meet the achievement of the income target, as well as balancing the economic target indicators with socio-environmental factors.

There are four key dimensions which are targeted to be realised in the *Post-2015 Agenda of Sustainable Development Goals* of the United Nations (2015a): inclusive economic and social development, environmental sustainability, peace, and security. The United Nations (2015a) has established a set of eleven thematic consultations on, respectively: conflict and fragility, education, environmental sustainability, governance, growth and employment, health, hunger, food and nutrition, inequalities, population dynamics, energy and water. In addition, social protection has become a top priority of the *Post-2015 Agenda of Sustainable Development Goals* of the United Nations (2015a).

In this context, the integration of Indonesia's *Master Plan* (MP3EI) (2011) with local cultures, peoples and communities, in relation with socio-environmental factors, is needed in order to become feasible and effective. Since the improvement and equality of education is one of the priority areas which UNESCO (2011) would like to achieve worldwide, the role of advanced Master programmes, such as the *Integrated Microfinance Management* (IMM) Master Programme at the Faculty of Economics and Business of Universitas Padjadjaran in Bandung, becomes important as a promising initiative in Indonesia's system of higher education.

The 'bottom-up' approach as a major strategy in the M.IMM curriculum of integrating indigenous knowledge and institutions for poverty reduction and sustainable development should further be promoted and implemented on a nationwide scale. There are hundreds of ethno-cultural groups in Indonesia with many local geniuses capable of preserving traditional knowledge and institutions, and making a difference to local communities in attaining the sustainability of socio-economic and environmental development for many future generations. Notwithstanding, the rich local experience and wisdom accumulated over so many ages should be documented comprehensively and discussed intensively throughout all levels of the education system, particularly in higher education. These and related efforts, for which time is pressing, would bring Indonesia to the position of realising its targeted share of the *Post-2015 Agenda of Sustainable Development Goals* of the United Nations (2015a).

---

[6]See http://www.indonesia-investments.com/projects/government-development-plans/masterplan-for-acceleration-and-expansion-of-indonesias-economic-development-mp3ei/item306.

References

Alkire, S., Conconi, A., & Seth, S. (2014). *Multidimensional poverty index 2014: Brief methodological note and results*. The Oxford Poverty and Human Development Initiative (OPHI). Oxford Department of International Development, University of Oxford.

Bateman, M. (2010). *Why doesn't microfinance work? The destructive rise of local neo-liberalism*. London: Zed Books.

Bologna Declaration. (1999). *Joint declaration of the European Ministers of Education*. Bologna: University of Bologna. http://www.cepes.ro/services/inf_sources/on_line/bologna.pdf.

Indonesia Law No. 20 (2003) *National educational system*. Article 65, Chapter XVIII. Jakarta: Ministry of National Education.

Directorate-General of Higher Education (DGHE). (2003). *Higher education long term strategy 2003–2010 (HELTS)*. Jakarta: Ministry of National Education.

Duvendack, M., Palmer-Jones, R., Copestake, J. G., Hooper, L., Loke, Y., & Rao, N. (2011). *What is the evidence of the impact of microfinance on the well-being of poor people? Systematic review*. London: EPPI-Centre, Social Science Research Unit, Institute of Education, University of London.

Master Plan *Percepatan Pembangunan Ekonomi Indonesia* 2011–2025 (MP3EI 2011) ('Acceleration and Expansion of Indonesia Economic Development 2011–2025'). Jakarta: Government of Indonesia.

M.IMM Master Programme. (2016). *Internal data of students of IMM study program*. Faculty of Economics and Business: Universitas Padjadjaran.

Roodman, D. (2011). *Due diligence: An impertinent inquiry into microfinance*. Washington, D.C.: Centre for Global Development.

Slikkerveer, L. J. (2007). *Integrated microfinance management and health communication in Indonesia*. Cleveringa Lecture Jakarta. Trisakti School of Management & Sekar Manggis Foundation. Leiden: LEAD, Leiden University.

Slikkerveer, L. J. (Ed.). (2012). *Handbook for lecturers and tutors of the new master course on integrated microfinance management for poverty reduction and sustainable development in Indonesia (IMM)*. Leiden/Bandung: LEAD/UL/UNPAD/MAICH/GEMA PKM.

Slikkerveer, M. (2016). Personal Photo Collection.

Slikkerveer, L. J., & Ambaretnani, P. (2010a). *Societal needs assessment for IMM in Indonesia*. Leiden: LEAD, Leiden University.

Slikkerveer, L. J., & Ambaretnani, P. (2010b). *Educational needs assessment for IMM in Indonesia*. Leiden: LEAD, Leiden University.

Slikkerveer, L. J., & Ismawan, B. (2012). Introduction to integrated finance management in Indonesia. In: Slikkerveer, L. J. (Ed.), *Integrated microfinance management for poverty reduction and sustainable development in Indonesia*. Bandung: LEAD-UL/UNPAD/MAICH/GEMA PKM.

Tsui, Kai-yuen. (2002). Multidimensional poverty indices. *Social Choice and Welfare, 19*(1), 69–93.

UNESCO. (2011, November 26). *Asia-Pacific regional convention on the recognition of qualifications in higher education*, Tokyo.

United Nations (UN). (2015a). *Post-2015 agenda and the sustainable development goals (SDGS)*. New York: United Nations.

United Nations (UN). (2015b). *Population and vital statistics report 2015*. Department of Economic and Social Affairs. Population Division. New York: United Nations. https://unstats.un.org/unsd/demographic/products/vitstats.

United Nations Development Programme (UNDP). (1996). *Human development report: Growth for human development?*. New York: Oxford University Press.

United Nations Development Programme (UNDP). (2011). *Human development report: Sustainability and equity: A better future for all*. New York: Oxford University Press.

United Nations Development Programme (UNDP). (2013). *Human development report: The rise of the South: Human progress in a diverse world*. New York: Oxford University Press.

United Nations Development Programme (UNDP). (2014). *Human development report: Sustaining human progress-reducing vulnerabilities and building resilience*. New York: Oxford University Press.

United Nations Development Programme (UNDP). (2015). *Human development report: Work for human development*. New York: Oxford University Press.

United Nations Development Programme (UNDP). (2016). *Asia-Pacific human development report: Shaping the future: How changing demographics can power human development*. New York: UNDP.

United Nations Statistics Division (UNSD). (2016). *World statistics pocketbook*. New York: UNSD.

Nury Effendi is the Dean of the Faculty of Economics and Business of Universitas Padjadjaran in Indonesia. He obtained his Ph.D. from the University of Oklahoma (USA) with a specialization in Financial Economics. Besides teaching and research, he is a Member of the Advisory Board of Bank Jabar and the Indonesian Association of Economics Graduates (ISEI). He is also Senior Lecturer in the Course on *Integrated Microfinance Management* (IMM), Bandung.

Asep Mulyana is Lecturer at the Department of Management and Business of the Faculty of Economics and Business of Universitas Padjadjaran, Bandung, Indonesia. He received his Ph.D. Degree from Bogor Agricultural University (IPB), Indonesia, and is a specialist in the field of the Management of Cooperatives & Small-Medium Enterprises. He was the Coordinator of the Course on *Integrated Microfinance Management* (IMM) in Bandung from 2012–2015.

Kurniawan Saefullah is Lecturer of the Faculty of Economics and Business, Universitas Padjadjaran, Bandung, Indonesia since 1998. He graduated from Universitas Padjadjaran and the International Islamic University, Malaysia. He has published several books and articles with his colleagues, including Introduction to Management and Introduction to Business. In 2009, he joined the International Project on Integrated Microfinance Management (IMM) for Poverty Reduction and Empowerment in Indonesia. As a Ph.D. Researcher at the Leiden Ethnosystems and Development Programme (LEAD), Faculty of Science of Leiden University, The Netherlands, he is finalising his Ph.D. Dissertation on 'Gintingan in Subang: An Indigenous Institution for Sustainable Community-Based Development in the Sunda Region of West Java, Indonesia' under the supervision of Prof. Dr. L. J. Slikkerveer. The focus of his research is on the important role of the local cosmology of Tri Tangtu in the process of sustainable community development, while he also developed a model of integration of community-support institutions on the basis of the Indigenous Knowledge Systems (IKS) in combination with the IMM and ICMD approaches, such as financial, medical, educational, communication and social services through the indigenous institutions of the local communities in the Sunda Region of West Java.

Epilogue

Despite an overall process of steady, albeit rather uneven worldwide economic growth over the past decades, estimated at 3.1% in 2015 and projected at 3.6% for 2017, and increasing endeavours by not only international organisations and agencies, such as the World Bank (WB) and the United Nations (UN), but also by national governments, MFIs and NGOs, poverty remains a huge worldwide problem, posing a serious global challenge to the well-being, health and peace of humankind in the next few decades (*cf.* IMF 2016).

According to the World Bank, today 836 million people are still living in extreme poverty, which is equal to about 1 in 6 people earning below the international poverty line of $1.25 per day. Remarkably, the vast majority of the global poor are living in rural areas, often poorly educated and mostly employed in the unstable agricultural sector, while more than 50% are under 18 years of age (*cf.* World Bank 2016). Moreover, in many developing countries, the gap between the rich and the poor continues to widen each day as a reflection of increased inequality in income and consumption, further hampering the reduction of poverty, particularly in the rural isolated areas. In addition, during the last few years economic growth has fallen back worldwide during the period of time known as the Great Recession, having a direct impact on the fragile situation of the poor.

Notwithstanding certain progress which has recently been reported in reducing poverty worldwide, the number of people globally living in extreme poverty still remains unacceptably high, and given current global forecasts of economic growth, poverty reduction may not be fast enough to realise the *Sustainable Development Goals* (SDGs) of the United Nations (2015a, b) of ending extreme poverty by 2030. Several international reports indicate that over the past decades progress has been made in poverty reduction; the first *Millennium Development Goal* (MDG) of the United Nations (2015a) to cut the poverty rate in 1990 in half by 2015—that is after 25 years—was realised ahead of its schedule. There is, however, general consensus that the highest priority should be given to a drastic reduction of the number of people still living in extreme poverty. The majority of these deprived people mainly

L. J. Slikkerveer et al. (eds.), *Integrated Community-Managed Development*, Cooperative Management, https://doi.org/10.1007/978-3-030-05423-6

inhabit two large geographic regions of the planet: Southern Asia and Sub-Saharan Africa (*cf.* UNDP 2017).

The international development agenda for the 21st century, led by the United Nations' *Sustainable Development Goals* (SDGs) and the World Bank Group's *Mission of a World Free of Poverty*, encompasses two major directions in the current efforts to reduce poverty within one generation: the *financial-economic approach* promoted by the World Bank, and the *human development approach* supported by the United Nations. As described in Chap. 4, the first approach is well-reflected in the annual publication since 1978 by the World Bank of their *World Development Reports*, providing a comprehensive overview on the economic dimension of development, in which each year a particular aspect of development is highlighted as a reflection of the progress and experience in international world development. These Reports also elaborate on the efforts to attain poverty reduction by the implementation of their financial-economic 'World Development Approach', involving mainly the application of monetary measures through the provision of financial support and services of loans and credit to developing countries.

In the course of the 1980s, the World Bank started to support the initially successful '*microcredit model*' which had originally been introduced by Mohammad Yunus and the Grameen Bank of Bangladesh to help the poor through the provision of largely subsidised financial services to groups of clients, such as micro-loans, micro-insurance, micro-savings and micro-deposits. Soon a growing number of Microfinance Institutions (MFIs) started to provide these credit services throughout the developing world. However, the shift in the late 1990s from *microcredit* to *microfinance*, supported by the WB and the IMF, introduced a new era of commercialisation and expansion of banking institutions with increased costs and interest rates, often at the expense of their low-income clients. A growing number of authors including Karnani (2007), Bateman (2010), and Roodman (2011) show that the neo-liberal process of commercial banking has eventually even weakened the position of the poor, who soon started to refer to microfinance as the 'poverty trap'. In its recent assessment of the state of affairs, the *Global Monitoring Report* of the World Bank (2015–2016) further expounds its twin goals of ending extreme poverty by 2030 and promoting shared prosperity, measured as income growth of the bottom 40% in the developing world, which, however, is still much worse off when it comes to access to non-financial services including education, health, communication and sanitation.

At the same time, the United Nations and its Agencies started to focus on realising poverty reduction through their 'Human Development Approach' of people-centered support for the poor, as it transpires through the publication of its annual *Human Development Reports* since 1990. Already in the Report of 1997, entitled *Human Development to Eradicate Poverty*, a strategy to alleviate poverty through sustainable development is made rather explicit, involving a number of activities, such as empowering men and women to ensure their participation in decisions which affect their lives, investing in human development such as health and education, and affirming that the eradication of absolute poverty is not only feasible and affordable, but also morally imperative (*cf* UNDP 1997). As regards the particularly vulnerable position of indigenous people, the Report also acknowledges that in many parts of the world, disparities in income and human poverty affect the indigenous people disproportionately, as they are in general poorer than most other groups in developing countries.

In its latest Report, entitled *Human Development for Everyone*, UNDP (2016) argues that caring for those left out of development requires a four-pronged policy strategy at the national level, including reaching those left out, using universal policies of inclusive growth, pursuing measures for groups with special needs, making human development resilient, and empowering those left out. Although the Report reaffirms that the indigenous peoples are particularly susceptible to poverty, its recommendations do not surpass mere general statements, such as that: "*measures are needed for some groups with special needs—including the indigenous peoples—in order to ensure that human development reaches everyone, and that recognition of the special identity and status of these marginalized groups is necessary*" (*cf.* UNDP 2016).

Several evaluative studies in the field of poverty reduction and development are pointing to a general lack of local peoples' active participation, involvement and contribution to outside development-oriented interventions of policies, programmes and projects, supported by the WB, UN, MFIs, NGOs, Governments or private organisations, forming a major obstacle to the realisation of poverty reduction and sustainable development. Both the WB and the UN as leading organisations in the worldwide struggle against poverty are showing their attention and interest in these important aspects for the realisation of their related goals and objectives, but a focused exploration, design and implementation of adequate and effective participatory strategies tends to remain missing in the overall policies. Indeed, the WB in its latest Report on *Poverty and Shared Prosperity* (2016) recognises the need to pay special attention to inequality in income and consumption, which specifically impair the poor, but the proposed policies remain focused on macro-economic and central fiscal policies with sector-specific interventions and public investments, instead of paying extra attention to the primary need for active involvement of the target groups, *i.e.* the local poor and low-income families at the community level.

Similarly, the latest Human Development Report of the United Nations, entitled *Human Development for Everyone* (2016), provides an informative overview of the current state of the 17 Sustainable Development Goals, based on a new global indicator framework conducive to recommendations in various sectors in order to achieve poverty reduction worldwide by the year 2030, but again, the implied strategies seem largely to continue focusing on initiatives and actions from the 'top down' to the local peoples and communities, which has shown to produce so many adverse results in the past. In spite of recent experience and a cumulative body of knowledge on promising 'bottom-up' approaches of participatory development, such prolonged centralist orientation in realising the SDGs on poverty reduction still seems prevalent in many international circles, as exemplified by a recent statement of Amina Mohammed, Deputy Secretary-General of the UN (2016), on strategies to end poverty, reduce inequality and tackle climate change: "*When the centre of government functions effectively, collective expertise from across the public sector can be mobilized and brought to bear on the most pressing decisions confronting a country.*"

While the WB continues to largely seek financial-economic solutions for the complicated financial and non-financial problems of poverty and inequality among different population groups around the globe, the UN recognises the crucial role of the human dimension in poverty eradication, but fails to implement strategies which are commensurate with the local culture in terms of indigenous knowledge, beliefs, practices and institutions which have enabled local people and their communities to live and survive over many centuries in their predominantly rural environments.

Meanwhile, as described in several Chapters in this Volume, a group of scientists and experts in the applied field of global socio-economic development started to reverse the international development paradigm of the 1970s and 1980s of 'development from the top' to 'development from the bottom', based on both the growing academic interest in *ethnoscience* and the accumulating body of knowledge from the field concerning the crucial, albeit largely ignored role of indigenous knowledge systems and institutions in the achievement of sustainable community development. The indigenous people constitute an important factor to be taken into account as currently, there are more than 370 million self-identified indigenous peoples around the world in about 70 countries. In Latin America more than 400 groups have been documented, while in Asia and the Pacific an estimated 705 indigenous peoples have been identified. In total, the indigenous people account for around 5% of the world's population, but 15% of them are living in poverty (*cf.* UNDESA 2009, 2016; IFAD 2012).

Several studies started to document that, on the one hand, the indigenous peoples possess centuries-old unique knowledge, wisdom and experience about living in harmony with their bio-cultural world, but on the other hand, they face severe deprivations caused by political, economic and social exclusion, rendering them today among the most vulnerable and poor population groups around the world.

As mentioned above, a growing number of case studies from around the globe were brought together in the CIKARD/LEAD/CIRAN Global IKS Network, encompassing about 35 IK-Centers worldwide, documenting the promising results of the 'bottom up' approach in development. The recognition, revitalisation and functionalisation of these indigenous systems of knowledge, beliefs, practices and institutions in this approach in many sectors of the development process, known as the 'Cultural Dimension of Development', has not only encouraged the active participation and involvement of local people in development programmes and projects, but also soon gained recognition and appreciation of national and international organisations, including UNESCO, UNEP, the EU and the World Culture Forum (*cf.* Warren et al. 1995).

As indicated above, the *Leiden Ethnosystems and Development Programme* (LEAD) of the Faculty of Science at Leiden University in The Netherlands, as one of the three Global IK Centers, conducted development-related research in the field of *neo-ethnoscience* on 'Indigenous Knowledge Systems and Development' (IKS & D) in East Africa, Southeast Asia and the Mediterranean Region, supporting the positive results of the shift in the development paradigm towards the 'bottom-up' approach in several sectors of the developing world, including health care, education, economics, and communication.

In its endeavours to further implement the 'Cultural Dimension of Development' within the context of poverty reduction and sustainable development, with particular attention for indigenous peoples, LEAD focused on the development of alternative strategies to increase the active participation and involvement of local people and communities in outside interventions of development policies, programmes and projects. In this way, new methodologies and models have been designed to engender community-centered integration of local and global systems of knowledge and practices which contribute directly to the realisation of poverty reduction and sustainable community development, promoted by the SDGs of the United Nations (UN 2015b).

As described in this Volume, new methodologies such as the *Leiden Ethnosystems Approach*, the integration models of indigenous and modern knowledge and institutions, including the *Indigenous Knowledge Systems Integration Model* (IKSIM) and the evidence-based *Integrated Poverty Impact Analysis* (IPIA), have been designed with a view to providing a contribution to the realisation of the SDGs of the United Nations by the year 2030. In these activities, special attention has been given to strategising the often disregarded indigenous institutions in order to strengthen their shared objectives of reaching poverty reduction and sustainable community development through a holistic framework of increased processes of knowledge integration, communication, decision-making, participation, cooperation and capacity building at the community level.

Meanwhile, over the past decades, Indonesia has provided an outstanding example in South East Asia of the theoretical, methodological and practical setting for the design and implementation of the alternative, more people-oriented development strategies, where LEAD's counterparts, including Universitas Padjadjaran in Bandung, have taken up the challenge to jointly develop the alternative approaches and models to sustainable community development with a view to promoting empowerment and fighting poverty at the community level throughout the country.

In spite of a steady economic growth of more than 5% over the past years, the poverty rates of well above 10% of the population of Indonesia still continue to persist, unfavourably standing out against the lower poverty rates elsewhere in the South-East Asian Region. Unlike the general expectation that poverty tends to reduce under conditions of economic growth, poverty levels remain at 12% and above since 2013 throughout the country. As indicated in several Chapters of this Volume, the undesirable poverty profile of the country is mainly determined by a large segment of the population of 40% being classified as 'poor' on the basis of the international poverty line of $2 a day. As Aji (2015) documents, about 68 million Indonesians have been classified as 'near–poor' in 2014, while some 55% of the households who were classified as 'poor' in the same year were *not* 'poor' a year earlier, indicating the high risk of falling back into poverty. Moreover, the pace of poverty reduction is slowing down as, from 2006 to 2010, the yearly poverty incidence declined by 1.2%, while from 2011 to 2014, the yearly rate of decline decreased to 0.5%. According to Aji (2015): "*Increasing difficulty in reaching the remaining poor and rising disparity in economic growth have contributed to the slowdown.*" While the majority of the poor people in Indonesia are living in the rural areas, a complicating factor is the vast geographical differences between the provinces of the country. The highest poverty incidence has been recorded in 2014 in the provinces of Papua (27.8%), West Papua (26.3%), East Nusa Tenggara (19.6%), Maluku (18.4%), Gorontalo (17.4%), and West Nusa Tenggara (17.1%), surpassing East Java, with a poverty incidence of 12.3%. In addition, increased inequality in income and consumption since 2000 has further hampered the process of poverty reduction, increasing the gap between the rich and the poor.

In order to turn the tide, the Government of Indonesia has recently launched several *Poverty Reduction Programmes* with the support of the World Bank, as described in some preceding Chapters. In the National Medium–Term Development Plan 2010–2014 (RPJMN 2010–2014), the government's poverty reduction strategy comprised three clusters: social assistance, community empowerment, and microenterprise empowerment, which focused on providing access to non–collateralized credit for microenterprises. To help design and oversee these programs, a national team for accelerating poverty reduction was established, chaired by the Indonesian Vice President. An evaluation of these programmes carried out by the SMERU Research Institute in Jakarta in 2010, however, showed mixed results, pointing to difficulties in targeting the government assistance to the target groups.

Similarly, the government recently launched the Indonesian Village Law No. 6/2014 (2014) (*Undang-Undang Tentang Desa*) as a means of institutionalising community-driven development as a national policy. In 2015, the Village Law underwent a transition from a development programme to a legal institution, securing the annual block-grant funding to villages under the PNPM-Rural programme which has been replaced by the transfer of funds as mandated by the new Village Law. Although this new strategy certainly contributes to the increase of financial resources at the community level, the main challenges to its successful implementation are the unwanted creation of new opportunities for increased corruption, collusion, and nepotism, paralleled by continued disinterest and non-participation by the non-elitist target groups of the poor in the local communities. Again, such a centralist 'top-down' approach continues to ignore the local peoples' indigenous knowledge and institutions, grounded in their traditional cosmologies, which as a 'Missing Link' in the policies obstructs the active participation of the local people in the related activities, ranging from the conceptualisation of the policy objectives, to the design and decision-making, and the final implementation at the community level.

Embarking on the premise that education is a key concept to realise these objectives, LEAD further developed, together with its counterparts in the above-mentioned target regions of East Africa (NMK), South East Asia (UNPAD) and the Mediterranean Region (CIHEAM-MAICh), advanced Master Courses in *Ethnomedical Knowledge Systems* (EKS) and *Integrated Microfinance Management* (IMM), while this Volume seeks to pave the way for a similar post-graduate training in *Integrated Community-Managed Development* (ICMD) with a view to providing new cadres of *integrated microfinance managers* and *integrated community-based managers* to guide and manage both the provision of financial and non-financial services as well as to implement the newly-developed alternative strategies of IKS-based poverty reduction (*cf.* Slikkerveer 2012). In this way, a contribution is provided to Indonesia's prioritised endeavour to reduce poverty at the community level, and as such also to the realisation of the SDGs of the UN by the year 2030.

It is hoped that the new community-oriented strategies elaborated in this Volume will not only be further developed by the Indonesian system of higher education, but also be implemented by the Indonesian government as a nationwide programme of community-based poverty reduction and sustainable development throughout Indonesia and elsewhere around the globe.

References

Aji, P. (2015). Summary of Indonesia's poverty analysis. *ADB Papers on Indonesia*, No. 04. Manila: Asian Development Bank.

Bateman, M. (2010). *Why doesn't microfinance work? The destructive rise of local neoliberalism.* London, New York: Zed Books.

International Fund for Agricultural Development) (IFAD). (2012). Indigenous peoples: Valuing, respecting and supporting diversity. Rome: IFAD. www.ifad.org/documents/10180/0f2e8980-09bc-45d6-b43b-8518a64962b3.

International Monetary Fund (IMF). (2016). World economic outlook (WEO). Washington, D.C.: International monetary fund.

Karnani, K. (2007). Microfinance misses its mark. Stanford social innovation review: Informing and inspiring leaders of social change. Summer 2007.

Roodman, D. (2011). *Due diligence: An impertinent inquiry into microfinance.* Washington, D.C: Center for Global Development.

Slikkerveer, L. J. (Ed.). (2012). *Handbook for lecturers and tutors of the new master course on integrated microfinance management for poverty reduction and sustainable development in Indonesia (IMM).* Leiden/Bandung: LEAD/UL/UNPAD/MAICH/GEMA PKM.

Suryahadi, A., et al. (2010). *Review of the government's poverty reduction strategies, policies, and programs in Indonesia.* Jakarta: The SMERU Research Institute.

United Nations (UN). (2015). *Millennium development goals report.* New York: United Nations.

United Nations (UN). (2015). *Post-2015 agenda for sustainable development and sustainable development goals (MDGs).* New York: United Nations.

United Nations (UN). (2016). Sustainable development goals: 17 goals to transform our world. New York: United Nations. http://www.un.org/sustainabledevelopment/blog/2017/07.

United Nations Department of Economic and Social Affairs (UNDESA). (2009). State of the world's indigenous peoples. Report ST/ESA/328. New York: UNDESA. www.un.org/esa/socdev/unpfii/documents/SOWIP/en/SOWIP_web.pdf.

United Nations Department of Economic and Social Affairs (UNDESA). (2016). Global sustainable development Report 2016. New York: UNDESA. https://sustainabledevelopment.un.org/content/documents/2328Global%20Sustainable%20development%20report%202016%20(final).pdf.

United Nations Development Programme (UNDP). (1997). *Human development report: Human development to eradicate poverty.* New York: Oxford University Press.

United Nations Development Programme (UNDP). (2016). *Human development report: Human development for everyone.* Nairobi: UNDP. http://hdr.undp.org/sites/default/files/2016_human_development_report.pdf.

L. J. Slikkerveer et al. (eds.), *Integrated Community-Managed Development*, Cooperative Management, https://doi.org/10.1007/978-3-030-05423-6

United Nations Development Programme (UNDP). (2017). Sustainable Development Goals in Action. Nairobi: UNDP. http://www.undp.org/content/undp/en/home/sustainable-development-goals.html.

United States Agency for International Development (USAID). (2013). *Getting to zero: A discussion paper on ending extreme poverty*. Washington, D.C: USAID.

Law, Indonesian. (2014). *Undang-Undang Republik Indonesia tahun 2014 tentang Desa (Indonesian Law on the Village)*. Jakarta: SEKNEG.

Warren, D. M., Slikkerveer, L. J., & Brokensha, D. (Eds.). (1995). *The cultural dimension of development: Indigenous knowledge systems. IT studies on indigenous knowledge and development*. London: Intermediate Technology Publications Ltd.

World Bank. (2015–2016). Global monitoring report: Development goals in an era of demographic change. Washington, D.C.: World Bank.

World Bank. (2016). *Poverty and shared prosperity: Taking on inequality*. Washington, D.C.: World Bank. http://www.worldbank.org/en/publication/poverty-and-shared-prosperity.

CPSIA information can be obtained
at www.ICGtesting.com
Printed in the USA
LVHW081034111120
671392LV00003B/19

9 783030 05422